The Scientific Voice

THE CONDUCT OF SCIENCE SERIES

Steve Fuller, Ph.D., *Editor*
Department of Communication
University of Durham, England

The Scientific Voice

Scott L. Montgomery

THE GUILFORD PRESS
New York   London

© 1996 The Guilford Press
A Division of Guilford Publications, Inc.
72 Spring Street, New York, NY 10012

Printed in the United States of America

This book is printed on acid-free paper.

Last digit is print number: 9 8 7 6 5 4 3 2 1

Library of Congress Cataloging-in-Publication Data

Montgomery, Scott L.
   The scientific voice / by Scott L. Montgomery.
     p. cm. —(The Conduct of science series)
   Includes bibliographical references and index.
   ISBN 1-57230-016-7. —ISBN 1-57230-019-1 (pbk.)
    1. Science—Miscellanea. 2. Science—History. 3. Science—Social
aspects—History. I. Title. II. Series.
   Q173.M83 1995
   500—dc20   95-43660
                                                      CIP

To Kay, who made this book inevitable

Le plus beau serait de penser
dans une forme qu'on aurait inventée.

[Most beautiful of all would be to think
in a form that one had invented.]

PAUL VALÉRY, *Tel Quel II*, 1943

Preface

Die Essays, die für dieses Buch bestimmt sind,
liegen vor mir, und ich frage mich . . . kann aus
ihnen eine neue Einheit, ein Buch enstehen?

[The essays destined for this book now lie before
me and I find myself wondering . . . can a new
unity, a "book" be produced from them?]

GEORG LUKÁCS,
"Über Wesen und Form des Essays: Ein Brief an
Leo Popper," *Die Seele und die Formen, 1911*

There is a well-known story about the aging Otto von Bismarck, the famed "Iron Chancellor" of Prussia. Shortly before his death in 1898, in a meeting that might be said to have marked an encounter between the centuries, he sat for a brief interview with a fledgling war correspondent. During a moment, perhaps, of journalistic torpor or naiveté, the reporter fell back on the obvious: "What, to your mind, Herr Chancellor," he asked, "is the determining fact of modern history?" Raising an eyebrow, Bismarck looked squarely at his guest and replied, not without a mixture of dismay and irony: "The fact that North America speaks English."

Language, we are often told, manifests the power of culture in one of its most conspicuous forms. But even more than politics or literature, sci-

vii

ence, in its modern guise, must be counted one of the great examples of this truth. That a vast amount of global scientific activity today, like North America, "speaks English" would seem proof enough. But perhaps not.

The essays in this book, in fact, arose from my own ignorance and dissatisfaction along these lines. Trained and employed as a scientist (geologist), yet with long-term interests in writing and language, I could not escape a curiosity about the nature of the discourse I produced and consumed for a living. My technical education had not prepared me to deal with such curiosity. Indeed, in many ways, it dissuaded me from pursuing it. Geology is a magnificent science; a great many phenomenologies of the world fall under its purview. It is unique in defining a realm all its own yet drawing within its borders the knowledge and discourse of so many other fields—physics, chemistry, botany, zoology, astronomy, various types of engineering, and more (geologists are at once true "experts" and hopeless "generalists"). Finally, it is also one of the few "hard" sciences still in touch with immediate experience of the world and also one of the few (biology is another) that boasts an enormously rich and varied literature, until recently prone to the highest orders of elegance. And yet, very little of this literature gains much attention among current-day geologic professionals, whose training, labor, and language seem to predispose them to have minimal concern for their own "field history," for the layers of evolution that exist within it. At the level of everyday use and awareness, geologists are largely deaf to the voices of history and culture that they themselves embody.

At some point, a late one I admit, I began to suspect that the machine was haunted. The writings collected here, most of them published during the past several years, reflect efforts to give flesh to the ghost of my suspicion. No doubt, there is much that remains naive and preliminary in them. That academic authors, for example, in a variety of disciplines—sociology, history, literary criticism, and rhetoric, among others—have been examining scientific language for well over a decade is something of which I was personally unaware until rather recently. This means that my own "discovery" has been largely a private one. By "discovery" I mean a precipitated awareness that within scientific discourse—this language that I had so often handled, like most scientists, as a piece of technology—there is to be found another world, or rather a grouping of worlds, a clamoring of voices from ages past and ambitions contemporary that are no less an essential and guiding element of science than are facts, hypotheses, and theories. These are not exactly the terms employed by those who pursue analyses of science. But they give some idea as to the sensibility guiding the essays in this book. That scientists speak and write the science of today in variable dialect and with profound inheritance, like North America does English, is a focus of the inquiries and reflections presented in this book. But more than this, too, I perceive in this speech a living repository. Beneath its instrumental surface there teems an accumulated life, impossible to fossilize.

Many, no doubt, would shun this type of organic metaphor. They would accuse it of being too ideological. But the hyperconsciousness of such possibilities that so often seems to rule discourse studies today can have the effect of killing any impetus or ability for good writing. This leads me to a second confession of dissatisfaction. In the large and growing literature on scientific discourse, I have personally found a wealth of insights, many valuable analyses, and helpful, often cunning, manipulations. But there is also much that appears to me limited, in a number of ways. Much reading and diligent search have not dissuaded me from this: beyond a certain point, my own curiosity cannot be pacified or entirely persuaded by the existing methods aimed at analyzing science, language, and culture. There is too much that seems either left out or doted upon.

Generally speaking, writers are divided into two main camps: those who apply some form of critical or literary "theory" to science (structuralism, deconstructivism, Foucaultism, hermeneutics, feminist critique, etc.), and those who follow the ways and means of "discourse analysis" (mainly sociologists, students of rhetoric, and communication specialists). Most members of the first group have been proud outsiders: science remains for them a dark or a gleaming territory, resistant and mountainous, to be characterized from an intimidated distance, often with leveling intent. These writers, it seems to me, approach the wall, Science, throw up their hands in often fine and well-considered gestures, and then retreat, defeated by the very success of their labors. Members of the second group, scholars interested in the rhetoric of science, are often methodologists first and thinkers second. In my reading of this literature—admittedly incomplete, selective, and biased by "insider" experience—I have found many things of interest and of use, but also a great deal of repetition. A few ideas seem to be recycled endlessly; exciting insights can be passed over with a nod; small pieces of history are often chewed to puree. Moreover, a good portion of the writing itself has the aesthetic qualities of winter cement, something that seems particularly unfortunate given that language and its powers are exactly the issue. The explosion of analyses of technical discourse during the past two decades is truly a cultural phenomenon. A new vein has been found; a new industry is underway to mine it. But the rush betrays a quality of obedience, I think. Like all raw material refined into manufactured products, no small portion of the relevant work on scientific discourse has tended to remain within the bounds of easily marketable forms.

Discomfort and dissatisfaction, however, are among the best reasons for writing. My hope here, therefore, has been not to disavow or throw shadows on this literature but instead to try and add to it. I should say, too, that my goal has nothing to do with either weakening or reaffirming the idea of objective scientific knowledge. This, it seems to me, remains more an ethical question than an epistemological one. The attitude that sees more of social content in science, by definition, than knowledge of the physical world is interested in something other than scientific reality as it actually exists. Such an attitude has constructed a different "science," extracting certain inevitable cultural ingredients and leaving the "rest" behind. Indeed, whether one calls science the engine of modernism or a black box truth, it would never have gained the real-world power it now so obviously holds unless it had the capability for creating knowledge that comes as close to the "objective" as possible. To deny this truth is to doom one's own analysis to triviality. To be serious about knowledge and its contents and influences in society, one must first accept it as knowledge—tentative, probable, undergoing correction, generated through language, rich with matter specific to its time and place, but also real in terms of its capability for expressing truth. If this is not enough, we have all of modern technological reality, in all its workaday forms, to convince us.

The essays in this book are not aimed at challenging scientific knowledge. On the contrary, their aim is to look at the forms in which this knowledge has appeared and to say something about where they have come from and what they mean. This includes showing how science, in its "expression," has often acted to limit science as truth.

What follows comprises a diverse, ragged, sometimes tendentious, but hopefully original collection of pieces. An early version of Chapter 1 was published in 1987 in the journal *Science as Culture*. This polemical lament on the character of scientific discourse in general prompted a host of perceptions and ideas that, I felt, begged their own pursuit. The more insistent of these appear as the subjects of the succeeding chapters. Each was written as a separate project, each deals with a different aspect of technical language, and each involves a different approach and style. Some essays are analytic; others are impressionistic; still others verge on diatribe. All have ingredients of each of these, in varying amounts. I have tried in every case to remain engaged with my own, scattered curiosity; to avoid a single, unified view of science and its discourse; and to explore the social and cultural consequences of language. Several chapters include large amounts of scholarly material and seek to add to this material new facts, perspectives, or interpretations—in particular, those essays on the mapping and naming of the Moon (Chapter 4), on translation and Japanese science (Chapter 5), and on Freud (Chapter 6). Others, such as that on the history of scientific discourse (Chapter 2) and on the imagery of disease (Chapter 3), involve more opinion and perception than scholarship per se, but to different degrees.

In the end, if the results of this inquiry do not allow me to write my own science any differently (for how could they?), they have at least provided the consolation of a more knowing participation.

No doubt some readers will question whether or not these individual pieces, as fragments of an unfinished pursuit, together comprise a true whole. Some may even ask whether the term "essay" can be appropriately applied to them. Montaigne, it seems to me, is not to be invoked here (like a reflex); the world and its contents are larger today than in his time, and the essay, with its searching qualities of dissatisfaction, should be free to expand likewise, without being chained to some primal father. The critic Georg Lukács, too, is exemplary in having tried to perform exactly this type of enslavement, but in a manner more adapted to the present century. In his "Essence and Form of the Essay," which serves as the introduction to *Die Seele und die Formen* (The soul and forms, 1911), he offers an impassioned plea, a failed demand really, that the essay be looked upon as an artwork in its own right, as a self-contained creation with unparalleled power to impose new order on that which already exists. Every piece of writing, Lukács says, strives for plurality as much as for unity. The uniqueness of the essay lies in its ability to employ both, simultaneously, in the quest to reveal a domain of thought that would otherwise remain hidden and interior. Its critical power is to make the invisible visible, not (like other arts) to create new visibilities altogether. The failure here is in the irony of Lukács' reclamation: art is kept the high ideal, an unassailable measure to which the essay must ascend. When he speaks of this reclamation, moreover, Lukács employs a special term of his own making: *Kunstgattung*, "species of art." Science thus enters in the form of terminology; the purity of Art is besmirched, qualified, claimed by reference to non-Art.

The point is not how the essay can be made artistic, or how breakdowns can occur between prose and fiction, but how, brought to a high level, it can contain a good deal of the complexities, the ponderings, the excesses, and even the contradictions, that are common to thought itself. From Lukács we gain the suggestion that the essay, at its peak, should be a condensation of a much larger work. Ideas should spark within it, as if metal were struck at every turn. It should produce both aesthetic and intellectual effects. It should be reactive, opinionated, even humorous, stumbling, overly dense (containing its own humiliations), but also serious and committed in its goals, not glib. One of Lukács's better insights is his perception that scholarship too often suffers from its own codes of literary behavior: though frequently dealing with fundamental problems of life and thought—how human beings have made the world they inhabit and conducted themselves within it—the scholar must often pretend his or her work is really about other things, about books, images, history, literature, and so on.

I have tried not to make this "mistake," at least not too often. Nor would I pretend that the various pieces in this book represent anything more than a groping toward the type of ideal that Lukács suggests. It is not necessary that one choose, finally, between the hopes of adding to the stock of common knowledge or to the stock of opinion about it.

Acknowledgments

My sincere appreciation goes first to two editors who, early on, found value in my work: Les Levidow of *Science as Culture* and Stephen-Paul Martin of *Central Park*. Earlier versions of several chapters (1, 3, and 7) appeared in both these publications. To these supporters, I extend my warmest thanks. A portion of Chapter 3 appeared in *Cultural Critique* (No. 25, 1993). My gratitude goes to Jonathan Kahana, for his diligence, and to Oxford University Press for permission to reuse this material.

Ewen Whitaker provided invaluable assistance on matters related to the history of lunar cartography, as well as most of the illustrations in Chapter 4. His generosity and enthusiasm, as well as his diplomatic corrections, are here acknowledged with deep thanks by a "fellow traveler."

Albert van Helden also read portions of the manuscript, offered valuable advice, and has saved me the embarrassment of certain errors.

Peter Wissoker and Rowena Howells of The Guilford Press, whose professionalism has been a constant encouragement, helped see this work through its various incarnations. I here express my deep appreciation for their help, patience, and interest.

The writing of this book was otherwise largely a personal affair. Assistance on translations from the Latin was very generously given by Doug Machle and Professors Alain Gowing and Allen Sass of the University of Washington. Unless otherwise noted, all other translations are my own. I have also received much help from several friends and colleagues, especially Jeff Eaton, Steve Fuller, Dorothy Nelkin, and above all, Dominic DiBernardi, who was present at the beginning. Finally, my love and gratitude go to my wife Marilyn and my son Kyle, who wrote more of this than they may ever know.

Contents

The Scientific Voice

1

The Cult of Jargon

Reflections on Language in Science

᚛᚛᚛᚛᚛᚛᚛᚛᚛᚛᚛

The everyday difficulty is to use words "pure and simple," without getting entangled in their emotional lives . . . the scientist is to a large extent freed from this temptation. He knows very well the danger of using words.

—JOHN WOLFENDEN,
"The Gap—and the Bridge," 1963

I

The T-DNA transfer process of *Agrobacterium tumefaciens* is activated by the induction of the expression of the Ti plasmid virulence loci by plant signal molecules such as acetosyringone. The *vir* gene products act in trans to mobilize the T-DNA element from the bacterial Ti plasmid. The T-DNA is bounded by 25-base pair direct repeat sequences . . .

To read a passage like this is to feel like a child again—suddenly bathed in a rich, warm fluid of sound, able to mimic, to splash in its

1

bubbly surface, utterly free of any adult responsibility for meaning. Put more harshly, to anyone outside the field of molecular biology, this paragraph (the first part of a recent abstract in a prominent journal) is likely to seem as distant and inaccessible as the frozen surface of Pluto, capable of communicating as much total sense as if it were written in Linear B or ancient Chinese.

Incomprehension of this kind doesn't make us feel stupid or incompetent, only unprivileged, closed out, ignorant, and—most of all—innocent. But if this is true, if such writing does make us feel excluded from a certain grown-up world of truth and truth telling, if it therefore reminds us of how ordinary our knowledge is, then it has had at least some degree of success.

II

Scientific writing, in its broadest sense, is quite likely the most triumphant, the most imitated, the most universal form of human discourse ever developed "after Babel." During the past 100 years, it has risen to a glorified preeminence over all other styles of written communication, having become the model of authority and presumed accuracy to which nearly all forms of expression have increasingly turned for "advice." As an enormous library of individual tongues that have adopted a single style of truth telling, "the common language of science" (as Einstein called it) has evolved to a level where it seems as fully absolute, independent, self-justifying, and unassailable as the facts it claims to transmit. Indeed, it would be hard—perhaps impossible—to deny the impression that here lies the grand master narrative of modernism, ideally suited to its content. What sort of faith, then, might we say seems to beat at the heart of this discourse? Simply this: that language can be made a form of technology, a device able to contain and transfer knowledge *without touching it.*

There seems something magnificent and fecund in this. Elevated pragmatism, supranational community: are not these, too, the flesh and spirit of this common narrative? A life of the mind that carries on far above the self-interested caprice of the sociopolitical moment, cleansed of moral or emotional diversions, purified of climatic influence—such would appear to be the rhetorical promise. If one works within this discourse, one seldom feels (today) any need or necessity to question its demands or its effects. Instead, one is urged by its codes and conventions to become another worker in the mill of hypothesis and data, to accept (as one has been taught) a role as its servant, its agent more than its author. On occasion, certainly, there are chances for invention and whimsy: a literary borrowing here, a sensational or impressionistic turn of phrase there (e.g., "white dwarf star," "hormonal permissiveness"). For a brief time, these may even become the source of debate or discussion, momentarily disrupting, as it were, an otherwise eminent lack of linguistic self-reflection among scientists.

Yet very soon, and of necessity, such terms lose their special charge. Familiarity bred of standardized usage causes their metaphorical foam to subside; they become identified strictly with their referents. The grand narrative closes over them, blending all once more into what appears to be a serene and infinite sea of epic denotation. Here is a voice, then, rising from this surface, that would qualify itself not merely as that of efficient use and commonality, but of epistemological solidity, power, and—above all—presence. It is a voice, in the end, that seems more than any other to embody and carry forth the modernist project of expressing absolute truth, objectivity, and therefore progress, without any end in sight, without any change in step, with immanent goals that appear to defy any chance or legitimacy for plurality.

In saying all this, however, one poses the beginning of a discussion, not its end. No one—not history itself—can at this point deny the real-life status, utility, and material power of this common language. It would be a grave and misguided error to blithely call technical discourse "just another set of narratives" that needs to view itself more "modestly" (Eagleton, 1987). Nor does it make much difference whether or not this discourse has become the subject of detailed professional analysis (it has). Academic curiosity notwithstanding, no formal study of "scientific communication" can hope to have any effect upon its institutionalized place and power unless such study leads directly to a revolution in cultural sensibility. As it is, one lives today encased, surrounded, and frequently controlled by the real-world products and dictates of scientific discourse, by the entire universe of technology that it makes possible, by various forms of technorationality that have emerged from within it. To call for one or another type of "modesty" is to avoid what science has become in Western culture; indeed, it is to call for ignorance.

But there is more, too. For neither can there be any doubt as to the unparalleled influence this discourse has had with regard to knowledge as a whole, in all disciplines. Clearly, the reified axioms of "evidence," "documentation," and "proof" that were institutionalized within it at an early point have since profoundly changed the nature of writing in every area of scholarship. One sees this at all levels throughout the realm of professional expression—in the triumph of the journal and therefore of the journal article; in modes of citation and referencing; in ideas concerning "research" and "theory"; in the straightforward explanatory style that has been adopted by most academic fields as the principal approach to writing; in a general structure of narrative that opens with an "introduction," moves on to the main text, and then ends in a "conclusions" section; in the demand for authors, whether in literature or history, to "support" and qualify their ideas by quoted evidence (to treat such ideas as hypotheses, in other words); and perhaps most of all, in the more subtle requirement that any nonfiction work of prose construct a logical, step-by-step type of argument that seeks to induct its readers into a clean, measured type of rational thinking, thereby providing both the form of and apologia for "truth." In all these things, modern professional writing follows the lead of science. What is sought, above all, is to construct a world of readerly experience where logic and evenhandedness rule. Such an image reflects back a coherent representation of modernist subjectivity itself, of the author as an embodiment of strictly controlled reflection and uncompromising intellectual honesty. This, of course, is a fiction (both writing and reading are always a chaotic affair); but scholarship—nay, authorship itself—demands such an idealization. It, too, is part of what is called knowledge.

Thus, a second question: What might the present period of history, saturated as it is with professional languages, be contributing most to the formal character and production of discourse as a whole? Or, more simply put, what is the central linguistic turn or drift of our own time? Clearly, this is not a grand and celebratory expansion of the vernacular. It is not an era for the likes of Dante, Murasaki, Rabelais, Luther, or Shakespeare. The colloquial realm of word making seems more and more to have become the responsibility of "marginal" groups in bourgeois society, for example, blacks, teenagers, and gangsters, those for whom language serves as a weapon of difference and identity. No, the most irreversible current in the discourse of power today would seem to be in exactly the opposite direction: call it jargonization, the weedy growth of local, specialized lexicons of expertise. Or more broadly, the established, even institutional tendency for any area of knowledge production, whether high or low, menial or mental, struggling or secure, to gain legitimacy through the building of an authoritative, insider voice—one whose status depends, at some point, on "exclusion."

Once again, science (and most of all, natural science) has been the historical model for this. Is it, therefore, the responsible agent as well? The answer must be no, at least in any strict sense. The uptake of scientific language as the most profound of linguistic prototypes is an enormously complex affair, related to the advent of science and technology as the bearers of truth and material power, and therefore to the Industrial Revolution, the spread of positivism, the belief in the "technological sublime," the rise of the university—in other words to the vast changes in Western society and institutions wrought during the last century.

Perhaps, if one desires a more metaphysical answer, the ultimate cause might be said to lie with the broader appetite in the West for appearances of certainty—for what Nicholas Schöffer (1978) has called our endless Occidental search for "new confidences," or what Roger Caillois (1973) has described as our need for "coherent adventures." An appetite, one might then say, that has always been at the heart of the Western concept of knowledge itself.

But while it may not be the cause, *sensu stricto*, scientific speech has plainly set the functional standards, nay, the very limits for jargonization. And by this I mean not so much the degree of specialization—for there have always been secret languages, the shibboleths of guilds, the arcana of cults, the argot of underground or renegade societies—but the detailed strategies that are involved, the very "how" of expression. And this, finally, prompts a still more central question: If the languages of knowledge have determined to seek in scientific discourse the end point, the very asymptote of their ideal success, if their historical momentum in other words is to become more and more "technical," even ecstatically so, what then is happening to them?

Here, Sartre (1949, p. 15) has given us a wonderful yet ominous phrase: "the word, which tears the writer of prose away from himself and throws him into the midst of the world." It is a violent image, but therefore a necessary one. How, we might ask, does the scientist, as writer (of the present, the future), become "torn" in this manner? How does s/he become engaged in the act of writing? What does s/he come to touch in the world through this writing? Toward what does her or his authorship hurl her or him?

III

It is my hope to try and tackle these questions in just such terms as Sartre sets out. Not as a scholar, concerned with forms of analysis, but rather as an author, of scientific and other works, who feels forced to speculate about "the prison house" of technical discourse and where history has placed such effort in the scheme of the present.

To answer Sartre's challenge, it seems to me one must deal with two different levels of representation. The first I have already introduced: it is the level of the master narrative, the primary codes and conventions that give scientific language its appearance of unity and totality, and that demand from its users many forms of strict obedience. It is a level that one might call "the social surface of technical discourse," that helps constitute the face of what is normally called science. The codes and conventions that go into the making of this face are real; they can be identified and discussed. But they are by no means the whole story. Beneath, between, and behind them lies a second topography of forms and meanings, one that in fact belies all final unity. Indeed, this is an area where recent scholarship has had interesting things to say, either in revealing or suggesting how a close reading will point up local failures of logic, conflicting interpretations, unneeded redundancy, literary tricks and turns—realities that all underscore the sense that the "scientific" remains unequal (or at the least, unfaithful) to its own promise of totality. This expands still further when one looks deeper into the actual terms that scientists have chosen for their jargon, their often ripely metaphoric qualities, their connections therefore to other areas of sociocultural meaning.

Scientific language can thus be viewed as a realm of great coherence and chaotic heterogeneity at one and the same time. There is no final truth to it, no single scheme for emptying it of meaning. Any such meaning will always be relative to the type of reading(s) performed upon it, whether this be for the purpose of single interpretation, as in the doing of science, or for diverse ones, as is sometimes the case with historical study. Neither approach, however, can ever hope to exhaust what this discourse "has to say"; neither should be believed when it makes claim to so grand a result (relativism is not the point; the derivation of meaning is always a social, and often a professional, process, based on practices of use). Put differently, technical discourse might be said to contain a mixture of modernist standards and postmodern[1] realities. While the former allow for its scientific substance, and dictate a great deal of the experience of authorship, the latter cannot be ignored, for they prove in the end that the writing and reading of this language is a densely cultural (one might say human) activity in which many fascinating and crucial influences, conflicts, and ambitions have been deposited.

To begin, then, it makes sense to examine the modernist level first, how it is built. And here one must look at three primary aspects to scientific discourse.[2] They are as follows:

Creation of Insiders and Outsiders

First, and most evident, there is the aspect I began this chapter with, that of exclusion. Technical language sets up a barrier between those who can speak and understand and those who cannot.

Socially, professionally, and even personally there is advantage to this. Such a discourse, after all, "exalts, reassures all the subjects *inside*, rejects and offends those *outside*" (Barthes, 1973, p.122). The speaking world is sundered in two. Within, the context is one of literacy and praxis, in effect a tremendous cosmic chattering of knowledge, broken and divided into a thousand dialects, all speaking at once, yet full of diverse exchange and cross-intelligibility too. Outside, however, where the nonscientist resides, so many dialects merely ring as so many variations of incomprehensibility. The context here, in other words, is not merely illiteracy but passivity. Indeed, to the untrained ear, eye, and hand, this discourse can only seem less a federation of collaborative voices than a vast and monolithic castle of impenetrable speech, one that would appear to justify talk of "science" in the singular (whereas the "arts" and the "humanities" might always, if only by implication, remain plural). Thus one conclusion (to be applied to all imitators): whatever benefits this language can offer to functional communication, these must be weighed against an intractable ability to make knowledge inaccessible (not secret).

Such inaccessibility alone has the power to intimidate. Even literary critics would be forced to admit the discriminatory power that has been granted technical language in our society. Able to close out, to awe, even to jam all other forms of speech with which it comes in contact, this discourse is always fencing an outline around the territory (for truth) and lower social status (relative significance) of its competitors, for which it even has its own term of secular inability: "lay speech." This shutting out, in fact, is something everyone—including scientists themselves— have experienced and know well: here, for example, where one stands mute before a physician's explanation of an illness or treatment (one's body becoming a voiceless thing); there, coming across an article on a topic in an unfamiliar field; or elsewhere, witnessing, perhaps, the humiliation of an argument at the hands of a technical phrase, a question defeated by an answer filled with jargon.

There is, too, the ability of this discourse to defamiliarize and thereby to mystify the ordinary. A single word can sometimes be enough to shift the context of one's own entire experience, to cause the "scientific" to rush in and take over (in pedagogic guise). On a simple walk in the woods, for example, it is enough to say: "Not *snail*, but its 'real' name, *gastropod*; not *rain cloud*, but *cumulonimbus*." Light? Poison? Heartbeat? But by the demands of truth, these should be quanta, toxin, systole/diastole. The potential for mystification, here, based on the ontologic powers invested in naming, takes an ancient and yet very specific form. In pharaonic Egypt, knowing the name of a thing or a person meant having some degree of possession over them (which is one reason why priests and rulers commonly had several names, at least one of which was secret). Being named by science, the world is suddenly possessed by the "scientific," rendered under the sign of "phenomena." In a sense, it is taken away from us and placed in a glass case, turned over to expert hands, to those who really know. We are mystified by the very fact that science seems to have named everything already. What is there left, linguistically, for us to discover?

These, and a hundred other encounters like them, occur on a daily basis. They are inevitable social effects of the nature and status of technical speech, part and parcel of the "microphysics of power" that it has acquired, in Michel Foucault's well-known phrase. In the face of such power, outsiders usually have little choice but to remain dazzled. Take, for example, George Steiner, who writes, "At seminal levels of metaphor, of myth, of laughter, where the arts and the worn scaffolding of philosophic systems fail us, science is active" (1984, p. 440). Or take Roland Barthes, who remarks of a biology lecture, "I was fascinated by this scientific language—euphoric for me precisely to the degree that I had no resistance to offer it" (1985, p. 102). Such fascination, with its underlying sense of humility or humiliation, has perhaps aided in making technical discourse the subject of full-fledged academic study in recent years, in the style of historical, sociological, and literary analysis (one way, it would seem, that the humanities have sought a degree of revenge).

More often, however, exclusion begets a kind of intellectual fatalism or useful despair for those interested in studying the sciences. Many examples could be given of this; the tendency to examine the formal structures and sociology of technical writing (e.g., how it is constructed by workers in a particular laboratory) without ever dealing with its ideas or the etymology of its terms is merely one glaring instance. The general problem, meanwhile, is implied by a more extreme example, shown by those writers who have gone so far as to suggest—without humor—that in studying scientific work, one ought to "adopt the stance of the anthropologist coming into contact with a strange tribe" (Harre, 1981, p. v). The language of science is the tongue of foreigners, equally exotic whether spoken in the native hut of the laboratory or the villages and cliff-dwellings of the professional meeting. In a sense, however, such timidity has its understandable side, as I have been saying. It is expressive of the distance inherent in the limited literacy that technical writing so often creates. Yet stunned or euphemized agreement to this distance does not seem an adequate response.

How foreign, then, is foreign? How deep does this inaccessibility go? Writers on language such as Roman Jakobson and A. J. Greimas have helped make it clear that an important aspect to any jargon is its ability to define the active character of knowledge, that is, to make it in some part functionally inseparable from terminology. In everyday terms, despite the wishful claims of educators and science writers, technical language is not merely the hardened and removable skin of scientific learning, but instead its very substance, the flesh of its erudition. Fundamental concepts, general theories, propositions—these can always, to some degree, be made amenable to paraphrase. Depending on the field, they can sometimes be rendered almost entirely into ordinary language. But such accommodation has its limit, and this limit is soon reached. What we call "data" and "evidence," especially if of recent vintage, very rarely can be translated so simply—not at least without a return to a degree-zero plane of explanation, and thus a formidable, step-by-step process of building up a reader's literacy through language acquisition. To wring the *full* meaning out of even single phrases—for example, "purine auxotrophy subjected to an IVET selection system"—would require paragraphs, perhaps pages of explication. What effort might then be needed to translate the detailed sense of the entire opening quotation to this essay? The scale of necessary education can only seem daunting.

The "foreign" adjective thus has its degree of justice. Scientific knowledge is both expressed in and, in part, *conceived* through a discourse that has undergone a tremendous compression. As any technical dictionary will show, it is a form of speech made superheavy by modes of shorthand condensation, by substitution, redefinition, fusional reduction of terms, and by the continual adding on of new and more precision-oriented nomenclature. No small part of this nomenclature is still, today, loyally coined out of Greek or Latin roots. Two centuries ago, when these languages formed the core of the required curriculum in Western education, there was a certain general readability to such jargon and, correlatively, an undeniable logic to the invention of new terms and names on the basis of these dead tongues. Today, however, words like *glycolipid* or *hypidomorph* have a measure of historical distance added in to their linguistic removal from the ordinary.

From a writer's point of view, the formidable density of technical prose has also come from a different source altogether. Simply put, it has grown not merely from terminological compression, but from a gradual, progressive elimination over time of any lighter, personal, or (for lack of a better word) "literary" modes of expression. We know this today: "science" and "literature" are, for the most part, famously separate in their linguistic domains (however much they might borrow, now and then, from each other). The long-term and by now clichéd battle between these two domains in the 20th century is at least as much a matter of language as it is epistemology or "culture." The result for science, indeed, is a discourse that commonly strives to be as un-literary as it can be, that trades off being "expressive" for a chance to approach the "performative,"[3] that does not merely try to document or articulate knowledge but to transact it as well, in its most naked, skeletal form.

This suggests the plight of science writing (which I will discuss more fully below). At this point in history, toning down the use of technical terminology in scientific discourse invariably means the elimination of detail and subtlety. Details in science, however, are not embellishment; they are information, facts, points of logic, twists of theory, and the like, and their deletion means, without exception, loss of knowledge. Explanations to the uninitiated, told in ordinary (media) speech, are forced to fall back on various strategies of generalization—for example, interpretive description, analogy, simile, or selective summary—none of which can avoid sacrificing information for accessibility, all of which involve choices guided by particular ideological agendas that frequently count among the standard tools employed by professional science writers (often automatically, as it were), whose job it is to make their subject more friendly and domestic, to reconstitute it, often enough, into such stereotypic and otherwise unexamined categories as the "breakthrough," the "cure," "research frontiers," and the like.

In most attempts at low-level popularization, there is an effort either to remove as many technical terms as possible, leaving only a few scattered remnants here and there, or to retain such jargon in the form of undigested narrative chunks. Either way, the effect is the same: scientific language becomes like a series of remnant islands, standing eminent in a shallow discursive bay. Such popularization, so often distinguished either by its routine and self-annihilating promise of utopia ("the most complex problems of the universe told in simple, everyday language") or by one or another type of sensational use ("miracle grass to act as erosion-control hedge"), is nearly always on the side of mystification.

Scientists and scientific language always hold the key to vast and unknown empires.

What about when technical terms are themselves adopted for metaphoric uses, whether by popularizers or others? Here, other interesting events occur. This type of usage is commonly aimed at creating some sort of literary effect, often dramatic: "The modern metropolis, with its recombinant structures . . . "; "The deficit, our economic black hole . . . "; "The entropic character of modern love." Again, there is an attempt to feed on the prestige, apparent precision, and broader suggestiveness of technical language, which is served up as if it were the semantic food of the gods. At the same time, such language is put toward a saleable aesthetic of colored effects and glittering spectacles that, in the end, makes science seem that much more dazzling and mythic. These types of popularizing, then, tend to retain, and indeed often even improve, the raised, distant character of scientific discourse. Given the ways in which technical jargon appears in such writing, inaccessibility is hardly overcome or tempered. On the contrary, it provides a kind of valued source for status and legitimacy, one that can be colonized or fed upon for its ultimate, final-word authority.

And yet, there is also something revealed in this type of usage that might otherwise remain hidden. Such usage, that is, can itself be used as a lens to look back into the interior content of scientific discourse. Through its exaggerated images, popularization can force to the surface certain beliefs or ideologies that exist like latent dogma within technical language. Employing "recombinant" to describe the urban environment, for example, immediately calls upon visions of the city as involved in processes of self-assembly, design, centralized planning, all of which make use of given materials, stored in some type of "memory." This image of making and remaking according to specialized, centralized codes is a metaphorical elaboration of nothing less than the new biological view of the body, inaugurated since the unraveling of the DNA structure, a view that conceives this body as a vast computer system and which might be amply expressed (if a bit fatalistically) by such well-known statements as this one by François Jacob: "Biologists no longer study life today . . . instead they investigate the algorithms of the living world (1970, p. 299)."

What, to follow another example, of linking black holes to the deficit? What beliefs emerge here? The reading one might offer is not difficult or circumlocutive. Implied is the idea of a coming "disaster" in the economy, a sudden or imminent collapse that will be wreaked by implacable "laws" of cosmic dimensions, well beyond the limits of human control. Posited is a vast natural catastrophe, to which human beings can only respond as observers (victimized or not). Such, then, is contemporary cosmology in micro, based on the advent of a violent universe since Einstein—a cosmos not merely relative, but endlessly convulsive, cataclysmic, glamorously stimulated, teeming with gigantia, in a word, a universe *interested* in its own witness. Such projected beliefs, moreover, reflecting simultaneous excitement and unease (of grand proportions), extend even well beyond the expanding bounds of cosmology. At least one well-known philosopher of science, Gerald Holton, has taken special note of this:

> It struck me some time ago that since the turn of the century the terminology entering physical science often had a thematic root contrary to the older (and persisting) themata of hierarchy, continuity, and order. That is, the newer conceptions, perhaps corresponding to the characteristic style of our turbulent age, tend to be characterized by the antithetical thema of distintegration, discontinuity . . . dislocation, indeterminancy, uncertainty, . . . strangeness . . . negative states, forbidden transitions, particle annihilation. It is indeed the terminology of a restless, even violent, world. (Holton, 1986, p. 240)

Holton ends his passage here, perhaps a little puzzled, even slightly overwhelmed at the larger suggestions of his observation. He is on the verge of a different reading of scientific language altogether, not merely a historical one but a more deeply cultural one. The changeover in ideologies that he notes surely cannot be ignored for its broader implications. Indeed, physical science would seem now to carry a "themata" of its own century of historical context, one not merely full of disorder or breakage, but of threat, loss, and most of all death.

What this says, finally, is that the question of accessibility is far more complicated than might first appear. Scientific language, in many of its parts, is not really so "foreign" as our anthropological guide might hope to convince us. The terms that make it up and the rationalities embedded within it can hardly be considered as objective or as distant from ordinary cultural experience as is commonly believed. A great portion of this language—which, after all, *is* language and must therefore share in all the large realities of what languages are and do—can be looked into for many "external" contents, many connections to the "outside."

The "Who" of Technical Speech

A second modernist feature of scientific discourse, seeking to mitigate such plural contents, can be seen in how it chooses to speak. For what does one ordinarily hear?

Scientist or not, one hears the persona of univocity, unbroken statement, the single voice of the scientific style. But how achieved? How constructed? For the most part, through a series of grammatical and syntactic strategies that attempt to depersonalize, to objectify all premises, such that they seem to achieve the plane of ahistorical essence: "Recent advances have shown . . . "; "Analyses were performed . . . "; "The data, therefore, indicate . . . ." The narrative is driven by objects, whether these be phenomena, procedures, earlier studies, evidence, or whatever. These are the subjects that perform the crucial action, that absorb the responsibility for what happens. One therefore encounters a type of speech that appears to remove from itself any overt traces of both a private and a social process. There are no obvious signs of place in a specific culture. One is not allowed to say, for example, "When I sat down at my lab that rainy Friday morning, low on funds, sick with fever . . . ," only: "It was determined that. . . ." And no doubt the honesty of such an opening as: "With the money being poured into Star Wars–related research, we quickly switched priorities . . . ," or "Barely a step ahead of the Harvard group, we knew we had to . . ."—if used in place of "The present study examined . . ." or the like—would utterly demolish the context of the scientific altogether.

There are other strategies, too. A range of first-order, nonnegotiable grammatical codes also serve as conventions for the blank face of the narrative persona. Denotative statement, for example, by which nearly every sentence is meant to read like a simple declarative, is achieved by the banishment of anything overtly resembling literary technique, any concern for sound, rhythm, flourish, and so on. There is a minimal use of pronouns and a habitual reliance on transitive verbs, which acts in part to reduce the possibilities for syntactic complexity. Vernacular modifiers, especially those expressive of emotion, are either missing altogether or reduced to a few programmatic outbursts, utterly tamed by the demands of standard usage ("remarkable," "unprecedented," etc.). There is a reflexive use of the passive tense, often combined with a pleonastic repetition of important terms ("Areas of fracture density are shown on the accompanying fracture map; fracture strike has been quantitatively measured and plotted . . .").

As readers, we are therefore asked to perceive in every field of study only the purged voice of a coherent, harmonious, and successful investigation—an entirely opposite persona, one might note, to the theme pointed up by Holton above. What seems to appear is Truth, not a claim for it; the Scientist, not a particular individual; Data, not writing. We are not supposed to hear the local effort and struggle of a personal writer or group of writers, a research team with its complex hierarchy and competitive status, nor the larger determinants of scientific production itself, with its varied interests, its political motives and connections, its social locations, cultural priorities, its sexy or crisis topics.

At this point, one might feel the need to disagree. Are there not certain aspects of any scientific text that clearly show something relevant to the context of writing? For example, what of citations and references; what of the frequent listing of grant agencies; what of acknowledgments made to other colleagues; and so forth? Might not these types of structural ingredients offer a glimpse into a more sociopolitical world? The answer is both yes and no. Within a journal article, certainly, there occurs the frequent, punctuated use of citations. Here, it might appear, is evidence both of individual agreement and of conflict. Here are researchers embedding each other in their work, drawing on each other's prestige even as they try to establish or further their own. There are attempts at persuasion, at borrowing, stealing, or even diminishing the "reputational capital" of others, often at crucial points (e.g., of weakness) in an argument or hypothesis, where it needs some rescue support. All true enough—and yet, in a sense, irrelevant. For the broader and more constant image put forth in the overwhelming majority of articles or books, an image itself based on the appropriating of such "reputational capital" and to which an entire field of inquiry (called citation analysis) has come to be devoted, is that of a harmonious and supportive community ruled by unified respect, loyalty, and collegiality. Yes, there are conflicts, disagreements, even battles on this plane; indeed, errors both trivial and most foul are often discerned. But, in the end, all this is evened out by the code that demands each author not dwell on such things too much, on accusation itself, but instead direct his or her speech toward the reestablishment of communal truth and advancement. In those cases where an author is singled out for correction, for what is termed "negative reference"—for example, "The data of Bartle et al. appear erroneous for the following reasons . . ."—the conventions of scientific authorship require that treatment avoid any and all implications of true deviancy. The standing of such authors *as scientists* is never called into question. This is true, moreover, even in cases of fraud or theft (as a number of cases in recent years have shown). The offending party might be impugned, even openly, as a bad scientist, as having breached a code of ethics, or worse, but s/he is never stripped of the stripes of "scientist" altogether. Such cases, moreover, are the rare exception. Far more commonly, parties involved in a dispute are made to appear as if they are merely on the other side of a gentlemanly debate, or, more seriously, like lambs who have strayed from the flock. Their sins have been committed in good faith and without blood ("A more likely interpretation of such results is . . ."). This may be stretched at times, in the ordinary literature; the scratch of accusation, of implied inadequacy can be made ("These data seem misinterpreted due to a failure in . . ."). But somehow, at the last, the effort is a corrective one, only very seldom dismissive. Conventions of professional politeness are upheld ("We therefore offer what we believe is a better analysis of these same data . . ."). For, in the end, to overly demean or attack another scientist has less the meaning of airing dirty laundry than it does of casting doubt on the field itself.

Scientists, of course, know all this well enough and often comment on it in private. A particularly vicious and irreconcilable conflict between famous researchers, involving decades of work, dozens of student careers, aggressive competition for resources, and the like will most often appear in the published literature as a more or less gentle debate or disputation, a "forum for open discussion." Those on the inside of such debates know the code: a few calm words can connote paragraphs of near-invective. But the public surface, the professional face of science, never ruffles, thus providing one of the more profound yet ordinary fictions of this discourse. There is little place in technical writing for the overt and daily humiliations that take place in professional science—to bring these to the surface, with sometimes self-congratulatory abandon, has been the task of the mass media in recent decades.

Citations, then, provide a vision of Alexander Pope's famous dictum: "All discord, harmony not understood." While allowing and admitting conflict (what some might call the postmodern, deunified side to science), they nonetheless channel such conflict into a professional rhetoric that makes it appear familial. What, then, about other elements of the technical article that seem to speak of a social dimension? The common article today, true enough, has a complicated layout. By no means can it be called a simple text, stripped of all external references. On its title page, for example, it normally includes some mention of social locations, and therefore, if only by implication, of political/economic forces. The institutional affiliations for each author are usually provided. Granting agencies are named and thanked. Gratitude may also be expressed—in standard, passionless tones, of course—to those who have helped review the article or supply the author(s) with information. Dates of initial submission to the relevant journal and of final acceptance are commonly given. Such are among the formal elements that might be said to offer, now and then, a keyhole view into how and where a particular article came to be written.

Yet it is important to note the structural position and status of these elements. They are typically exiled to the bottom of the page, or to the very end of the article, and invariably appear in much smaller type than that of any other part of the text. Sociologically, they are truly "the fine print." But the message is clear: they constitute something of reduced significance, a minor footnote providing a peripheral (and mostly internal) accountability, something that occurs both outside and beneath the more important content offered, the science.

To make things still more evident, one might consider the title page, gateway to this science. This, too, is built to reveal a detailed hierarchy of significance. At the top, for example, there appears the name of the article, in bold, headline type. Next below comes the author's name or their names, followed by an abstract, often again in bold; and finally, the main body of the text. Depending on the journal, authors' names may appear in the smallest print of all in these title-page elements or they may be printed in a type size just a bit smaller than that for the title. There is never a "by" in front of them. Each name usually has a superscript, a number or a letter, that refers to that particular author's institutional affiliation, given at the bottom of the page. He or she is thus part of the text, part of an institution, again part of science.

Every article closes with a list of references. Almost never, even in the longest examples, are any of these from a nonscientific source. Indeed, few are from fields not directly related to the research or commentary at hand. The vision this implies of what knowledge is needed to perform good science comes back to a type of pure, sealed professionalism, with no need or interest regarding outside nutrients. The references themselves, furthermore, often appear in the form of a standardized code—for example, *J. Vac. Sci. Tech.* **255**, 2203—frequently intelligible only to insiders (if there were a legitimate way to abbreviate people's names, this too would probably be done for authors). The total image is therefore not merely one of shared interest or prestige among workers. Nor is the view we get from this confined to an apparent sign of originality, if very few references are given, or a mixture of thoroughness and timidity, if a great many are supplied. The reference list is instead akin to a documentary of a local athenaeum. It is a roll-call of select workers, all of whom are laboring toward common goals, each adding his or her precisely shaped brick to the walls of intellectual community. It is this, then, that closes the gate at the end of each entrance into science.

🖫🖫

Some of these perceptions have been noted before, in various ways, not the least by scientists themselves. Nearly 30 years ago, a BBC broadcast by Nobel laureate Peter Medawar caused some stir by asking (in its very title) "Is the Scientific Paper a Fraud?" and then going on to answer in the affirmative (see Medawar, 1990). At the time, Medawar blew a whistle (and he knew it). He noted that the standards for writing a technical article put forth a false image of scientific labor, one that proposed the researcher following through a smooth Baconian process of inductive information gathering, objective observation, and final conclusion making. Research, said Medawar, is intensely biased by theory, and is in any case a messy business, full of groping and guessing. "Proof"—for a result or a discovery—is very often something that is added later on, a kind of constructed, whodunit logic imposed in part by the standard structure of the scientific article. Medawar's conclusion was that a great deal of the human reality of doing science is cleaned away in the act of writing, in the creation of science as a finished product. Medawar's article attracted strong negative responses from scientists at the time. Their professional image was at stake. Of course, what he said was obvious to any insider; his real crime was to bring such understanding into the open, to expose the truth that science, too, depended on the writing of fiction for its embedded authority.

Another example along these same lines, perhaps the most famous of all, is offered by James Watson's *The Double Helix* (first published 1968). Despite everything else that might be said about this publishing milestone, it should be noted that the author himself freely (if somewhat disingenuously) admits, almost as a footnote, how the whole Sturm und Drang surrounding this famous episode of discovery, with its multiple levels of personal, institutional, and international rivalry between British and American researchers, winds up in a published result whose deadpan beginning can only be seen as a type of oblivion: "We wish to suggest a structure for the salt of deoxyribose nucleic acid (DNA)" (see Watson, 1980, p. 237). *The Double Helix*, in a sense, can only be viewed as an attempt to reclaim—in heightened and dramatic form—what is obliterated in this single line.[4]

Since the late 1960s, meanwhile, many academic authors, mainly sociologists, have gone much further in documenting this type of discrepancy as a process. Studies of this kind, for example, have traveled no small distance to show the enormous amount of activity, influence, thought, and interaction that is progressively erased or rendered irrelevant during the actual writing of particular articles (e.g., Knorr-Cetina, 1981; Latour and Woolgar, 1979; Latour, 1987). Among such lost elements studied are many pointed up by Medawar and Watson: guesswork and false starts; abandoned paths of logic; conversations and battles among participants and others; actual moments of insight; the pressure of deadlines; any and all emotional states or problems; and, of course, the detailed trials and tribulations of research itself.

Some of this, it seems to me, is less an issue of scientific discourse being "fraudulent" than of it being a form of authorship. Leaving out many aspects of the unsuccessful or the unexpected, eliminating the messiness, the trial-and-error qualities of the rough draft, rearranging and rewriting the whole into a coherent story, upholding, in other words, an image of the writer as eminently rational and in control—this is all simply, yet profoundly, an integral part of composition itself. All writing, whether of novels, poems, musical scores, or technical articles, seeks to idealize its author in this way. A certain confessional dimension is lacking, purposely left out, so that a sacred idol can remain standing: the Author, a fictional perfection (see Foucault, 1979). No less than science, literature in its published form usually provides scant evidence of the social, political, economic, and personal forces that have led to its creation. What novel admits, openly, that it is trying to take advantage of a particular public sentiment? What realm of "art" today will acknowledge its tendency to pander to given tastes or to copy styles and subjects already proved successful (how easy it would be to give examples!). Where science seeks to provide a heroic objectvity, literature intends a heroic subjectivity. Both, however, desire something of the heroic. And yet, at the same time, who invades the novelist's study, what ethnographers, sociologists, or anthropologists follow every movement of his/her pen or cursor, as if observing a "foreign" species? The literary biographer is not of this type: s/he is more of an archaeologist, whose province is the kingdom of the dead. The difference, of course, is that science has long claimed a truth in which none of the things of politics and culture matter. Moreover, its knowledge, by virtue of its very material power in the world, deserves far more to be a contested and analyzed territory.

Much of the analysis that seeks to put back the many lost realities of scientific labor therefore seems valuable, important, necessary. To document what is missing from the scientific article, how it is built and how digested by its audience, forms a sizeable portion of what is now known as "science and technology studies" (STS) and is doubtless a relatively new and helpful way to comprehend the labor of scientists (who are, after all, producers and power brokers of tremendous import in the realm of knowledge). Today this work seeks to go beyond exposing many everyday "secrets" familiar to most scientists (if only intuitively), and to search out other more detailed and more general aspects showing how science is actually performed and how its knowledge is continually being constructed, shaped, and revised. In one writer's summary, it has "sought to open the black box, to reveal the uncertainties, the negotiations, the dilemmas and controversies that inform, not exceptionally but as a matter of course, the very making of 'science'" (Webster, 1991, p. 15).

To a degree, however, there is irony in this. That such studies have had to be done—that there has grown up an entire field of literary biography dedicated to revealing the social aspects of technical discourse—can be said to reflect the overall historical power of this speech, which for so long dictated a reading of it as objective and neutral, without important external content. Breaking down the doors of common sense has meant defeating this power in some measure, returning to scientific work its most evident aspects as a labor of human beings. But this act of saviorship seems in large part a limited and self-referential one. Written by scholars, for other scholars, without any real public intent, it cannot but often remain parasitic on its nemesis, which always looms in the midground as a negative yet required master of ceremonies. Eternally invoked, as just seen, is the old black box model, a vision in which many scientists no longer believe, yet which is posed as the "standard version" that must, time and again, be revealed as fraud. As a spur to inquiry and to proclamations of new insight, it must never be defeated entirely or allowed to retreat. One wonders, is there room for self-exhaustion in this type of scholarship? Perhaps the curtain can only be pulled back and the exclamation "Aha!" delivered so many times before the audience begins to feel restless, moves on, demands a refund, or else a new act. Perhaps this is one reason why some strains of thought in this area have moved toward an anxious and unnecessary extreme, claiming, for example (as the constructivists often do), that there is no such thing as "truth" or "nature" at all, or that the natural world plays no central, adjudicating role in the building of scientific forms of knowledge. For a host of professionals in many fields, say, for example, geology or marine biology, who spend their professional lives in the field, touching, handling, smelling, collecting natural materials, this type of statement can only sound like drawing-room silliness, a matter of intellectual cognac that evaporates if not drunk quickly enough (and in sufficient quantities). In fact, it is part of a distinct prejudice among many constructivists, directly inherited from positivistic notions, that all real science takes place in the laboratory, through experiment. In such settings, "nature" might seem an abstraction, a matter of language and language making only. But the whole of science is enormously larger than this. Indeed, it would be fascinating to examine exactly how different the idea of "nature" or "natural phenomena" is distributed among a range of different fields. But this type of study is not what the STS bandwagon calls for.

As it happens, the black box image *is* one that remains current for a fair majority of the public, for media science writers to some degree, for government officials, and for science teachers as well. More, it is a model whose historical power has been entirely public: this is the meaning of its modernism. Belief in an impartial, objective science, or at least in its possibility, still exists in most quarters of society today. People today may well understand that scientists fight and scratch among themselves, that they are not always models of honesty and integrity, even that they are sometimes locked in battles for fame. But somehow all this still remains detached from the idea of science itself, as knowledge. To the degree that this belief helped socialize the scientist's reading of his/her own language in the past, it has continued to do so for the larger image of science among everyone else too. To open the box, in almost any sense, is therefore a corresponding act of historical and public significance. It seems a fair question to ask of the new scholarship, therefore, if some part of its goal is not so oriented. Is it indeed for society in general that this work is conceived and written? Is it to non-STS insiders that its results are directed? Unfortunately, the answer is no. Popularization might be taken on as a topic, ripe for study, but it is almost never engaged in, as an act of commitment. As yet, most of the understanding that has come out of academic analyses of technical discourse has no real advocacy, no social role beyond simply generating more work for scholars, carving niches and filling them. However much it may lay claim to interdisciplinarianism, it remains intrepidly disciplinarian in the larger sense, prey to what one might call the timid roar of the thesis. Like scientific articles, it seems written only for other professionals, other producers who are also the sole consumers. Too often, it is not merely parasitic on the body scientific but imitative of its insidership as well. This is why, too, I believe, that it is often on the prowl for what are called essentialist explanations—the final nature of "science" and the "scientific"—which, in the end, only return us to the modernist surface with which we started.[5]

The "Who" of Authorship

One final way of rephrasing Sartre's question is to ask who the author, as author, might be in this discourse, what type of persona s/he must assume. This seems either a very simple question to answer, or else a very difficult one.

In technical writing, after all, the personal dimension to authorship is reduced to a minimum. One may speak of literary strategies, rhetorical moves, or whatever, but at the level of actual experience these things are constrained by conventions that occasion little excitement or aesthetic pleasure in an individual author. As a personal act, this type of composition demands a kind of death of the self, a literary annihilation. This is because, when taken in total, the codes involved seek to hold under severe pressure the biographical (emotional, experiential) writer and thereby condense around a spectacular refusal of human subjectivity. The fundamental tactic, as Barthes once said, is to make content everything, expression nothing, to make all as instrumental as possible, the output of a generic producer. As author, the scientist must largely become "an anonymity, an absence, a blank space."[6] And in truth, any point at which there emerges something resembling a truly personal or literary style in a technical article is commonly considered to be a point of *failure*, when required standards are transgressed and "scientific" discourse begins to break down. Among the scientific community, the personal excites a degree of suspicion, even discomfort or disdain.

We might then ask: What about when this language is spoken, say during a lecture or at a conference? Isn't this where a return of the singular person must inevitably occur? Ordinarily, yes. But this is a discourse which is first and last a *written* one; in speaking it, an individual is not necessarily speaking a language of his/her own. At the extreme, they are merely providing a local face to a voice that requires no specific body, a voice both communal and recitative.

If one remains true to the demands of such expression, one cannot help but be its transitive messenger or vessel. One cannot avoid practicing a repressive sobriety, suspending or minimizing many of the natural components of personal speech itself: play, humor, exaggeration, diversion, excitement, anger, any of which are possible only when the codes of this language are warped, broken, or subjected to some sort of relieving interruption. Whether writer or speaker, the requirements are the same: one "is debarred more by convention than necessity from using any of the devices that lead to graciousness in [expression]" (Savory, 1967, p. 57).

In the universe of science, language is very rarely an object for inquiry or experiment. Or when it is, the invention tends to begin and end in terminology alone, in the coining of new names or vocabulary, never extending its desire to the "how" of writing itself. On the whole, language remains instead a secondary medium, an application employed according to rules and laws that seek to give it the lithic poise of a solved equation, the finality of a function with all its variables plugged in. Even the most startling or revolutionary discoveries—those uncovering the existence of a new fundamental particle, active volcanism on a Jovian moon, a grand theory such as that of plate tectonics—must, without fail, be draped in the same drab coat as if they were the most ordinary and uneventful presentations of reworked data. At the most, one is offered a language that seems caged beneath layers of understatement. Watson and Crick, in their seminal paper on DNA, felt bound to confess no more than: "this structure has novel features which are of considerable biological interest" (Watson, 1980, p. 237). However coy this declaration may have been, the irony of understatement here loses force by the very fact that it depends strictly on obedience to a given code of oblivion and erasure. To tell truths of enormous power and import without excitement, fear, or wonder: such is the standard.

Certain thoughtful writers, Stephen Toulmin for one, have maintained that during the past two decades contemporary science has become more and more subjectivized, that is, less dependent on the mythology of the detached and neutral observer, more willing to admit the truth of probable rather than final knowledge. Yet if this is so, it presents an interesting problem. This is because, as any perusal of technical books and articles written over the past half-century would show, the language of scientific writing has tended, on the whole, to travel the exact opposite path. Though often qualified, as in the past, by the use of the conditional tense, if clauses, and the like, it has more broadly continued to eliminate the pronoun "I" from most writing, to increase the abundance and density of its jargon, and to flatten its pitch down to a near-monotone. Less subjectivized, it has also become more univariant, more detached from the writer *as writer*, less tolerant of articulatory modes from other areas of human endeavor. If today much of science appears Einsteinian in sensibility, it nonetheless remains Newtonian in expression. As it stands now, the old authoritarian determinisms, the twice-told tales of neutrality, dispassion, "finality without end" (Kant) all remain hardened like resistant fossils within this genre of writing.

One might well wonder at certain general effects of this. For example, a great deal of space has been devoted to bemoaning the fact that a majority of scientists and engineers do not like to write and do not, from any point of view, write well. Many are said to actually fear and hate it. Most seem to feel that writing is a necessary chore, at best. There is, in fact, an entire postwar pedagogical tradition devoted to the attempted solution of this problem, to convincing the average scientist that "writing skills" are both essential and admirable. Many colleges, universities, and other educational organizations now have regular courses oriented toward trying to teach technical professionals "how to write clearly and effectively." Always, therefore, a central question hovers in the background. One of the more popular texts, in fact, felt impelled to bring it up:

> I ask you, as I have often asked myself, why it is that so many scientists, while capable of brilliant performance in the laboratory, write papers that would be given failing marks in a seventh-grade composition class. . . . Unfortunately, I do not know . . . and I doubt that anyone does. Perhaps there are no "answers." (Day, 1979, pp. 157–158)

But of course there are answers, as indeed there must be. And one of the most important seems to me as purloined as the proverbial letter in Poe's tale (a literary oversight, after all). For, given everything I have mentioned above, given what technical writing has become in terms of style, one is not a little tempted to ask, in turn, if a bit perversely: Why is all this "failure" surprising? How, in fact, could it be otherwise? Since this form of writing is neither particularly interested in language (in an expressive sense) nor in the writer (in an individual sense), how can it be expected to motivate linguistic interest or expertise? How can it not be a chore for so many, a task for even the most competent?

If this seems simplistic consider: there can be little doubt (and there should never be) that lab and field work can be exciting, interesting, sometimes full of unexpected turns and developments, demanding of routine ingenuity and off-the-cuff inventiveness. Very rarely is it the straightforward cookbook activity so often taught (to obvious detriment) at the undergraduate level. It requires, so to speak, the direct and immediate involvement of a responsive, thinking, deciding, and therefore emotive self, without whose presence it would very often break down altogether or at best yield unimportant or unusable results. "Brilliant performance" in this setting is always a personal thing; it represents one area where the individual can truly *add* him-herself to the "scientific." And yet, as I have said, writing up such work has little room for any of this; indeed, as an act, it involves exactly the opposite movement, a systematic banishment of one's personal existence. In truth, as the supposed culmination of all labor, this act of writing makes for a sad denouement to the possibilities for rich intellectual experience everywhere else.

An old adage suggests that clumsiness comes from lack of pleasure. But the irony here is that scientists are double victims: at once discouraged from the pleasures of writing, yet, at the same time, damned for the inevitable result. "To learn to write well," says our author above, "you should read good writing. Read your professional journals, but also read Shakespeare." The absurdity of this type of advice, whatever its stereotypic appeal, can only add insult to accusation. (If Shakespeare has any direct comportment with technical writing today, it is certainly not in the form of literary example or teacher, but instead something closer to that of a historical beginning point for a long, negative evolution.)

In summary, I have dealt (only glancingly, to be sure) with three central tactics of scientific discourse that help compose its "modernist" face: its ability to split the speaking world, its erasure of origins and influences, and its repression of the individual writer. Such things help define both the purified, decultured appearance of scientific language and the reasons for its intimidating power. If it appears fascinating to some—among them not a few of the more sophisticated literary critics and writers on language—this is because of a particular seduction to this power, and the pleasures of being overwhelmed and uninvolved. Such fascination, I would say, comes to those who are not users, who are relieved from the responsibilities of use. It is the flip side to the ethnographic/sociological view, which is equally entranced by such power but intent on its submission to another, to "analysis" and "method." For both types of writers, there is no reason—unless, of course, one accepts the global influence of this discourse as a linguistic model and its real-world force as a carrier of technological knowledge—to reject the spell and spectacle of its incantatory magic or to make this magic the subject of a broader critique and awareness among the public. To either group, it is essential that scientific language retain its thrill, its quality of being a universe apart.

So this, too, is perhaps why one should be suspicious. From an insider's point of view, after all, the writing and reading of technical language—as it demands to be read and written—is anything but spectacular or enchanting. If it is far more than a mere practical shorthand, a clean and efficient passing of tokens between specialists, this still does not make it aesthetically and personally an exciting form of language of which to be an author.

IV

According to the Russian theorist, Mikhail Bakhtin, all forms of discourse need to be thought of as *dialogic*. This means that they can only exist in a relationship of dialogue, both with a reader (intended or not) and with other discourses on related topics, past and future, spoken and written. Writing, in other words, never occurs in isolation, but is always a kind of dynamic affiliation that changes through time. According to Bakhtin, "Discourse lives, as it were, beyond itself . . . ; if we detach ourselves completely from this [connection], all we have left is the naked corpse of the word, from which we can learn nothing at all about the social situation or the fate of a given word in life" (1981, p. 292). This basic principle, so central to Russian structuralism, was used by Bahktin to develop a scheme for delineating how natural science and the humanities differ in their approach to language.

It is worth examining this scheme in terms of its broad principles for two seemingly opposite reasons. On the one hand, it highlights the importance of looking back at some of the general historical changes that have taken place within scientific writing over the past several centuries, especially how it has evolved and what it has lost. On the other hand, it proposes an understanding of technical language that is seriously and interestingly flawed, one that gives particular integrity to the standard black box view. This makes it useful; as a model for important error, Bakhtin's scheme can therefore be used as a lens for focusing critique.

According to Bakhtin's scheme, the human sciences carry on a far more constant and endemic dialogue with culture than do the natural sciences. This is because their objects of study are other forms of discourse, such as documents, literature, art, music, criticism, and so on. They are concerned with other "speaking selves," that is, "not just man, but man as a producer of texts" (Todorov, 1984, p. 17). Accordingly, the dialogic of the humanities can be called full, as it involves direct commerce with other types of speech and other voices. Natural science, however, Bakhtin wants to identify as requiring a true thing for inquiry, a nontext. The goal, linguistically, is to *give* speech to such an object, to discover it as a reason for language. There thus appears to be only one speaker involved in this inquiry: the scientist, who is both reader and writer. No other voices are allowed full entrance or standing: discourse becomes *monological* (science talks to itself and to no one else). Put

differently, in the words of J. F. Lyotard, "Scientific knowledge requires
that one language game be retained and all others excluded. . . . [It is] in
this way set apart from the language games that combine to form the
social bond . . . being no longer a direct and shared component of the
bond" (1979, p. 25).

True as this type of isolation might appear today, it could not be
assumed in the past. Before the end of the 19th century, other discourses
were a regular and fully overt part of scientific speech. Works of classical philosophy, quotations from the Bible, the words of artists or poets,
passages of personal reflection, of aphoristic wit or rhetorical humility—all these and many other forms as well were routinely employed in
technical papers. The works of Bacon, Galileo, Leibnitz, Newton, Buffon,
Humboldt—to name only a very few of the more well-known scientists
of the day—are simply full of such elements, indicating that the idea of
"science" itself was far more universal in terms of expression, far less
inhibited in terms of its willingness to borrow from other forms of discourse than it has since become. Such were the days when scientists were
known as "natural philosophers"; this title alone suggests the wide range
of different expressive means that a science writer might ordinarily draw
upon. Indeed, at the time of the Renaissance, when Copernicus composed his famous *De revolutionibus* (1543), differences between science,
philosophy, and literature hardly existed at the level of basic discourse.
It was common practice for both scientists and humanists to address a
reader directly, and to persuade through the standard use of citations
from classical *auctores*:

> Why should we hesitate to grant [the Earth's surface] a motion, natu-
> ral and corresponding to its form. . . . And why are we not willing
> to acknowledge that the appearance of a daily revolution belongs to
> the heavens, its actuality to the earth? The relation is similar to that
> of which Virgil's Aeneas says: "We sail out of the harbor, and the
> countries and cities recede." (see Shapley, Rapport, and Wright, 1958,
> p. 57)

A full 300 years later, and this practice is still widespread—though admittedly less a required element now, a holdover from medieval Scholasticism, and more an option employed for added effect and expressivness,
as shown in the work of such scientists as Charles Lyell, Georges Cuvier,
Pierre Laplace, and Thomas Huxley. Moreover, in a number of scientific
fields, persuasive techniques today often consigned to the realm of literature were employed in the domain of technical expression. Travel
writing, the conscious employment of metaphor, and dramatic effect were

common among these, as shown by the following example from Alexander von Humboldt's "Essay on New Spain" (1810):

> Descending from the central flat toward the coasts of the Pacific ocean, a vast plain extends from the hills of Aguasarco to the villages of Toipa and Patalan, equally celebrated for their fine cotton plantations. Between the picachos del Mortero and the cerras de las Cuevas and de Cuiche, this plain is only from 750 to 800 meters above the level of the sea. Basaltic hills rise in the midst of a country, in which porphyry with a base of greenstone predominates. Their summits are crowned with oaks always in verdure. . . . This beautiful vegetation forms a singular contrast with the arid plain, which has been laid waste by volcanic fire. (quoted in Mather and Mason, 1939, p. 179)

Use of personal, even of poetic, language in the midst of a scientific contribution was still an expected discursive phenomenon in many fields during the 19th century. On the evidence of their works, authors of science did not view themselves as a linguistic species separated from the writers of literature and nonfiction elsewhere. Darwin, for example, like many of his contemporaries, wrote scientific prose that could be read by a wide-ranging audience *and* that was technically legitimate in the eyes of fellow geologists, zoologists and botanists. Or better put, he wrote at a time when literature and science borrowed equally from each other's storehouse of persuasion, for both were partners in the larger field of Authorship and Readership.

🔳🔳

The record therefore stands clear: scientific discourse has only very recently driven all obvious nontechnical types of speech and authorship from its literary door, in the process largely giving up the love of writing for its own sake. The process has taken a long time (I will examine it briefly in a moment), but its effects can hardly be denied.

One such effect, seldom if ever mentioned in this context, yet of striking and momentous import, stands out with regard to the writings of Sigmund Freud. Freudian language has long posed a problem for scholars: is it scientific or something else, "technical" or "literary"? In large part, however, the question revolves on the standard translation of Freud's works into English. Bruno Bettelheim's pointed essay, *Freud and Man's Soul* (1982), was one of the first to make clear and public the outlines of this enormous effort at reconstitution, to point up what almost any third-year German student knows: that an elegant and reflective style, modeled on Goethe, praised by Thomas Mann and

Hermann Hesse, has been carefully recast into a cold, arid, jargon-filled prose—a strategic recasting of Freud's original, creative discourse into the very "scientific" model it was apparently intended to exceed and possibly overcome. Expertly deleted or revised in the English translation are such elements of Freud's style as his extensive use of wordplay, his sardonic humor and complexity of syntax, many of his frequent classical allusions, his nuanced changes of meaning for single terms—in short, everything that contributes toward the sophistication and introspection of his style, its intended grace. Most telling of all, perhaps, is the utter loss of much everyday language, which Freud very consciously employed: to mention only the most obvious examples, *das Ich* and *das Es*—the "I" and the "it"—changed into the Latinated "ego" and "id." Politically loaded terms such as *Massenpsychologie* ("mass psychology," or "psychology of the masses"), were neutralized into "group psychology." The religious connotations of *Seele* ("soul") were erased into "mind" or "psyche." And so on.

Why was this done? What was its purpose? Few writers, until quite recently, dared speculate on the matter, let alone make even the most obvious of connections. And this despite the well-known aggressiveness of Ernest Jones to possess Freud for the English-speaking world and to oversee his penetration into existing psychiatric and psychological circles. But one sees the answer today through its institutional result. Psychoanalysis in England and the United States models itself as a science, and has done so for the last 70 years. Until a few years ago, to become an analyst one was required to take a medical degree, to complete a residency, and to become a doctor in the fullest sense of training. Was Freud really a scientist? Is psychoanalysis indeed a technical discipline, comprising truths derived and tested from nature? Such questions are endlessly debated and never answered. But in the world of daily practice, of institutional power and reality, they were long ago decided. The Strachey English edition of Freud's total *oeuvre*, translated in the 1950s and 60s, was intended to finalize a Freud who had started to be re-invented decades earlier. In large part, Strachey simply adopted and expanded upon a technical vocabulary and style of translation established by Jones during the 1910s and 20s, during a time when an enormous preference existed for positivistic styles of knowledge in the English-speaking countries. The literary Freud, meanwhile—the Freud who clearly had the greatest influence on continental Europe until much later—was brutally suppressed, deleted in effect; a schizoid reality has been established around him. In the end, one is therefore urged to ask, as any good psychoanalyst does, just who this "Freud" is with whom we have been possessed for so many years? How is "he" related to that true self, that authorial self, the man who lived and wrote during the infantile stages of the current century? How has he too, as ideological creation, carried into our heads certain visions of "the scientific observer," particularly over the human subject and his/her chance for reform? These questions (and others) have been taken up in detail in Chapter 6.

V

Reading a random sample of scientific discourse written during the last century and a half one becomes witness to a process of distillation, in which various genres of literary expression are successively labeled non-scientific and banished. It is like a grand effort of decontamination, in which everything that does not finally contribute toward a passionless and purely factual "science" becomes volatilized.

In the 1830s, for example, it was perfectly expected that an author such as Charles Lyell might speak of ethical and philosophical matters in his famous *Principles of Geology*:

> In an early stage of advancement, when a great number of natural appearances are unintelligible, an eclipse, an earthquake, a flood, or the approach of a comet . . . [they are all] regarded as prodigies. The same delusion prevails as to moral phenomena, and many of these are ascribed to the intervention of demons, ghosts, witches, and other immaterial and supernatural agents. By degrees, many of the enigmas of the moral and physical world are explained, and, instead of being due to extrinsic and irregular causes, they are found to depend on fixed and invariable laws. (1990, I, p. 75)

By the turn of the century, with the full professionalization of science and its shift from the writing of books to that of journal articles, this type of reflection is largely absent, if not entirely gone, from scientific discourse. As a scientist, one no longer writes of moral matters or offers analogies for philosophical reflection. But while these things are now long gone, the forms of their eloquence and the declarations of their reasoning are not:

> The final results of my investigations in the Durness-Eriboll region during August last seem to me to indicate most distinctly the probable truth of the theory which has long appeared to myself to be the

> only possible solution of the Highland difficulty. I believe that we
> have in the . . . Highlands of Scotland a portion of an old mountain
> system, formed of a complex of rock formations of very different
> geological ages. These have been crushed and crumpled together by
> excessive lateral pressure, locally inverted, profoundly dislocated,
> and partially metamorphosed. This mountain range . . . is of such
> vast geological antiquity, that all its superior portions have long since
> been removed by denudation, so that, as a general rule, only its inte-
> rior and most complicated portions are preserved for us. (Charles
> Lapworth, "The Secret of the Highlands," 1883 quoted in Mather
> and Mason, 1939, p. 579)

Like Lyell, Lapworth the author is deeply within his own text as a living
"I." He feels no hesitation about telling his reader about his *belief*. One
might note that his paper, like Lyell's *Principles*, was a pathbreaking
work in the history of geology, and thus his literary skill, which sacrifices nothing of clarity for expressive fullness, nothing of information
for accessibility, seems all the more accomplished in the light of present-
day technical prose. For what might we expect from a prestigious geo-
scientist writing today? Surely, in its most accessible form, something of
this order:

> Cycles of relative change of sea level on a global scale are evident
> throughout Phanerozoic time (Vail et al., 1977). The evidence is based
> on the facts that many regional cycles determined on different conti-
> nents are simultaneous, and that the relative magnitudes of the changes
> generally are similar. Because global cycles are records of geotectonic,
> glacial, and other large-scale processes, they reflect major events of
> Phanerozoic history. (Vail and Mitchum, 1979, p.469)

What remains? A speech almost totally anonymous, like a proteinous
extract from the *logos* itself. Today, the "I" of the past, where allowed
to return, is relegated to preliminary and inessential remarks, either in a
preface or else a brief introduction, and this in books, not articles (e.g.,
a commemorative volume of some sort). There are exceptions, to be
sure. Major episodes of conceptual shift within a discipline, as during
the introduction of the theory of plate tectonics during the late 1960s
and early 1970s, can sometimes open a breach, allowing for surface expressions of personal belief or opinion. Particularly heated rivalries or
debates within the literature may do the same, where one sometimes
hears a private voice raised in celebration, contradiction, or perhaps
even chastisement:

"Homo sapiens" . . . must see facts as part of a pattern, of a story, a myth. In modern scientific jargon, myth-making is called model-building, and although current myths (models) consist of more factual information than those of old, still they contain too much imagination. This applies particularly to the concepts of mountain building. (Laubscher, 1969, p. 551)

In the appropriate context, then, scientists recognize such writing as a sort of welcome deviance, a forgiveable or even appealing truancy. A moment of interest is returned to language. There is the potential for pleasure, and it is a pleasure that comes directly from the awareness of a difference from the norm. Endemic sobrieties are lifted; one is drawn in, urged to respond, right there, on the spot, being touched by writing as a consciously personal act, as conscious of itself.

But this has its message as well. For it is through just such a *coupure* that one realizes (again) how reflexive the standard scientific style has become—and, sadly enough, how little of the scientific life has a place within its overt forms of expression. With respect to this human side—which is as full of passion, partisanship, theft, inspiration, suffering, in a word, creative desire, as any endeavor of invention or discovery must be—there is but little vocabulary and less space. Historically and linguistically, these have all been forced to go underground, as it were, to become visible not at the level of comment, of statement or opinion, but rather (as I will discuss in a moment) in other, largely hidden, aspects of narrative and meaning.

One would hope, as Jeremy Bernstein has written, that "it is now a cliché that scientists are human beings" (1982, p. 251), and therefore that it has always been false and demeaning to assume that the aesthetic experience is an alien thing to scientific work on any level or at any scale. But one should be frank: if a majority of scientists and engineers still reflexively downplay, diminish, or simply don't discuss the aesthetic sensibility in their work, if they prefer to abide by the age-old image of "the scientist," with its implication of passionless reason and logic, is it not at least in some part because the discourse in which they find themselves immersed, and which they must endlessly consume, metabolize, and produce for the duration of their professional lives, has qualities within it that embarrass this kind of expression at every turn? One need not believe in the Whorf theory (the idea of linguist Benjamin Whorf that language determines thought) to find common sense in such a question. Dispassion, neutrality, detachment, intellectual purity, objectivity—are these not the myths, after all, that would demand an expert who is immune or irrelevant to psychology?

VI

The victory of jargon over the past two centuries has come about for a variety of reasons. It is often said to have gone hand in hand with cognitive removal in the natural sciences. As the realm of human experience has been left behind, the need to invent or modify language has grown, and with it, the necessity to compress speech so that there is constantly new room for an ever increasing number of terms.

During the last 100 years, there has come as well what can only be called a loss of linguistic sophistication at many levels of speech and writing, what Steiner calls a "retreat from the word" (1984, p. 283). The relationship is a difficult and multilayered one, at least partly related to growing shifts in what Michel Foucault has termed the "archaeology of knowledge," in this case largely due to the professionalization of knowledge and the uptake of scientific discourse as a model form of truthful communication. Whatever the detailed reasons, written language has been simplified and reduced in many ways. Recalling the past eloquence of science itself, one can only regret that "literature is alone today in bearing the responsibility for language" (Barthes, 1973, p. 120).

A further lament, then: since technical discourse has given up much of its own experimentalism, great chances have been missed. Given, for example, the infinite richness of the world's mythologies, it can only seem bland and unimaginative (not to mention imperialistic) to have so rigidly colonized the solar system with the Greco-Roman tradition. Until the very recent images sent back by the *Voyager* spacecraft—images which, gratefully one now thinks, gave rise to an exploded demand for new names on many of the Jovian and Saturnian satellites—one would have been fully justified in asking why there is no "room" for the gods and heroes of China, Japan, or India in the skies (science fiction having continually proved itself more imaginative on this score)?

And in truth, the largely latinated surfaces of the Moon, Mars, and now Venus are almost totally unintelligible and unpronounceable for most people—and especially non-Western peoples—hopelessly distant and cold to the literate imagination of the vast majority of humanity. *Arsia Mons, Vallis Marineris, Mare Aestatis, Taruntius, Palus Epidemiarum*—who, we should ask, are these names intended for? For whom will they suggest the striking geographies of the planets within and beyond the orbit of our own? If anything should be a "polyphony," a memorable diversity of voices, shouldn't it be the heavens, in which all the world's cultures have, since the earliest ages, found their frightened and beautified reflection?

The point, however, should not be pushed too far. For such a polyphony has recently come to exist, though in selective fashion. New discoveries concerning the surface features of other objects in the solar system, particularly from *Voyager*, have led to some striking and, in part, welcome changes in this area. Features recently identified and mapped on moons such as Io, Ganymede, Callisto, Enceladus, and Rhea reveal a wealth of names taken from the pantheons of Arabian, ancient Egyptian, Persian, and—to a lesser degree—Japanese and Chinese mythologies. To the craters of Mercury have been applied the names of famous writers, musicians, and artists: Balzac, Basho, Beethoven, Goya, Hawthorne, Eitoku, Liang K'ai, Mark Twain, Sayat-Nova, Surikov, Valmiki, Phidias, to name but a few among more than a hundred.

And yet, despite everything, one cannot help but note certain limitations inherent in the new naming schemes. For example, only dead and ancient religions are utilized (the Occidental idea of the "archaeologic"). Those bodies named after specific texts (e.g. the Saturnian moons, Mimas, Japetus, and Enceladus, named after Malory's *Morte d'Arthur*, the *Song of Roland,* and *A Thousand and One Nights*) utilize only those of Occidental origin or of traditional fame in the west. The six major satellites of Uranus are all from Shakespeare's *A Midsummer Night's Dream* (Miranda, Ariel, Umbriel, Titania, Oberon, and Puck). In some sense, therefore, a grand departure from the past is not to be found here. The birth of new names, and the responsibility for it, turns back at some point to a celebration of a single cultural vocabulary.

VII

For all these reasons, then, one feels that beneath its poise of rational dignity, scientific discourse has come to embody a certain distrust and fear of language. "The danger of using words" for this discourse, after all, lies in their *innate* ambiguity, their natural ability for carrying more than one meaning at the same time. If we were to say (with Lacan, and many feminist writers) that all discourse is charged with gender standards and representations, we would have to say that the speech of science is inclusively, even ideally male. We might even call it the voice of a

Protestant deity (in his more calm and demanding moments) whose desire for order, ascetic restraint, bounded community, and obedience is absolute.

It cannot be an utter surprise that this discourse has other effects, even on those who have taken it to task. One of these is a particular kind of anti-intellectualism. For example, regarding questions of language, we find even so excellent a writer on science as Peter Medawar proclaiming in a recent work: "I have long thought that nearly all books on nearly all subjects—especially the philosophical—are much too long" (1984, p. xi). Here, in short, is the backhand slice of Occam's razor: everything (worth saying) can somehow be distilled into an alembic of elegant candor. Length therefore becomes equal to waffle and verbiage; brevity, the soul of truth ("wit" no longer having a place). Here we are, then, dragged back into the grade school classroom, our pants pulled down, threatened with the "clear and concise" ruler wielded with schoolmarm zealotry. Here, too, is a result of taking the standards embodied in scientific speech too literally, too empirically, too much at their word.

⑤⑤.

I have mentioned, somewhat figuratively, the gender aspect to scientific discourse. It is interesting in this respect to note that when the topic of words and their "native ambiguity" has been brought up by various scientist-authors during the present century (and it often has), there frequently occurs a turn to overt sexual imagery. John Wolfenden, for one, remarks that, when it comes to words, the scientist "seeks to free himself from their unconscious blandishments by inventing for his own purpose a form of expression which is innocent of these seductions" (1963, p. 21).

Such thought and terminology go back to the origins of modern science in the 17th century, to Francis Bacon, Thomas Sprat, Robert Boyle, and John Wilkins (see Chapter 2). For these writers, too, most of whom helped found the Royal Society of London, one of the first modern scientific organizations, words were not merely objects or signs but living things; they have "lives," in Wolfenden's terminology, and in Sprat's, "vicious" abilities to "snatch away" the rewards of knowledge. The fundamental Baconian plan was an attempt to deanimate this living power, to "kill" it such that scientific language would more closely approximate the "primitive purity" of "things." But having "lives," words also have sex: "swellings of style" in Sprat's phrasing, "seductions" in Wolfenden's. Either way, if not properly controlled, they will almost certainly "turn elusive," run away into "loose implication," or "fall" into the chaotic distractions of "feminine generalization," into a permissive and wanton display of "soft" emotionalism. The scientist must therefore have no "intercourse" with this type of language; he must protect himself from its "vanities" by creating a sterile or eunuch discourse, a speech that allows him the impossible stance of being both "innocent of seduction" and at the same time wholly aware of its forms and its "intentions." It is in this sense that he must, in short, become the godlike inventor of a linguistic Eden, from which everything tainted by Eve—everything identified, one can only surmise, with culture—has been banished.

But aside from its obvious sexism, there is a distinct quality of self-denial in this type of attitude, in the Baconian program as a whole. If words, *by their nature*, are seductive, dangerous, full of emotional life, how can it be possible to employ them in "pure and simple" fashion? How can they be denaturalized,, how stripped of their inherent powers yet still left capable of performing more focused and disciplined work? Can redefinition alone, and what is called "restricted usage," shake from them everything that is theirs, everything that maintains them as shared cultural property, that makes them perform certain social acts, everything in effect that makes them words? Or, to make use of sexual imagery again, if at the start they are androgynous, combining both "male" and "female" qualities, then isn't it only through one or another form of repression that the former can be made so preeminent? And if so, is it not then possible that a return of the repressed may occur at any moment, demanding as a result one or another "illicit" reading?

VIII

It is here that one enters the gates of a second level of representation within this discourse. The "modernist" view that we have been concerned with so far, the view of a totally univocal, selfless, exclusionary discourse, embodied in a range of conventions for writing, needs to be amended by certain "postmodernist" ideas on how various new ways of reading this language erode such a view, from the inside out as it were. The modernist view, that is, though based on aspects of expression that are only too real, does not define the entire reality or significance of this discourse. Such a view (dominant until now) sees technical language as entirely "scientific," as self-referential to an extreme degree. This is basically the view of Jürgen Habermas (e.g., 1971), when he describes this discourse as the inscription of "a model community." And indeed, in writing it, a scientist is largely forced to become a kind of modernist actor. S/he, as I have said, must become a productive agent for standards of expression that give a face to the promise of objectivity, value-neutrality, consensual truth, and so on. A so-called postmodern view, on the other hand, might begin by opposing Habermas's notion with an idea derived from Benedict Anderson (1983), that of "imaginary communities." Communities, says Anderson, exist largely on the basis of "the style in which they are imagined"—in this case, the style of the scientific (a kind of "nationalism?"), embodied in the modernist surface of technical discourse. But this quality of the "imaginary" has another dimension as well: what is made to seem whole and harmonious, universal and bounded, is actually in some sense a mirage, for this discourse also inscribes conflict, struggle, broken identity, and failed logic. A postmodern reading would then be involved in looking through the promise of a "model community," in showing its apocryphal nature, demonstrating that the repressions of authorship (which, again, are quite real) do not in fact create a writing that is pure, neutral, seamless, or unburdened by cultural/political contents.

At the same time, the fundamental program for a postmodern reading of scientific texts would itself be heterogeneous, and only partly formed at any time. Its aim would be to open up science—any single text—to a multiplicity of readings, many of which would not at all be interested in uncovering one or another essence of this category, "science," or in trying to reduce it to just another ordinary type of speech, but rather in revealing it as a social, literary, aesthetic, and political realm of expression whose borders with the rest of culture are diffuse, unclear, and changeable. In broad attitude, a postmodern view

> would involve a more sceptical attitude toward the concept of science itself. High energy physics and meteorology, molecular genetics and natural history, petrochemistry and cosmology are interestingly different from and complexly related to one another (nor do difference and complexity vanish *within* such domain boundaries). The boundaries of "science" in any case cannot be readily located by identifying such domains: technology and material culture . . . the recruitment of allies and resources, science classes, museums, advertisements and popular culture, may be as intimately a part of "science" as theories, observations, and the reasoning which connects them. The idea that there is a "natural world" for natural science to be about, *entirely* distinct from the ways human beings as knowers and agents interact with it, must similarly be abandoned. (Rouse, 1991, p. 161)

The key word here, I think, is "entirely." Natural science is about two things at once: condensing perceptions of the material inverse into words, and, inevitably, inscribing the human world of perception into the same products, this knowledge.

What follows is a brief summary of several approaches one might include within the postmodern program. Some of these derive from French structuralist and poststructuralist ideas (those of Barthes, Derrida, Foucault, and Serres most of all), from reader-response criticism (whose leading theorist is Roman Ingarden), and from speech–act theory (which sees linguistic communication as a series of "moves" or "games"). Several of the approaches I discuss have been promoted during the past decade or so by linguists, discourse analysts, sociologists, and others who have turned to the study of scientific texts (not always with results that depart from the modernist view). Other approaches I offer as possibilities for new readings.

The Gaming Table

One type of "illicit" reading of technical language takes the form of stylistic analysis. This means a very detailed look at just how this discourse works, step by step; how it seeks (succeeds or fails) to persuade; what social interactions it holds incarnate; what it does as a series of observable acts, carried out within scientific texts. This approach, or, more accurately, series of approaches, views an article, for example, as a progression of encounters; it is said to engage in certain language games with regard to framing the reader's response. Examples might include the use of certain words or constructions to make claims for legitimacy, for example, through uniqueness: "Previous work has failed to note . . . "; "A novel feature of these data. . . ." Or, in contrast, acceptability: "Our results agree closely with . . ."; "Such observations do not contradict. . . ." Or, alternately, value: "To date, a major hindrance to studies in this area has been . . . "; "A fundamental problem in developmental biology remains. . . ."

Other linguistic moves might involve certain types of caution or "politeness" (Myers, 1988): "The data presented here can be taken to suggest . . ."; "If those these conclusions are correct it would seem likely that. . . ." Or else, as I indicated earlier, strains of soft argument: "We find this model unacceptable for the following reasons. . . ." On the other hand, there may be requests for sudden changes in focus and logic: "It might be asked at this point whether . . ."; "We should not assume from the foregoing that. . . ." Or else realignments: "It is not the purpose of this paper . . . "; "A full review of these topics is beyond the scope of this article. . . ."

The use of questions in technical writing, though somewhat rare, is a distinct move to break the narrative flow, to grab attention: "Is such a hypothesis valid?"; "How are such results to be interpreted?" Questions can also be used to halt one line of reasoning and begin another, or simply to pose a kind of rhetorical end point and make claim to future progress: "Why does this occur? We do not yet have sufficient data to resolve current difficulties in this area. . . ." Beyond this, the strategic placement of various conjunctions—*if, however, despite, thus, therefore*—is also used to procure or maintain the reader's interest, or else to move the reader between graded positions of distance vis-à-vis an accepted theory or result: "If we accept . . . "; "This implies, however . . . "; "Thus, it becomes necessary to conclude. . . ."

Verb tenses, meanwhile, can be manipulated in complex ways within scientific texts, particularly with regard to their invocations of time. They often act, for example, to shift the narrative between different temporal frames, and to invest each of these frames with a specific function regarding truth or fact. This is especially true in those fields where "the past" in various forms is itself the focus of analysis. Such disciplines would include evolutionary biology, geology, and archaeology, where the "past" is constantly being made part of "the present." An example might be the following:

> *New* geologic data from *recent* wells on the West Florida Shelf and magnetic anomaly data *indicate* that pre-Mesozoic basement terrains on opposite sides of the Florida Lineament *were* contiguous *prior* to Triassic-Jurassic volcanism and *exhibit* only minimal lateral offset across the Florida Lineament *at present*. (Christenson, 1990, p. 100, italics added)

Even in such an abbreviated quote one finds no less than four different zones of "the present" being represented ("new," "recent," "pre-Mesozoic terrains indicate/exhibit," "at present"), and two zones of "the past" ("terrains . . . were"; "prior to . . . volcanism"), with the latter being contained within the former. At a still more detailed level, the procuring of information, of data, is related to two of these time zones ("new," "recent"), whereas one zone of "the past"—"pre-Mesozoic basement terrains"—acts both as the object of one present ("indicate") and the subject of another ("exhibit"). Finally, one sees that it is only in these two subfields of "the present" that the actual investment of "fact" occurs, and in such a way that the data becomes itself the demonstrative subject of this truth (the scientist again being deleted as the maker of knowledge). The uses of time in scientific writing, in other words, can act in strategic ways to position "fact" at particular moments in the narrative, and to make it seem the product of a "past" made "present" by the very existence of "data."

Reference citations, too as I noted earlier on, can be studied in terms of where and how they appear in a text, what they sometimes can reveal about consent and conflict (I am less interested here in the discipline of citation analysis, which too often makes the claim of being a "true science of science," than in some of the uses to which its approach can be put). An example might be the tendency of some authors to jam the narrative full of a large number of references whenever a "risky" statement is made, beyond a certain level of generalization. This has the effect of invoking the image of collegial support and community—and, in a larger sense, making it seem as if ideas are the creators of such community. Alternately, an author may choose to do the exact opposite, to generalize without support, thereby claiming such ideas entirely as his/her own, perhaps dancing along the margins of "accountability." More often one encounters a compromise between these two positions, thus a text in which the "individual" has a complex and unstable position vis-à-vis such things as authorship, originality, and the discipline itself. The work of citation within the narrative, that is, constantly moves back and forth between differing degrees of collegiality and of independence, between pleas for legitimacy on the grounds of similarity and those of difference. The "individual," as author (single or collective) might be said to alternately disappear and reappear, or else to merge into and emerge out of a presumed collective.

Finally, while patterns of citation can be valuable to map out, and no doubt have more than a little to say about the sociology of developing "frontiers," they need to be given a larger historical context. It is very striking to see, for example, how the actual practice of citing other authors has grown quantitatively over the past century. In 1900, it was common for an article to carry no more than three or four references to other work. By 1950, this had increased substantially; today it is not unusual for reference lists to run two or more pages in length. What might *this* mean with regard to changing standards of "contribution," "originality," and "documentation"? The realm of citation analysis might itself be expanded to answer such questions.

🖩🖩

Such analyses, however, as I have implied before, will have their limits. If too enraptured with methodology, they can become tiresome, endlessly repetitive, and can be used to belabor unimportant, common-place conclusions. Again, as currently practiced, they often serve the cause of the modernist view of science, being firmly rooted in a desire to "discover" something universal about this enterprise (see note 5). And in another sense, by viewing a scientific text as a series of choices or rhetorical functions, as "moves" roughly equivalent to those of a chess game, this approach can project too much rationality onto the writing process itself, too full a conscious reasoning. It can attribute far more intention and narrative ingenuity to an author than is ever justified, making everything seem the outcome of a literal choice.[7]

No doubt the more interesting view such analyses offer is to show just *how* messy and politically driven scientific effort can be. It is not merely that they can prove the ideology inherent in the image of the model community of Science, bound by the adhesives of agreement, con-formity, and verification. A detailed look at the rhetoric of science gets us closer to something almost at the other end of the spectrum. What it can show us is conflict at many levels of work and thought; territory staked and being staked; an enterprise that is often near the verge of flying apart into thousands of urgent pieces, each with its own special plea for a "revolutionary" stance.

This, moreover, brings up another aspect of technical discourse, a rather obvious one (for writers). To a truly rigorous eye, measured against the terms of strictly reasoned argument, scientific articles in any field are almost never without a great number of significant faults and fail-ures, again at many levels. The idea that a "good" research paper ap-proaches the realm of seamless logic is absurd. Nearly all contain ex-amples of sloppy or poor writing (indeed, this is exactly what most tech-nical writing manuals say), or else important inconsistencies, missing transitions, gaps, irrelevancies, and the like. The vast majority of scien-tific authors have a very haphazard or incomplete sense of narrative structure, a flaw often disguised by the use of devices like subheads and titles, but nonetheless easy to find if one attempts to follow even simple trains of thought in detail. One finds, for example, that the rationale for deciding paragraph breaks is often intuitive and unpredictable. Infor-mation from cited references is referred to both in the present and the past tense. The order of significant observations or data can be casual or slapdash. The whole, that is, very frequently has the feel of only partly sorted material, an attempt to "get everything in" or, given the realities of status and competition, to "just get it down on paper."

For someone used to reading technical literature on a daily basis, it is clear that authors very frequently introduce ideas, personal biases, and speculations under cover of night, so to speak, by means of one or another rhetorical cloak: "In this vein, it seems appropriate to consider . . . "; "Some workers have suggested . . . "; "The possibility exists that . . . ." There may be even constructions that rise from the floor of credibility like a house of cards: "If one accepts . . . then it becomes conceivable that . . . and when we consider. . . . Thus it can be concluded. . . ."

All this exists and can be easily found in the most sophisticated journal or textbook, despite the presumed stringencies of style sheets and peer review. The point is not to condemn such lapses, or to claim scientific discourse as being no more "privileged" than any other (as I have said, a false idea in any case, given obvious cultural realities). Presumably, one could make a field of study out of these defects, employing them toward some happy finding with regard to the nature of the writing process in science. Vain hopes aside, it seems more profitable to me to comprehend them in other terms, not merely as revealing discontinuity and slippery logic, but as speaking for a narrative space more open than it is closed, more porous to an author's demands for status, for heuristic interpretations, for airing pet grievances, or promoting personal theories or even whims—all by employing rhetorical tropes that are acceptable, part of the literary ethic. What would ordinarily be called "mistakes," then, might be said to exist *because* of these needs for expression, *because* there is no other place or way to satisfy them. And indeed, when one takes the time to read and compare articles written a century ago with those written today, a very interesting trend becomes clear: namely, that a much higher degree of narrative coherence and logic was then the rule, when scientists were still trained to be writers, that the loss of literary technique since that time has brought loss of this, too.

Image as Text

Another approach to reading technical discourse would be to look in detail at its graphic materials, its images. Illustrations present a different text for analysis. They are not merely visual, as representations, but are visualizations of certain norms and standards. They can therefore be read in many ways: in terms of their epistemological function, the types of truth they tell (e.g., Hoffmann and Laszlo, 1989); their iconography, or pictorial effect and formalism (Bastide, 1985); their history of development (e.g., Rudwick, 1976); or the changing ideological vision they might embody with regard to a specific subject or manner of conception (e.g., Gilman, 1988; Myers, 1988). Finally, they too might be read as "rhetoric," specifically, therefore as containing internal contradictions, multiple and competing messages at various levels of meaning. An image, no less than a textual passage, always carries a certain degree of risk or instability. There is always the chance that it might be perceived in new ways, productive or critical, not intended by its maker.

One thing, however, should not be overlooked: visual materials are one place where the aesthetic dimension to scientific work and thought rise to the surface—where aesthetic experiences and responses, where elegance and pleasure as parts of the scientific, become possible at the level of expression. This might be said to help demolish one side of the modernist argument surrounding technical language, the argument that it disregards all emotion for the sake of elevating "sense."

Images in science are enormously diverse and have been so for sometime (Robin, 1992). They play an extremely varied role among the many different disciplines and can be seen to combine a range of functions even within single articles. In fields such as engineering, climatology, or cosmology, where a great variety of such materials are routinely employed, and are always undergoing some degree of innovation and experimentation, there often exists an interesting kind of tension between text and illustration: it is often as if two universes of readerly experience lay side by side. The former, plodding ahead in its denotative way, can be relieved and countermanded by the latter's visual appeal, its grace, ingenuity, and even spectacle. Certainly, in recent years, the conflict has had occasion to increase a good deal, due on the one hand to the continued tendency toward further jargonization and flatness in writing, and on the other to the introduction of many new and sophisticated visual materials, from electron microscopy to closeup satellite pictures of the planets. During the last decade, many journals have introduced color into their illustrations. In the 19th century, when publishers lavished considerable expense on scientific works they felt might sell well, color illustrations were common. The rise of the journal helped end this by 1900; now the situation has reversed. This is particularly the case in geology, where brilliantly tinted maps, cross-sections, and other materials regularly appear in books and articles. Within the space of the text, these materials often take on a life of their own; there can be a quality of wildness or distraction to them that stems from excessive variety or sheer number, from an intensity of color and a range in visual language that go beyond the needs of the information itself. When this occurs, the text no longer can be said to cohere, but instead threatens to break apart the subject itself—whether a mountain range, continent, or marine process—into a series of aesthetic flourishes, trials, or even fetishes.

Locating cultural bias and ideology in scientific illustration, meanwhile, has been especially focused on medicine (e.g., Gilman, 1988; Haraway, 1989). This is one area where norms relating to the conception and treatment of human beings are quite literally open to view, laid out upon the visual table. The path for this type of reading was largely begun by Foucault, in his two books *Madness and Civilization* (1965) and *Birth of the Clinic* (1973). It is a valuable path, to be sure, but for a broader postmodern reading of scientific discourse, a somewhat obvious one. To date, most critics have left untouched the realm of physical science—the forms by which physics and chemistry conceive matter, astronomy and cosmology the universe, and so forth. Each of these—and the visual languages they employ—might be considered ripe for analysis. Moreover, the historical role of image making might itself be examined in terms of how it contributed to the formation of each discipline as a "science," in the sense that the "scientific," as understood by our culture, has denoted certain approaches to representation (and not others), "ways of world-making," in Nelson Goodman's felicitous phrase, that might be said to have taken their place within the larger context of Western image making as a whole. The power of the telescope and microscope to usher in a new era of scientific and general representation should be clear in this regard. Geology, as another particular example, emerged as a science just at the time when certain ideas of "landscape," "travel," and "nature experience" reached their peak. As a means of *seeing* the Earth, both literally and symbolically, of investigating and studying its near-surface forms and their internal contents, geologic science was born in the midst of powerful cultural predispositions. These included the romantic fascination with landscape, said to be full of the sense and quality of "hidden force"; pastoral longings brought on by the effects of rapid industrialization and urbanization; the fashionable appetite for "natural history" and "nature sketching" among the middle classes; the tendency for artists, poets, and even musicians to seek impressions out of doors and then return to some inner chamber, there to compose a final "work" or "study"; technological developments such as the invention of lithography (1798) and photography (1827); the expanded tendency of the new age of "exploration" and geography, aimed at cataloging the reality of distant regions (e.g., Alexander von Humboldt); and finally, the growing tendency, fed by industrial expansion, to view colonial empires in terms of their "natural resources," such as coal, minerals, quarry stone, and the like. In all these (and other) aspects of social experience during the early 19th century, one finds an enormous greed for visualizing different aspects of the Earth, for combining the touristic and the documentary, the painterly and the verbal, the personal and the objective. Properly understood, geology and its evolving language of illustration need to be placed within this kind of context, related to the larger role of the image and its uses during this period. No so-called internalist view—which would interpret this language as developing purely within the confines of natural history or, perhaps, the technology of illustration itself—could possibly do justice to the emergence of a science with so many "external" connections, both obvious and subtle.[8]

Postmodern readings in this area, therefore, have a great deal to choose from, a wide field of play. Any visual language, after all, is also a form of visualized culture and history. With this as a starting point, many entrances into the wider contents of technical discourse can be made.

One Voice, Many Tongues

There is a very practical reason why scientific discourse fails in its bid to become fully universal in every way: it remains divided among the many separate languages of the contemporary world, each of which impresses a quality of difference upon it. In what way or manner is science in French different from that in German or English or Russian or Hindi? What cognitive, structural, or semantic changes take place when it is transferred from one language to another?

These are questions of intimidating complexity. In large part, they remain unanswerable without a true and final cognitive model of language acquisition and use, something yet elusive and in its beginning stages (Gardner, 1985). Yet it might be interesting nonetheless to examine the basic forms in which science occurs in different languages, especially a non-Western, ideographic language such as Chinese or Japanese (see Chapter 5). Here, in fact, the basic conventions of authorial restraint are clearly similar to those in English (though with interesting amendments): objects perform the main action, and there are few allowances for confession, or for literary grace. Yet beyond this surface similarity, differences abound. The nonsymbolic pictorial nature of the written language, for example, makes it possible for much of the most arcane terminology to be "read" by ordinary people (who understand the individual ideographs). Thus, the level of exclusion is not the same as in Western languages and needs to be understood in other ways (e.g., that of social access, people's attitudes toward science, etc.). Moreover, as detailed in Chapter 5, the origins of scientific terminology in Japanese are far more variable than anything seen in the West: there are native coinings; ancient names and terms handed down from Chinese; terms that have been semantically translated from other languages into Japanese; terms that have been adopted phonetically (by means of a separate alphabet, generally reserved for foreign words) from a range of Occidental tongues—Dutch, French, German, Latin, as well as English, each of which dominated specific fields during the past 100 years, when Japanese science came of age. Shifting influence among these different languages has been related, in part, to political as well as cultural and technical factors. Technical nomenclature in Japanese is a historical hybrid, full of evidence of Japan's past fidelities.

What all this suggests is that each nationalized portion of technical discourse— particularly those portions (languages) in which science writing is a relatively recent innovation—could be studied according to how it has evolved, how it continues to change, how the choices made by its practitioners reveal important things about the spreading and inscribing of cultural power, even how the power of translation manifests itself in the history of the modern world.

No Word Is Innocent

Finally, one comes back to broader linguistic ground, to words themselves. Whatever the apologists for Bacon might ultimately say, the "native ambiguity" of ordinary language is not the point. Pure denotation and self-referentiality, isolation from the rest of culture: these are not possible in science for a somewhat different reason. Language, after all, cannot be deprived of its *meaningful* connotative associations and urges; frames of reference always leak. There are always possibilities for readership that can call these urges back.

Even, for example, in the opening quotation to this chapter, the untrained eye and ear strike on the words "induction," "virulence," "mobilize," and "bounded." What they perceive is a *military* presence, not merely a scientific one. Still more revealing would be a look, say, through any recent textbook in immunology. This would inevitably turn up a rank-and-file nomenclature among which would occur such terms as "killer T-cells," "suppressor T-cells," "natural killer cells," "T-cell proliferation," "target cell," "invasin," to name but a few. Far from being "pure and simple," these terms are literally saturated with cultural attitude. They bleed with our societal terror of disease (illness as biofascism), our anger at it, and our intent to deal with it not through visions of growth or healing but rather those of *war*. No argument is needed: here is not a dispassionate or disinterested language, but one that is intensely didactic, affective, engaged with the politics of this cen-

tury at many levels (see Chapter 3). Indeed, this area of medical discourse seems the very opposite to the garden of verbal delights promised
in science not so very long ago (see John Wolfenden's quote at the beginning of this chapter, for example). That such promises are still often
made and accepted today can therefore be seen to participate directly in
furthering a cultural ideology of no small importance.

Similar readings, of a more subtle nature perhaps, could be made
in any professional language of science. "Metaphor is never innocent,"
Derrida states, "it orients research and fixes results" (1978, p. 17).
Moreover, a text is never a single, impermeable structure of meaning.
Semantically, it wanders, skids, calls upon other regions. Put differently,
"a text is not a line of words releasing a single 'theological' meaning . . .
but a multi-dimensional space in which a variety of writings, none of
them original, blend and clash" (Barthes, 1977, p. 147). Other structures, other voices, other contexts of meaning are always there, waiting,
as it were, for emancipation.

When scientists select or create such a word from the existing semantic mass, they often want to temper it with a ferric precision, to iron
out all connotative demands or suggestions. The scientist wants his term
to be a purely technical sign, with no index of meanings. But this cannot
occur, even locally, by what is called "restricted usage," for as I have
tried to make clear, words can never be orphaned from their extended
"family of meanings" (Wittgenstein). Perceiving any word as exclusively
technical—any word, above all, that has long been used in other forms
of discourse—can only come about from a socialized practice of reading, one that relies in the end on a kind of trained blindness.

What this suggests, is that we could read technical discourse in this
analytical way, too, searching for the conceits within it. Why do this?
Many reasons, but most of all to show where and how relations might
exist between the denotative (scientific) and the connotative (sociocultural) contents of knowledge and where the ideological intentions of
each coincide. In the language of immunology, with its biomilitary vocabulary, one can easily see what exists everywhere else in scientific
speech: words or phrases whose drifting figurative energy cracks the
crystalline spell of univocal coherence. And more, a language that reveals not a coincidental, but rather a *systematic* relation between technical conception and a deeper (in this case, terrorized) societal sensibility
that envelops it.

Surely, then, a detailed historic tracing of scientific lexicons—a
sociolinguistic excavation, as it were, showing where crucial terminologies have come into being and what forces and connections have acted

to guide their selection—would tell us much about the true and changing place of science and its discourse within culture. Surely, too, it would
show a deeper possibility of connection between science and the arts,
again at the level of cultural influence. For such an approach to reading
technical discourse does nothing if not prove the untruth of W. H. Auden's
well-known aphorism: "Scientific knowledge is not reciprocal like artistic knowledge: what the scientist knows cannot know him."

🖅🖅

The above summary covering possibilities for a "postmodern" reading of technical language is by no means complete or at all comprehensive. I've attempted to cover only a sample of approaches, no more. Some
of these seem more powerful to me (as a scientist) than do others; in
particular, the tracing of lexicons, of metaphorical conceits, of the role
of translation, and that of illustration could all have much to offer by
way of showing the processes in which technical knowledge emerges out
of cultural needs and sensibilities. In any case, the types of readings I've
tried to describe indicate that no single approach can hope to encom-
pass the whole field.

Some other approaches could be mentioned. Those related to literary theory come immediately to mind. Yet these, too, have their limits. If
we take the view of narratology, for example, any scientific text can be
shown to tell a number of different "stories," each with its own formal
elements (often in confused order). To the semiotician, on the other hand,
the interrelationship between visual and written text, as well as the functioning of signs within each, will be far too variable (and important) to
be exhausted by a single interpretive scheme. Reader-response criticism
also fails to fully capture "how this discourse works," either because
there are simply too many "readers" embodied in the average text, or
because the "historically specific conditions of reception" are entirely
obvious, that is, they are those of the modernist surface itself. And finally, as we have seen, stylistic analysis can suffer the fate of simply
piling up "observations" about "narrative acts," and then sifting among
them for some hint or whiff of *principium scientia*.

As a program, repudiating science's claim to a discourse that is
pure, objective and logical has only begun. Statements like those of
Todorov, Lyotard, and all others who see science and its language as
being "held apart from the social bond," are the mark that the old myths
have not really been challenged, at least not very deeply. As we have just
seen, it takes little to show that this language, and therefore its knowledge, do in fact "know" us quite intimately, as of course they must, that

they are not so much embedded in society, like an encrusted gem, but are simply part of it, like an element in an alloy. Like any human knowledge, no matter what its form or style, scientific knowledge is engaged in dialogue with the culture that makes it possible. Therefore, our task is to unlearn these myths (which all return to notions of separation and purity), to discover a means to explode them. One such means is to ignore the protective walls of "context," and to read this discourse in new ways, on the more open plains of praxis.

IX

Where does all this leave what is called "science writing"? With something of an unenviable task, to be sure. This form of writing exists, after all, because of what scientific discourse has historically become, because of the density of its jargon and the inaccessibility of much of its content (realities that no "postmodern" reading can ever mollify or change). Surely, the very reason for science writing and its critical role in any modern democracy (for this must be acknowledged) has more than a little to do with the exclusionary principle inherent in contemporary technical language. Moreover, given the present-day character, demands, and evolutionary momentum of scientific discourse, this role will hardly decline in importance during the near future.

Not merely a source or mediator, science writing is also a creator of social reality, both for those within and those without science. It must act as a political broker and agenda maker, somehow endorsing the elite status (significance) of scientific knowledge and at the same time the public's "right to know." Such is why the great popularizers, from Playfair to Asimov, have so often been grand apologists and skilled diplomats. It is also one reason why they have so often employed the language of superlatives and measured sensationalism: such language flatters both sides.

More immediately and practically, however, science writing has seen fit to choose among three main tactics: those of (1) interpreter, (2) promoter (pro or con), or (3) observer. The first of these seeks neutrality; the second, attention (science as spectacle); the third desires, most of all, some form of evaluation.

The first two of these strategies tend to merge, even become indistinguishable, in the mainstream mass media. They thus define the bulk of what the general public is exposed to, on a consistent and self-rein-forcing basis. As one who has studied the matter in depth, Dorothy Nelkin speaks of the "homogeneity" of contemporary science journalism, its tendency to "focus on the same issues, use the same sources of information, and interpret the material in similar terms" (1987, p. 9). Her final conclusion, in fact, is anything but sanguine:

> Too often science in the press is more a subject for consumption than for public scrutiny, more a source of entertainment than of information. Too often science is presented as an arcane activity outside and above the sphere of normal human understanding, and therefore beyond our control. Too often the coverage is promotional and uncritical, encouraging apathy, a sense of impotence, and the ubiquitous tendency to defer to expertise. (1987, p. 173)

To some extent, this might have been predicted. The attempt to merely "interpret," to restate science in simpler words, is both insufficient and impossible. This is not only because neutral explanations (of often nonneutral knowledge) are unobtainable. It is also because, even from the outset, the "what you need to know" approach could only leave knowledge in a vacuum, on a planet without an atmosphere. Such a tactic mimics one aspect of scientific writing, that of trying to make everything appear pure and disconnected, untouched by any "external" influences. A science writer who assumes the interpretive stance would like to make this same claim, pretending no involvement in the historical and institutional conditions for knowledge, nor in its social implications. In effect, such a writer seeks to shield himself from any real responsibility as a writer, as someone with the power to expand (not merely regenerate) public sensibility. Everything is represented as if it occurred deep within the hallowed halls of science. Nor are such writers ever willing to admit that they are translators more than interpreters, that is, that they themselves *do* something to this knowledge, by remaking its language often in accord with cultural bias (I will offer some general examples in a moment).

What, then, of the marketing approach? Clearly—whether the form taken is that of salesmanship or warning—this too serves to elevate the privileged status of science and its discourse. Here, in fact, we get what we have most come to expect from our media, a spectacular science that is either drenched in marvels, breakthroughs, geniuses, mysteries, and so on, or one that threatens us with the specter of "monstrous" or "unknown" consequences. This world of Manichaean superlatives—so often a kitsch world of thoughtless effects—is exactly what most people have become comfortable with, not merely from newspapers, but from periodicals such as *National Geographic* or *Reader's Digest*, from the greater share of TV documentaries, and from nearly all books that promise us such things as "the most complex problems of the universe told in simple, everyday language." This is a world whose terrain and inhabitants seem utterly fixed. Nothing occurs, at a fundamental level, beyond the endless recycling of a gee-whiz terrain ruled by crusaders "racing against time," by lovable eccentrics seeking the "wonders of the cosmos," by mandarins of brilliance uncovering the "secrets of life," on the one hand, or, on the other, a dark, brooding landscape roamed by forbidding forces and substances that have been "unleashed upon mankind," by wizards and their apprentices, and by oracles gothic in their tone, eager to tell of an "uncertain" or "profoundly altered" future.

Examples for this type of science writing can be found almost anywhere. I offer only a simple instance:

> A turning point has been reached in the study of life. A turning point of such consequence that it may make its mark not just in the history of science but perhaps even in the evolution of life itself. . . . For the last 25 years molecular biology, the study of DNA, has made the fastest progress of any branch of science. A mass of pure knowledge has been accumulated. . . . The whole gene pool of the planet, the product of three billion years of evolution, is at our disposal. The key to the living kingdom has been put into our hands. (Wade, 1977, p. 1)

Much could be said about what is to be found in such a passage (e.g., exactly how does one determine that one field has made "faster progress" than another?). Yet it is enough to remark how nearly everything I have just discussed shows up here with a naked face. It might be noted, too, that few researchers today would agree among themselves (in a nonpublic setting) with the sort of thing being said in this kind of quote, especially the last two sentences. By almost any major standard—say that of medical therapy, of new cures for chronic diseases—the gains made during the 18 years since Wade wrote this have been, at best, modest and preliminary. No overnight "revolution" has taken place; cancer, M.S., diabetes, and now AIDS, as well as a dozen other widespread illnesses, are still very much with us and the claims now being made by experts with regard to their eradication are tinted only with tones of hope, not with the kind of absurd and irresponsible hubris reflected in Wade's words.

The interpreter and the marketer very often meet in mainstream science writing, as we have said. This would appear inevitable. The basic reason comes from democratic mythology itself. As an elitist form of knowledge, science must be retold to the public, and in such a way that it be made to appear under forms more familiar to that public. Science must "come down" and become a part of what is common, what is accountable. This does not mean that it loses status, not in the least; on the contrary, the terms under which popularization occurs are very often those derived from long-standing traditions (perhaps even rituals) for the manufacture of fame and notoriety, that is, traditions based on the creation of heroes, the proposing of mythic struggles, the power of miracles, and the like. Providing scientific knowledge with an everyday hagiography is especially evident in many films that portray the lives of famous scientists. In *The Story Louis Pasteur*, for example, perhaps the archetype for all such films, one sees Paul Muni battling in operatic isolation against the assembled intellectual and social forces of late-19th-century France. His final victory, as logically impossible given these (falsified) conditions as it is demeaning in its sentimentality, is a victory not for humanity but, as Pasteur himself says, "for science" (no phrase recurs so often throughout the film). One thus sees the enormous complexity of Pasteur's life and work recast into libretto, into the everyday aesthetic forms of a soap opera.

The same thing takes place in another way in a great deal of science writing. At the worst, whether in Sunday-supplement articles or in prime-time TV specials, one finds represented the image of an audience that is so hopeful of stupefying wonder, so lacking in any form of educational experience, that the natural world can only be explained in terms of: (1) carnival facts ("As many as 230 million such bacteria would fit on a single pencil dot . . ."); (2) the most banal metaphors, for example, those of the kitchen ("Once the batter of the earth congealed . . ."), human relations ("Particle interactions can be seen as a series of marriages and divorces"), or the neighborhood ("In the second decade of this century, we learned that our sun resides in the suburbs of a galaxy . . ."); (3) storybook imagery ("At that time, huge carnivorous monsters filled the seas . . ."); (4) military simile ("Our bodies are under constant attack from unseen enemies . . ."); and of course, (5) the sports scene ("Once the catalytic whistle goes off, various molecular teams go into motion . . .").

As Dorothy Nelkin (1987) indicates, much of the blame for the continued vitality of such stereotypes is due to two things. First, there are the practical, institutional constraints imposed by modern journalism. Here would be included the effect of deadlines, the small number of full-time science reporters, and the larger demand that news sell itself (too often as melodrama), or at least as a "good read." This, in turn, creates the requirement for a low common denominator of writing, so that the widest possible audience can be reached (newspapers being a capitalist product like any other). All of this can be said to encourage the use of images and language that are automatic, received, established. Second, we can convict the impetus among scientists themselves, much heightened in recent decades, to cultivate a positive public image by saying the "right" things at the "right" time, therefore fending off public criticism, advancing their individual careers, and gaining popularity for their own field as a whole (thereby lobbying in effect for increased federal funding). In contrast to scientists a generation ago, for whom publicity seeking was considered taboo, a form of conduct strictly condemned by professional norms, many of today's researchers publicly promote themselves and court publicity, for example, by slipping into the stereotypical "scientist" role—paternal, gently condescending, enthusiastic about "future advances"—whenever they are interviewed by journalists or other writers. Or else they decide to speak in the type of exaggerated, hyperbolic language already established for their kind: that is, in terms of "coming revolutions," of "unlocking mysteries," and the like.

The true tragedy in science writing of this kind is that it is wholly wedded to keeping its audience technically illiterate. It produces a "science" that is exactly what people have come to expect, one that is not merely distant but elevated, and for that matter, not merely elevated but unattainable, resistant, unknowable. It is no doubt one of the deeper failures of popular representation, both inside the classroom and out, that this "science" remains the most familiar we have, the one with which most people seem to feel comfortable.

⑤⑤

But to be completely fair, there may be another, even more basic problem. It is a problem that, from a more reflective or a sociolinguistic point of view, defines the deeper effort that science writing cannot avoid today, an effort that makes it both unenviable, as I have said, and profoundly challenging in the end. Fundamentally, the problem is this: to somehow discover or forge a stable plane between scientific and nonscientific speech. In effect, everything talked about regarding the codes for writing this discourse should help suggest the difficulties involved. Good science writing, if it hopes for some modicum of success, must somehow overcome as many insider–outsider strategies as it can, in particular the major ones I have tried to outline above. But still more, it needs to opt not for their destruction or erasure; this would mean too great a sacrifice of what science truly is. Rather, it should make use of them. To explain this, an analogy may be useful. Sociolinguists speak of what is known as "code switching" or "dialectic shifting"—displayed, for example, by urban blacks who have learned how to change from ghetto dialect to middle-class speech when the need arises, for example, during a job interview. Is it exaggerating things too much to say that in some sense a similar change in linguistic context is required to move between technical and ordinary speech, in trying to adapt a language of specific practice to one of universal consumption?

This, finally, brings us to the third type of science writing. If we are to admit some enviable alternative to the other approaches mentioned, how would it occur? What would it be? This I can say plainly: the best science writing has been able to observe, to evaluate, and to explain—but by way of diffraction, through an approach that one might call "postmodern" in sensibility. In brief, it has managed to integrate multiple discourses: history, politics, economics, cultural criticism, art, literature, philosophy, mythology, personal anecdote, fantasy, biography, and much more. Only by means of such integration, after all, can science be given back its actual place within the general culture. Told through and beside these other voices, that is (and here I mean something more than mere hagiography, chronological history, or intellectual scandal), science regains its living, its *ordinary* location in the physiology of culture. It becomes something knowable, something with a reality of connectedness.

A few classics in this area might include works such as A. S. Eddington's *The Nature of the Physical World*, G. de Santillana's *The Crime of Galileo*, and René Dubos's *The Mirage of Health*, as well as certain works by J. D. Bernal, Stephen J. Gould, Michel Serres, Robert Young, Richard Lewontin, Georges Canguilhem, and Gregory Bateson, to name but a few. In the writings of these authors, technical language does, of course, lose something of its active character; this is inevitable. But the loss is a measured and necessary one, required for the sake of new visibilities. In the great heyday of popularization, from the early to the mid-19th century, when the public hunger for science (and its many prophecies) seemed unquenchable, this type of approach was common, muddled, and in part unnecessary. It was common because science still remained merely one category of knowledge among many others; it was muddled because, in fact, the "scientific" was itself invested with all manner of heroic tropes from other discourses (especially literature); and it was often unnecessary because scientific language had not yet departed from the realm of general literacy (works by Humboldt, Cuvier, Darwin, and others were best-sellers in their day). Today, however, a very different set of historical realities exist, demanding (in my opinion) a different, more self-aware type of science writing. As elsewhere, these realities make for a certain irony: that the discourse of science, the language of truth, must be overcome in order that the truths of scientific thought and work be made apparent and accountable. A difficult task, to be sure. But then, considering the broader responsibilities involved and the breadth of understanding required to meet them, it perhaps should be the case that this writing be counted among our most demanding.

X

In the end, one is compelled to return to the linguistic universe within science itself. A common view holds that professional languages do not communicate with each other. They pass on the street, as it were, with little exchange beyond a perfunctory tip of the hat, a momentary comment about business or the weather. Once again, Barthes evokes a feeling of recognition: "The language of the *same* suffices us. . . . We lock ourselves into the [speech] of our own social, professional cell, and this sequestration has a neurotic value: it permits us to adapt ourselves as best we can to the fragmentation of our society" (1973, p. 116).

Familiar though they may be, these words are nonetheless overstated (as I have already intimated)—and particularly so in the case of science. Within the sciences after all, is not a single prison of separatist confinement, but rather a great dictionary or society of languages, each of which is not a hermetic and sealed volume but rather more like a living structure within a great body, rubbing up against its neighbors, stealing and exchanging nutrients along permeable edges. Or to adopt another image altogether, proposed by the philosopher Hilary Putnam, we might think

> not of a single boat but of a *fleet* of boats [whose inhabitants] are trying to reconstruct their own boat without modifying it so much [that it] sinks. . . . In addition, people are passing supplies and tools from one boat to another and shouting advice and encouragement (or discouragement) to each other. Finally, people sometimes decide they do not like the boat they are in and move to a different boat altogether. And sometimes a boat sinks or is abandoned. It is all a bit chaotic; but since it is a fleet, no one is ever totally out of signalling distance from all the other boats. We are not trapped in individual solipsistic hells . . . but invited to engage in a truly human

dialogue, one which combines collectivity with individual responsibility. (1981, p. 118)

The writer here was expressing a view of science as a whole, as a practice; it is, he says, neither monolith (heroic or otherwise) nor shattered column from the ruins of the Tower of Babel. It involves things and human beings, constructions and improvisations. Its complexity of interaction, moreover, without and within, is exactly what one sees in its discourse, if one peers below the surface. Constant neologistic barter, thievery, diversification, and mutual dependence—this is what one finds within scientific language at the level of actual use and invention. It is this linguistic activity that reflects, as well, what I have termed the second level of representation in this discourse, a level marked by all manner of struggle, disorder, claims making, creativity, love, and ambition.

It would seem, in fact, that one side of scientific research all but ensures the productive necessity of this neologistic activity. For as scales of investigation have reached first to the molecular and then to submolecular levels, so-called crossover phenomena (and thus terminology) have become more numerous, thereby revealing how the old borders that distinguished the primary fields of physics, chemistry, biology, and geology have begun to dissolve. In simple terms, our increasing ability to observe both atomic and universe-wide phenomena (microcosm and macrocosm) has meant that specialization leads to overlap as well as divergence. The early historical signs that this would one day occur on a regular basis, and would represent a definitive type of advancement, took place in the 19th century, with the creation of such hybrid fields as biochemistry (a term coined c. 1850) and physical chemistry (c. 1890). The process of intellectual and linguistic commingling has continued at an accelerating rate. Within the last few decades, a great host of new crossbreed disciplines have appeared, merging two, and sometimes even three different fields: biophysics, astrogeology, paleobiology, quantum chemistry, psychoneuroimmunology. A genealogical chart showing the patterns of such crossbreeding would no doubt provide a tool of high value to any historian or sociologist of science (one branch of citation analysis has attempted to do just this). There are many questions to be asked about such a process, how and where it occurs, and by what incentives and mechanisms. But the point I wish to make is merely a simple and obvious one, namely, that the very existence of such fields is ample evidence of an avid and expanding dialogic within science—a dialogic that has certain "internal" characteristics, certainly, but that is also more broadly motivated by "external" influences. Astrogeology, for example, is in part a direct result of a nationalist (and in part, military) program to "get there first," via NASA and the sending of space probes into both the inner and outer solar system. Psychoneuroimmunology, on the other hand, has come about because of broad public dissatisfaction with traditional medicine, and the consequent demand for a return to the study of how such things as "emotional state" and "mood" affect the body's conditions of health or illness.

One question of a more speculative kind might be asked. What kind of limit might exist to this type of dialogic in science? How far, and in what direction, might it eventually go? What might be said to define its final end point? One can envision two extremes: first, that every phenomenon or process, or at least those of identified importance, would become the subject of an entire attending field; second, and on the contrary, that growing overlap between fields will one day bring about the downfall of all divisions, thus fulfilling Bacon's early hope for *una scientia universalis*, a single grand community of knowledge and discourse.

Of these two possibilities, it would seem that the former has the better chance. Indeed, one already sees it in nascent form, in the growing tendency (due, in part, to the sheer accumulation of knowledge) to create entire subfields or "frontiers" focused on single entities: a subatomic particle or type of star; a species of dinosaur or fruitfly; a mineral or a metal. The great dream of a single unity for all science, which physics was presumed to one day bring about, proposed for example by philosophers like Rudolph Carnap and the Vienna positivists, appears today as another of the grand illusions of the past, a kind of statuesque nostalgia. Of course, one still finds this dream occasionally invoked: "as far as we can tell, it is in the area of elementary particles and fields . . . that we will find the ultimate laws of nature, the few simple general principles which determine why all of nature is the way it is" (Steven Weinberg; quote in Holton, 1978, p. 14). Old myths die hard, especially among the faithful with necessary ties to the material realities of the faith. For physics, this has meant the continued building of ever larger, more sophisticated, more expensive, and thus more politically invested forms of apparatus, starting with accelerators and culminating, most recently, in the Superconducting Supercollider (SSC), a $2 billion piece of equipment. The promise of "a final theory," written so insistently in the title of this apparatus, seems more desperate than ever, yet less willing to admit that the vision involved must be a social vision, not an "insider" one, just as it was in the days when science was itself seen as "the final source" for all secure and progressive knowledge.

The hope for such a theory rings hollow in another sense, too. Few today are naive enough to believe in a possible end to science. A glance at the contemporary scene is enough. It is no longer really tenable to propose such a point of discovery beyond which the whole future enterprise would be reduced to a sophisticated mop-up operation. In what way, after all, could the "ultimate laws" of subnuclear particles (should they exist) ever constitute truth for fields such as plant pathology or archaeology? Notwithstanding data or documentation on quarks—one of science's more recent borrowings from literature, in this case *Finnegan's Wake*—and their characteristic traits of "color," "flavor," "strangeness," "taste," and "charm" (surely physics could have done better than to make all matter in the universe seem so intrepidly middle class!), how would a theory of quantum electrodynamics ever be able to tell us anything meaningful about early hominid evolution in Africa, why bees prefer a certain type of flower, or ongoing continental collision in India? There no longer exists, in short, any fixed definition of the physical world, any one scale of reality into which all others can be collapsed. This "world," after all, is a result of continual and changing negotiation, not only between "things" and words, but between the impulse to know and the ways in which this impulse defines its objects.

🖫🖫

The more likely conclusion might be this: that science, like the universe itself, is moving by way of its varied dialogic away from unity toward an intricate series of constellations between various fields and subfields, toward the simultaneous trends of fragmentation and collision; discontinuity and junction; solitude and conviviality; consensus and competition. And it is in this sense that we might agree with Lyotard in one respect, that of viewing contemporary scientific work itself as an immense adversarial game: an unending attempt to create the conditions for new work, to find gaps or instabilities in existing structures, chances to make room for new knowledge, new kinds of scientific statements. From this perspective, the idea of "science" shifts entirely away from that of achieving consensus, toward the image of strife, a drivenness to look for nothing less than the possibility of "revolution" at every scale of inquiry. Given the general accuracy of this statement (which most scientists, I believe, would acknowledge), why is it so? A difficult question, to be sure, but one whose answer clearly lies not in any rhetorical agonistes between tradition and the individual, nor in any transhistorical, Hegelian theories about the "essence of scientific reasoning" (Karl Popper) or "the nature of scientific development" (Thomas Kuhn). Rather, one suspects the place to look would be: first, in the contemporary social structures of scientific work (those that *demand* competition, and shape it as well), and second, and perhaps more important, in the specific images and models scientists have absorbed with regard to their own personal careers, involving such things as success, achievement, "contribution," "advance," and the like—images, in other words, that are historical and cultural, that cannot escape some degree of grounding both in science education and, more broadly, in our current media-saturated culture, replete as it is with representations of "brilliance," professional and public glory, with the ballast of clichés so evident in much science writing and the rewards promised from their embodiment.

⑤⑤

Such images, in short, are part of science too. They are but one example of an array of realities—attitudes, beliefs, myths, methods of "internal" and "external" popularization, as well as the normally cited litany of resources, institutions, and technologies—that reveal how this category "science" is truly without fixed boundaries, how culture flows into it like a variable but insistent tide, and at the same time how "science" leaks outward through a hundred different channels, each with its own burden and flow.

I thus come back to an earlier point. For something of this agitation between inner and outer, between consensus and conflict, seems reflected in the tension that has grown up between the standardized scientific style on the one hand, with its voice of conformity without appeal, and the more relativistic, conflictual, and interdisciplinarian qualities that have come more and more to find a central place within technical understanding. At a certain level, as the "postmodern" readings above suggest, this tension must appear immoderate, almost neurotic. For on the one hand are the "modernist" ideals, the constraints of linguistic convention, professional behavior, of institutional attachment, and of what is to be counted as knowledge. These are the centripetal aspects, those that seem to contain and direct the whole, to aim it inward. Yet against these beat a near anarchy of other conventions and demands, of individual ambition and all that it implies, of competition at many levels and for different types of capital, material and symbolic, between researchers, research groups, institutions, nations, disciplines and subdisciplines, whose separate requirements are themselves often in conflict. These more centrifugal realities are not merely those of "social interest"; the individual still needs to be inserted back into the process and labor of science, and not merely as an "agent" but as a living, deciding participant. Purely formalist interpretations of this tension, in other words, will defeat its deeper complexity, will impose upon it an unfair brand of determinism.

The average scientist, meanwhile, does not feel him/herself to be torn apart by this tension. If s/he perceives it at all, it is usually in the form either of competing demands—teaching, grant writing, research, publication (sometimes called the "Four Horsemen" of contemporary university science)—or, perhaps, a momentary lament or regret for the dullness and impassivity of the scientific style. Some in positions of institutional power, it is true, value and encourage "rebellion" among younger researchers, hoping to provide motivation for advancement, for the overturning of given ideas. As described by Holton (1986), such mentors also rationalize this attitude through the belief that the social organization of science is there to channel any and all these "brutish energies" into "disciplined inquiry." But the point is that these same energies are themselves part of this social organization, as the desire to encourage them makes clear.

In the end, one wonders: how long and how far can scientific discourse go before being forced to change its rules? What point must be reached before those aspects of personal and professional need that lie "hidden" within this discourse are allowed again more overt forms of expression, those that are themselves more humble, more relativistic, more "interested"?

XI

Having adopted a speculative turn, I feel a need to pursue it toward a personal conclusion. Taking yet one more step back, the question of phenomenology rises up. What, one might ask, is the deeper need expressed within this discourse? What aboriginal longing does it contain? What is its intention toward the world? Or, ultimately and more simply, what does it *want*?

As it happens, this is not difficult to say: from its beginnings in the 16th century, modern (Western) scientific speech has taken on a single, breathtaking linguistic purpose: that of giving the material universe a voice.

To the ear of epic science, natural reality is a lush and resistant silence. It presents an infinite panorama of cosmologic secrecy and unspoken order. Sartre's words return at this point, for we see that the scientist, in his or her act of writing, is thrown into the world in two ways: first, into the realm of men and women (profession, knowledge, society, culture), and second, into the midst of this formidable and overwhelming hush, which nonetheless demands a script.

At base, this discourse does not want to end or to fill the world's ahuman silence, to appropriate or to master it, to shout into its darkness and make it erupt and echo with meaning. The desire that moves it is not only that which would interpret or translate the universe, but one that would speak *for* it—and it is this positing of its own voice for the world itself which has, possibly and most of all, led science from the time of Bacon to believe its signs must in some way approach the materiality of its presumed subjects. If the granitic solidity of this voice is in part illusory, it still contains a belief in the possibility for an elevation above time, beyond temporality and death. It holds within it the ideal of being always struck with the single note of a pure and clarion logic, and yet with a wan sobriety that, sooner or later, calls upon the idea (as expressed by Nicholas Schöffer): "*Ce qui sauve l'homme c'est sa diversité, ce qui le perd, c'est l'unanimité* (What saves man is his diversity; what fails him, unanimity") (1978, p. 235).

Of one thing, however, there can be little doubt. No other region of human discourse will continue to give birth to new language, new powers of naming, and new organizing metaphors as much as that of science. And it is exactly this that draws one toward another troubling notion. For despite its utility and ubiquity, this discourse, as it exists in the main, and for the writer most of all, must in some sense be seen as a less brilliant side to what is perhaps mankind's most infinite and productive—though often troubling—enterprise of intellectual exploration. Why? Not only because of what has to do directly with authority or power (this is a different problem). But because this discourse has within it standards that, whatever their efficiency, retain the capacity to dull the experience of being a day-to-day participant in this enterprise, and that argue formally against the human urge to generate speech in any personal sense, with differing degrees of expressive freedom.

In more practical terms, formal technical speech—the most preponderant and imitated that we have—has within it a standardized motive for aesthetic repression that depends for its justice on the verdict of convention, which, if overturned or disregarded sporadically or even systematically would not in the least lead to the downfall or trivialization of knowledge, but which, on the contrary, would return to it a degree of the range and humility that was once (and not long ago) a respected part of its voice. It is language, and today this means written language most of all, that can make knowledge convivial, celebratory, in a word, "festive" (Barthes).

Perhaps this will occur, at least locally. In image and in literary convention, the master narrative of science remains, more densely confident than ever. But in actual practice, in certain aspects of thought, and in the textual realities that lie beneath and within this narrative, one sees a different story altogether. The fall of the old *grands récits* in science—those that promise the unity of all knowledge, its inevitable benefit to mankind; the promise of pure objectivity; and of a truth beyond probability—now seems clear. If one prefers the idea of a "postmodern science" (rather than a series of postmodern readings of science), then one might focus on how this fall may have helped pave the way for the advent of concepts and terminologies (to quote Holton again) "contrary to the older (and persisting) themata of hierarchy, continuity, and order," those that give center stage to various principles of discontinuity and indeterminancy, and that codify or perhaps even articulate—often in very specific ways—something of "a restless, even violent, world." Certainly, there has been a return of "catastrophe" to many areas of scientific thought: the concept of sudden change, long denied for its religious connotations, has reentered almost every major field to date. One hears it in such ideas as the punctuated equilibrium theory of evolution; major extinctions caused by giant meteor impacts; chaos theory; black holes; continental breakup; and climatic shifts. This entrance of "crisis" into technical ideas does not at all mean that scientific explanation has given up its belief in causality, in mechanistic thinking. On the contrary, it has simply incorporated forms of disruption into its power of representation.

In such a view, "postmodern" science would have to be seen as both a continuation of, and a challenging amendment to, "modernist" science, which would have reigned up until roughly the late 1960s or so. But the point I want to end with is different. At the level of writing, this "postmodern" turn seems evident only with regard to certain areas of terminology, as I have said, those in which a greater tendency toward experiment and play are found. Perhaps this is a first step; perhaps this kind of trend will help encourage new styles of naming in other areas, and from there, a new demand for more adjectival types of expression. To my mind, it clearly marks the understanding among a growing number of workers that technical discourse can in fact be opened up, and that there is an increasing need to do so. The "postmodern" aspects of contemporary science may here be cracking through (not, to be sure, without their own ideologic contents). It remains to be seen just how this might continue, whether it will be satisfied with mere occasional wit and whimsy, or whether a change toward more heterogeneous styles of writing might actually become possible.

Yet for the interim, one can say with certainty of science what Flaubert once said of art: "Plasticity of style is not as large as the entire idea. . . . We have too many things and not enough forms."

Notes to Chapter 1

1. In what follows, I use the term "postmodern" in a relatively limited sense, most of all with regard to ideas of heterogeneity, internal multiformity, inconsistency, discontinuity, and the like. Though these form only a very small part of a much larger debate about the role and character of aesthetic realities, they seem of particular relevance to scientific discourse and thought. For a fair summary of the broader discussion as recently framed in art, literature, and architecture, see Harvey (1989; esp. chaps. 1–6).

2. The term "scientific discourse" refers in this chapter to the language used by the individual scientist as a practice; that is, as a means of professional communication.

3. "Performative" is a term introduced by the English philosopher J. L. Austin (see reference list) to indicate a type of language that actually performs an action simply by the fact of its having been spoken, for example, a marriage vow such as "I do," which seals the event, or a legal command or declaration, which begins a proceeding. By using it here, I mean to suggest that technical language seeks, most often, the level whereby saying a thing in a certain way makes it true.

4. Edward Yoxen offers an interesting study on the general influence of this (see his "Speaking Out about Competition," in Terry Shinn and Richard Whitley, eds., *Expository Science: Forms and Functions of Popularisation*, Boston: Reidel, pp. 163–181). It is Yoxen's notion that Watson had didactic motives, mainly "to create a new image of scientific dedication in an age of highly competitive endeavour, when one's own lapses could lose the race." In short, Watson's book revealed, endorsed, and promoted the doing of science as a type of capitalist intellectual entrepreneurship, locked in competition for "market share." This is a different image from what is standardized within technical discourse itself, an image that comports well with Medawar's emphasis on the groping, scrambling nature of much scientific work.

On another front altogether, however, it is interesting to note that Watson's book has been officially redesignated as "literature," thanks to its reissue in 1980 as a Norton Critical Edition (including text, commentary, reviews, and original papers). This act of placing the book in such a series on "the history of ideas," that is, might be said to represent a means of divorcing it from "real science," keeping it in the realm of a work *about* scientific work, not a direct product of such work, an ingredient in the public image making that is also an integral part of science and that helps define its larger public context.

5. This requires a degree of qualification and explanation. Perhaps I should say that I don't mean to try and dismiss this field of endeavor, which in fact involves, and has developed, a wide number of different and valuable approaches to the study of scientific discourse (stylistic analysis, semiotics, rhetoric, deconstruction, and so on). This work is no doubt an essential part of the larger effort in Western culture to defeat an older, naive, positivist view of science itself, especially as embodied in its discourse, a view that believed scientific thought to be above and beyond most worldly concerns and therefore unaccountable, both epistemologically and politically. The more critical and historically proven danger of this view was its influence in areas such as racial definition, intelligence, education, national destiny (e.g., as a matter of "organic evolution of societies"), in short, politics by many means— areas we can easily see today were deeply tinged by motives other than "scientific truth" but which, in fact, were widely viewed as coterminal with such truth at the time. Adopting the speech, the gestures, the claims of the "scientific," so to speak, was a means in the 19th and early 20th centuries to legitimize many programs of aggressive sensibility, with sometimes monstrous consequences. As recently pointed out by Daniel Chirot (1994), it remains in place in many parts of the non- (or less-) industrialized world today and continues to wreak havoc. "Science"—or rather the "scientific"— as the language of unchallengeable truth must not be allowed its former styles of prestige.

Little of this larger historical and political background, however, seems to be a motivating force behind discourse studies at present. I don't mean that such studies *must* begin from such a background in order to be significant. But it often seems as if they are lacking a certain central rationale, other than to try and get at the "nature" of science once more, or else to make use of technical language as a kind of final frontier for certain analytic methods, earlier honed in other areas.

As a scientist, however, interested in his own language, in what it means to write and read it, to study its history and its effects outside science per se, I find I have other problems with this work, too. One of these lies in some of its more basic assumptions, which, in part, it shares with textual analysis generally. It is frequently aimed less at the actual contents of scientific knowledge than at its forms (the two are by no means identical), at technical writing itself as a type of formalist construction produced by generic entities (scientists being commonly referred to as "agents" or "actors" that "perform" the work of various "interests"). Because of this tendency, it views science as a more or less grand unity, or shared plurality of unities. We are directed to see Science more than sciences; Chemistry and Biology, not chemistries and biologies—since, as every scientist knows, each field too comprises a multiplicity of procedures, rites, competing actions and methods, varied uses of the laboratory, the field, and so on.

This type of thinking, like that of analytic philosophy before it, continues to go its own way, in search of its own master narrative or narratives, looking for evidence on such grand (or local) universals as "the process of discovery," how "the scientific text functions," or the "operations" by which "scientific facts are constructed"—in short, one or another universal element or mechanistic principle in "the system." Such writing, in short, too often wants to generate truth texts about truth-telling texts. Obviously my own efforts here are not wholly removed from this sort of thing. But there is a difference between trying to characterize technical discourse and trying to pronounce upon its final "nature" or universal elements (in Chapter 5, on technical discourse in Japanese, I offer some comparisons to help show that such universals are false and impossible).

Because of this search for modernist narratives—wedded no doubt, in part, to some of the most basic notions of academic scholarship, itself influenced by science during an earlier period (e.g., the need to define a bounded subject, advance a field, generate fresh knowledge that both reveals and overcomes the limitations of the past)—one finds that most writing in this area partakes of a general blandness and flatness, a timidity of literary eloquence, that seems technical in style. It is comprised in the main of articles that have adopted a very similar (indeed, imitative) type of language, with very similar types of deletion: the very same kinds of realities that are missing from scientific discourse are absent here, too. At a certain level, strange ironies abound: one finds a writer making near-identical use of such conventions as citation, reference lists, "discussion" and "conclusions" sections, declarative argument, the gathering of "evidence," and so on, even as these things are being examined in terms of their "formal strategies" in scientific discourse!

To be fair, something of this has been noted by sociologists of science (see, e.g., Latour, 1981; Mulkay, Potter, and Yearley, 1983; Woolgar, 1988). Some writers, Bruno Latour in particular, have adopted styles that, while nonetheless still seeking some final "truth" about science, engage a far more interesting, playful type of language. The issue as a whole has even been given a name: the "reflexivity problem." This refers to the idea that any ideas or statements *about* science are themselves constructions, with their *own* sociopolitical contexts, and must be looked at by means of a "self-reflexive process." Clearly, though, there is some type of containment going on here: not only is the problem defined in terms of insider jargon, but the real immediate context, that of sociologic or rhetorical analysis itself, is never questioned or called to account. The old criticism of science as being a "self-policing" realm seems here repeated. "Reflexivity" too often qualifies as a kind of designated peer review. The larger implications involved, which have everything to do with what types of knowledge serve best as academic capital, and what types of roles such knowledge might have in society, are rarely approached. "Reflexivity," finally, does not seem to go an enormous distance toward a broader reflectivity. Certainly, as a concept, it doesn't do very much for the quality of most sociological language (indeed, as in the past, it is the discipline of history, i.e., history of science, where allowance is still the fullest for more interesting styles of authorship).

One more recent approach seems to me to offer a fertile alternative. Fuhrman and Oeler (1986), as an example (see also the essays in Pickering, 1992), speak of the need to critically examine the place of belief in society, vis-à-vis "science." The point might be not to consider scientists merely as "social agents," negotiating and calculating their merry way toward truth and status, but rather the very opposite—society as a scientific "agent," providing structures that frame and teach those beliefs. This would be one way to bring back in the historical-political role.

The most committed attempt to give the analysis of technical language a larger, mediating role between the academy and society is surely that of Fuller (1993), who argues that such analysis, like the power of discourse itself, must be viewed in political terms. As he notes, ideas in our own century have gained enormous (and sometimes terrible) force less because of their truth value than because of "the ability of their procedural languages . . . to get people from quite different walks of life to engage in projects of mutual interest" (p. xvi). Fuller is one of the very few, at present, to try and engage debate—indeed new disciplinary boundaries—in this area. His vision for studies of rhetoric generally, or "social epistemology" (as he calls it), is very high: its practitioners, he says, "employ methods that enable them to fathom both the 'inner workings' and the 'outer character' of science without having to be expert in the fields they study" (p. xii). This may well be true, but I am nagged by a wasp of doubt: method alone, no matter how embracing, has never been able to lift its practitioners out of their own setting. Inasmuch as the new rhetoricians will likely be academics first, and science policy consultants (or the like) second, they will be constrained by the everyday realities of the university, the true center of knowledge production in the Western world—whose structure, influence, economics, and discourse might very profitably be among their first subjects for inquiry.

6. In this sense, technical discourse would appear to satisfy, in specific, what a particular brand of literary theory (structuralism) tried to claim for all of writing. The "dead" or missing author, that is, appears as a principal character in the extreme formalism of one such as Michael Riffaterre, who writes that:

> The text works like a computer program designed to make us experience the unique. This uniqueness is what we call style. It has long been confused with the hypothetical individual termed the author; but, in point of fact, *style is the text itself.* (1983 p. 2)

The pseudoscientific aura to this type of statement is one that claims "fact" by disclaiming any interest in the day-to-day, contextual, human side of writing. It is, in a very literal way, the opposing complement to the older fallacy of giving total, subjective intentionality to the author for all possible contents of a work. There is a desire, at some level, to kill the author by giving language vampiric powers. This technical vision, as I hope to show below, can never serve as the *summa theologica* of writing itself, scientific or otherwise.

7. Often it seems forgotten, for example, that such terms as "function" and "move" are themselves metaphors in this context. In my own experience, it is obvious that different authors write in different ways, and that individual authors also write in different ways, depending on the subject, the type of article, the type of importance the writer wants to have attributed to it. This happens, too, just as with other types of narrative, in a sometimes haphazard fashion, by fits and starts, even in eccentric isolation (one researcher I know, for example, dictates while in the bathtub; another writes only at night, while alone; a third, meanwhile, has consciously adopted Anthony Trollope as his model, sitting down each day for exactly the same length of time, ending each session by abruptly stopping when the appointed hour sounds, whether in the middle of a thought, a sentence, or even a word). Moreover, this tends to change depending on what is being written about, what time of year it might be (influences of academic or work schedule), what journal is being aimed at, whether any deadlines exist, and so on. Differences between individual authors and research teams notwithstanding, I would imagine the writing process in science to be no less variable and without essence than in any other area of discourse, professional or otherwise. This does not, of course, diminish the realities of convention that must be obeyed, no matter what the specific circumstance of writing. But it does put a different contextual frame around the type of subtle changes, the individual "personalities" that one can find in technical discourse.

8. For these reasons I find Martin Rudwick's (1976) famous study in this area a bit too narrow and contained, despite its monumental amount of information and its many valuable insights.

References

Anderson, B. (1983). *Imagined Communities: Reflections on the Origin and Spread of Nationalism*. London: Routledge.

Austin, J. L. (1962). *How To Do Things with Words*. Cambridge, MA: Harvard University Press.

Bakhtin, M. M. (1981). M. Holquist (Ed.), *The Dialogic Imagination: Four Essays*. Austin: University of Texas Press.

Barthes, R. (1977). The Death of the Author. Transl. by S. Heath. In *Image–Music–Text* (pp. 142–148). New York: Hill & Wang.

Barthes, R. (1973). The Division of Languages. Transl. by R. Howard. In *The Rustle of Language* (pp. 111–124). New York: Farrar, Straus, Giroux.

Barthes, R. (1985). Day by Day with Roland Barthes. In M. Blonsky (Ed.), *On Signs* (pp. 98–117). Baltimore: Johns Hopkins University Press.

Bastide, F. (1985). Iconographie des Textes Scientifiques. *Culture Technique, 14*, 15–35.

Bernstein, J. (1982). *Science Observed*. New York: Basic Books.

Bettelheim, B. (1982). *Freud and Man's Soul*. New York: Knopf.

Bloomfield, L. (1970). Linguistic Aspects of Science. In C. F. Hockett (Ed.),

A Leonard Bloomfield Anthology (pp. 205–219). Chicago: University of Chicago Press.

Caillois, R. (1973). *Coherences Aventureuses.* Paris: Gallimard.

Chirot, D. (1994). *Tyranny in the Modern Era.* New York: Free Press.

Christenson, G. (1990). The Florida Lineament. *Transactions of the Gulf Coast Association of Geological Societies, 40,* 99–117.

Christianson, G. (1984). *In the Presence of the Creator: Isaac Newton and His Times.* New York: Free Press.

Day, R. A. (1979). *How to Write and Publish a Scientific Paper.* Philadelphia: ISI Press.

Derrida, J. (1978). *Writing and Difference.* Transl. by A. Bass. Chicago: University of Chicago Press.

Eagleton, T. (1987, February 20). Awakening from Modernity. *Times Literary Supplement.* (Quoted in Harvey, 1989 p. 9; see below)

Foucault, M. (1965). *Madness and Civilization.* Transl. by R. Howard. New York: Random House.

Foucault, M. (1970). *The Order of Things.* Transl. by A. Sheridan. New York: Pantheon.

Foucault, M. (1973). *The Birth of the Clinic.* Transl. by A. M. Sheridan Smith. London: Tavistock.

Foucault, M. (1979). What Is an Author? Transl. by J. Harari. In J. Harari (Ed.), *Textual Strategies: Perspectives in Post-Structuralist Criticism* (pp. 141–160). Ithaca, NY: Cornell University Press.

Fuhrman, H., and F. Oeler (1986). Discourse analysis. *Social Studies of Science, 16*(2), 300–315.

Fuller, S. (1993). *Philosophy, Rhetoric, and the End of Knowledge.* Madison: University of Wisconsin Press.

Gardner, H. (1985). *The Mind's New Science.* New York: Basic Books. (Epilogue, 1987)

Gilman, S. (1988). *Disease and Representation: Images of Illness from Madness to AIDS.* Ithaca, NY: Cornell University Press.

Goodman, N. (1979). *Ways of World Making.* Cambridge, MA: Harvard University Press.

Habermas, J. (1971). *Knowledge and Human Interests.* Transl. by T. McCarthy. Boston: Beacon Press.

Haraway, D. (1989). *Primate Visions: Gender, Race, and Nature in the World of Modern Science.* New York: Routledge, Chapman, and Hall.

Harre, R. (1981). Preface. In K. D. Knorr-Cetina, *The Manufacture of Knowledge: Toward a Constructivist and Contextual Theory of Science.* Oxford: Pergamon.

Harvey, D. (1989). *The Condition of Postmodernity.* Oxford: Basil Blackwell.

Hoffmann, R., and P. Laszlo (1989). Representation in Chemistry. *Diogenes, 147,* 23–51.

Holton, G. J. (1978). *The Scientific Imagination: Case Studies.* Cambridge: Cambridge University Press.

Holton, G. J. (1986). *The Advancement of Science, and Its Burdens.* New York: Cambridge University Press.

Knorr-Cetina, K. (1981). *The Manufacture of Knowledge: Toward a Constructivist and Econtextual Theory of Science.* Oxford: Pergamon.

Jacob, F. (1973). *The Logic of Life.* Transl. by B. Spillman. New York: Pantheon.

Kristeva, J. (1980). Word, Dialogue, and Novel. Transl. by T. Gora, A. Jardine, and L.S. Roudiez. In *Desire in Language* (pp. 64–91). New York: Columbia University Press.

Kronick, D. (1976). *A History of Scientific Periodicals: The Origins and Development of the Scientific and Technical Press, 1665–1790.* Metuchen, NJ: Scarecrow Press.

Lapworth, C. (1883). The Secret of the Highlands. In K. F. Mather and S. L. Mason (Eds.), *A Source Book in Geology, 1400–1900* (pp. 578–580). Cambridge, MA: Harvard University Press.

Latour, B. (1981). Insiders and Outsiders in the Sociology of Science; or, How Can We Foster Agnosticism? *Knowledge and Society, 3,* 199–216.

Latour, B., and S. Woolgar (1979). *Laboratory Life: The Construction of Scientific Facts.* Princeton, NJ: Princeton University Press.

Latour, B. (1987). *Science in Action.* Milton Keynes, England: Open University Press.

Laubscher, H. (1969). Mountain Building. *Tectonophysics, 7*(5/6), 551–563.

Levy-LeBlond, J. (1975). Ideology of/in Contemporary Physics. In H. Rose and S. Rose (Eds.), *Ideology of/in the Natural Sciences* (pp. 120–182), Boston: C.K. Hall.

Lyell, C. (1990). *Principles of Geology.* 3 vols. Chicago: University of Chicago Press.

Lyotard, J. F. (1984). *The Postmodern Condition: A Report on Knowledge.* Transl. by G. Bennington and B. Massumi. Minneapolis: University of Minnesota Press.

Mather, K. F., and S. L. Mason (1939). *A Source Book in Geology, 1400–1900.* Cambridge, MA: Harvard University Press.

Medawar, P. B. (1984). *The Limits of Science.* New York: Harper & Row.

Medawar, P. B. (1990). Is the Scientific Paper a Fraud? In *The Threat and the Glory.* New York (pp. 228–233). HarperCollins.

Mulkay, M. J., J. Potter, and S. Yearley (1983). Why an Analysis of Scientific Discourse is Needed. In K. Knorr-Cetina and M. J. Mulkay (Eds.), *Science Observed* (pp. 171–204). London: Sage.

Myers, G. (1988). Every Picture tells a Story: Illustrations in E. O. Wilson's *Sociobiology. Human Studies, 11,* 235–269.

Nelkin, D. (1987). *Selling Science: How the Press Covers Science and Technology*. San Francisco: W. H. Freeman.

Pickering, A. (Ed). (1992). *Science as Practice and Culture*. Chicago: University of Chicago Press.

Riffaterre, M. (1983). *Text Production*. Transl. by T. Lyons. New York: Columbia University Press.

Robin, H. (1992). *The Scientific Image*. New York: Harry N. Abrams.

Rouse, J. (1991). Philosophy of Science and the Persistent Narratives of Modernity. *Studies in the History and Philosophy of Science, 22*(1), 141–162.

Rudwick, M. (1976). The Emergence of a Visual Language for Geological Science, 1760–1840. *History of Science, 14*, 149–195.

Sartre, J-P. (1949). *What Is Literature?* Transl. by B. Frechtman. New York: Philosophical Library.

Savory, T. H. (1967). *The Language of Science*. London: Deutsch.

Schoffer, N. (1978). *Perturbation et Chronocratie*. Paris: Denoel.

Steiner, G. (1984). *George Steiner: A Reader*. New York: Oxford University Press.

Todorov, T. (1984). *Mikhail Bakhtin: The Dialogical Principle*. Transl. by W. Godzich. Minneapolis: University of Minnesota Press.

Toulmin, S. (1985). *The Return to Cosmology: Postmodern Science and the Theology of Nature*. Berkeley, CA: University of California Press.

Vail, P. R., and R. M. Mitchum, Jr. (1979). Global Cycles of Relative Changes of Sea Level from Seismic Stratigraphy. In J. S. Watkins, L. Montadert, and P. W. Dickerson (Eds.), *Geological and Geophyscial Investigations of Continental Margins* (pp. 469–473). Tulsa, OK: American Association of Petroleum Geologists.

Wade, N. (1977). *The Ultimate Experiment: Man-Made Evolution*. New York: Walker & Co.

Watson, J. D. (1980). *The Double Helix. Text, Commentary, Reviews, Original Papers*. New York: W. W. Norton.

Webster, A. (1991). *Science, Technology, and Society*. New Brunswick, NJ: Rutgers University Press.

Weinberg, S. (1992). *Dreams of a Final Theory*. New York: Pantheon.

Wolfenden, J. (1963). The Gap—and the Bridge. In *The Languages of Science* (pp. 19–34). New York: Fawcett.

Woolgar, S. (1988). *Knowledge and Reflexivity: New Frontiers in the Sociology of Knowledge*. London: Sage.

Yoxen, E. (1985). Speaking Out About Competition: An Essay on *The Double Helix* as Popularisation. In T. Shinn and R. Whitley (Eds.), *Expository Science: Forms and Functions of Popularisation. Sociology of the Sciences, Yearbook* (pp. 163–182). Dordrecht: D. Reidel.

2

In Equal Number of Words

Notes for a History of Scientific Discourse

🔲🔲🔲🔲🔲🔲🔲🔲🔲🔲

The greatest learning is to be seen in the greatest plainnesse.

—JOHN WILKINS, *Ecclesiates, or a Discourse Concerning the Gift of Preaching,* 1646

During the past two decades, with the rise of interest in language generally, the issue of discourse in science has attracted ever growing attention. Like a bright colored thread in an otherwise drab-tinted textile, the problem of language weaves its way through contemporary study in the sociology, history, philosophy, and cultural theory of scientific work and thought.

Different approaches to the study of scientific language are everywhere in evidence.[1] Replacing the old view of analytic philosophy, intent for the most part on discovering the laws of truth presumably inherent in technical expression, one now finds, as summarized recently by

Golinski (1990a) and Fuller (1993), that, like literature itself, scientific discourse is an object of analyses that draw their inspiration from hermeneutic and semantic approaches, from studies of metaphor and of onomastics, from detailed and comparative mapping of rhetorical strategies, from narratology and stylistics, even from reader-response criticism. Obviously, there is no single term to describe this work, varied as it is. One might instead say (to employ a phrase much in favor) that the new crop of investigators are out looking, in one form or another, for the patterns in the carpet that might offer some secret to the "literary technologies" of science.

This being said, however, enormous gaps remain. And one of the most striking, from any point of view, is the lack of a coherent history of scientific discourse. Curiosity demands an answer to certain questions: Where did this language come from? Given its enormous power today, commensurate with the power of scientific knowledge itself, how did it develop? What type of linguistic evolution and philosophy brought it to its present state? What stages did it go through, linked to what circumstances in Western culture?

In my own experience, few if any scientists have ever thought very much about these questions, let alone considered them as comprising a riddle whose solution might be of importance to the meaning of their work. This is not surprising, given the nature of today's scientific training. But no less is it true that scholars recently embarked on analyses of scientific language have failed to yet assemble the necessary materials or make the full-hearted attempt to construct such a history themselves. This, in part, may be due to the lack of an agreed-upon scheme for interpreting this language (see, e.g., Markus, 1987). Competitive variety among current approaches might just as easily act to postpone or discourage any single overview as to promise its eventual realization. Pieces, to be sure, are strewn throughout the relevant literature like fragments of a lost mosaic. There are now literally hundreds of studies of individual authors and their rhetorical strategies, of trends that developed during particular time periods, of the contributions or failed innovations one or another canonical author made toward the building of a truly scientific language, and so forth. Thus one discovers the suggestive shards for a number of different histories, perhaps a true beginning here and there. But however far and deep one may penetrate this literature, however broad its scope may seem, however reliant on the ideas of sophisticated theorists such as Michel Foucault, Hans-Georg Gadamer, Jacques Derrida, and Roman Ingarden, one will find revealed no single picture out of all this work and thought, no synthesis. At present, there exists no real history of scientific language, only a field of possibilities.

My attempt in this chapter is less to build such a history than to suggest certain elements necessary to it. The scholarly histories that remain to be written, particularly those that will seek after the "rhetorical functions" and "literary constructions" of technical discourse, comprise a formidable task. It is a task, moreover, that will reward some and not others. What I offer here are a few essential points that reveal themselves perhaps more to a different audience, to the writer of scientific language, with his/her daily experience of use. What has struck me most of all (as a user), time and again, is the change that has taken place over the centuries in what it means to be an author of scientific texts. What type of "author" was this, say 100 or 200 years ago? How was this author related to a certain type of reader, whether presumed or embedded in the text? How have both science writer and science reader evolved?

In what follows, I want to try to sketch out the contours of this authorship–readership relation at its inception—to speak about how the modern discourse of science began, the ambitions and freedoms it sought after, its inevitable loyalty, early on, to certain portions of the medieval literary past. I will then attempt to provide an outline of what happened to this discourse, the manner in which it developed, how it came, eventually (and only very recently), to achieve something of its original aims. I should say that it is not the dead space of pure texts, those academic objects built out of "strategies" and "moves," with which I am concerned. Nor is it the Author, as grand master or unconscious dupe. I am interested, instead, in scientific language as a major current in human expression during the modern era, a current that began by striving for openness and freedom, yet which centuries later came to close in upon itself, becoming accessible only to the trained and chosen few.

I

When did something like a technical vocabulary first make its appearance in English? The answer seems to lie in what today might be considered an unlikely source. Scientific language in English first appeared not in some major treatise from the hand of a thinker such as Roger Bacon or Robert Grosseteste, but instead in a small volume entitled *A Treatise on the Astrolabe* (dated 1391), written by none other than Geoffrey Chaucer. This text was composed by the poet to instruct his young son about the "new astronomy." Chaucer's *Treatise* contains the first known occurrence of many English words having to do with astronomical calculation and with navigation, for example, *azimuth, nadir, zenith, declination, ecliptic,* and *equinox.* Most of these were anglicized from the Latin, but a fair number were actually introduced from Arabic (indeed, Arabic names are used for several stars that now are known by Latin names, e.g., "Alhabor" for "Sirius"). Chaucer had evidently picked them up from various astronomical textbooks in use at the time and had simply vernacularized them for his own teaching. As he writes in his preface to "Lyte Lowys [Little Lewis] my sone," "sufise to thee these trewe conclusions in Englissh as wel as sufficith to these noble . . . Grekes these same conclusions in Grek; and to Arabiens in Arabik, and to Jewes in Ebrew, and to Latyn folk in Latyn; whiche Latyn folk had hem first out of othere dyverse langages" (1987, p. 662).

Such terms had actually been in use for almost two centuries prior to Chaucer's writing. They had entered European discourse along with the astrolabe itself (an instrument for determining position and for performing certain astronomical calculations) sometime in the 10th or 11th century. By the 1100s they had become common, made all the more so by a major period of translation that lasted more than 100 years and that brought into Occidental hands the overwhelming riches of Greek and Arabic science.[2] This episode proved the foundation and intellectual reservoir for the rise of the European universities in the late 12th century. By Chaucer's day, the astronomical terminology that had come from the works of Ptolemy and al-Sufi were an ordinary part of the technical vocabulary of Latin scholarship and science, and thus of university learning. What Chaucer did was to remain loyal to this larger context. His was not a scientific discourse aimed at inquiry, but instead a pedagogic one, embedded with certain terms of science. Yet his work, though small and ignored today, could be said to have marked a crucial moment, when an early form of scientific jargon was introduced into the common language, there to establish a foundation on which much could be built. To this degree, it seems hardly unfitting that a poet whose influence on the English language was no less profound than Shakespeare's should prove equally important to science.

🖫🖫

The true beginning of scientific writing, however, as a conscious *style,* occurred sometime later, in the 17th century. The primary mover is acknowledged to be Francis Bacon (1561–1626), whose critique of the past and whose program for the future were largely focused on language, both as structure and as method. It was Bacon's proclamation, first in *The Advancement of Learning* (1605) and then more loudly in the *Novum organum* (1620), that in order to progress beyond medieval sophistry, knowledge would require a new type of speech, a plain and unadorned style of writing capable of carrying the truth of the world in as direct a manner as possible. Previously, learning had fallen into the trap of favoring fine style at the expense of true inquiry; language had become a veil between the mind and the world. Scholasticism had allowed the power of words to "force and overrule the understanding, throw all into confusion, and lead men away into numberless empty controversies and idle fancies" (1857–1859, vol. 4; pp. 54–55). This power Bacon called a pagan darkness and a glittering coinage, labeling it an "idol of the Market-place . . . on account of the commerce and consort of men there" (vol. 4; p. 55). Natural philosophy (science), he argued, had failed to advance significantly over the ancient achievements of the Greeks and Romans because "sophistical and inactive" discourse had urged men to concentrate their powers on skill in rhetoric and not on physical realities and processes, on "words instead of things." What was now required was a simple, direct, and literal form of speech, and a terminology more tied to actual objects and related ideas, rather than to "fanciful" types of writing and "unskilled abstractions."

The Baconian view was therefore less a philosophy than an attitude toward language. On the one hand, Bacon expressed a profound distrust of words, of their ability to "react on the understanding" and to corrupt the mind. He held that the time-honored belief that reason would always triumph over words was false: language had its own power to defeat rational thought. On the other hand, Bacon also saw language as detached from the material world. It has no immanent link to truth, but must instead be brought under planned, systematic control, according to a particular vision. In order to become gold, it first had to be beaten into iron. It had to be conceived as a medium of exchange, or else as one of the natural philosopher's instruments. Bacon thus was not opposed to language per se, but was instead in favor of a particular kind of language (Vickers, 1985). The successful scientist would build a lexicon through a process of cataloguing the physical universe and its actions: names and simple descriptions would provide a base for higher concepts. Inquiry would be based on the witness of natural events and the casting of this private witness into a public speech. This speech, moreover, would be capable of conveying and reproducing the experience itself in a plain, direct manner. The scientist's language would therefore consist, at the highest level, of a series of tightly defined and linguistically discrete object-areas for study, for example, light, heat, motion, and magnetism. Once established at the level of terminology, these areas could then (and only then) become the sites of a progressive building up of observations, questionings, and conclusions. Study of them would advance because it would be rooted in material reality, not in words alone. Each new discovery would have its own name and would add to the great and expanding catalog of the world.

Bacon spoke as if the study of "experimental philosophy" was to begin with his work. Reading him, one gets the feeling that he sees himself as the pinch in the historical hourglass, the prophet of a new age about to emerge, full-blown, from his forehead. Yet, as one would expect, his Olympian ambitions drew heavily upon the material of other writers (Vickers, 1968). One example, seldom mentioned, was surely that of Richard Hakluyt, famed author of *Principall Navigations, Voyages, and Discoveries of the English Nation* (first issued in 1589, later enlarged into three volumes in 1598–1600), a work called the "epic of the English nation" and one of the most widely read and influential texts in the whole 16th century (Shakespeare was well acquainted with this work and drew some of his knowledge and scenery from it). Hakluyt was responsible for introducing a great deal of Renaissance geographic knowledge into England. He did so, moreover, by means of narratives that were intended to be very close to the speech of "plaine folk." This often took the form of firsthand accounts by various mariners (which Hakluyt transcribed from English sources or translated from sources in other languages), whose directness and simplicity offered the English reader something close to an empirical first look at many unknown parts of the Earth. Voyages, past and recent, were presented as if they were historical experiments of a sort, attempts to combine adventurous curiosity, personal gain, and increased wealth and prestige for England. Hakluyt's narratives also included descriptions of new phenomena— peoples, birds, flowers, trees, fish, animals, and so on—always reported in a straightforward, simple style. Narratively, the New World and the new era it proclaimed came to life through the medium of language. The effect could not have been lost on Bacon: the substance of Hakluyt's fame cannot be easily separated from this sensibility of opening up the English eye and mind to "the diversse realities" of the world.

Perhaps, however, the more important and direct influence on Bacon's philosophy came from his near-contemporary, William Gilbert (1544–1603), well-known London physician and author of *De magnete* (On the magnet), published in 1600, with which Bacon was intimately familiar. A second work of Gilbert's, *De mundo* (Concerning the world), was left unpublished at his death, but was edited by his half-brother and circulated in some form among court intellectuals such as Bacon and Thomas Harriot before finally being brought out in 1651 by an Amsterdam publisher (Kelley, 1965).

As a number of scholars have pointed out, Bacon was familiar with both these works of Gilbert. Kelley, who has studied the matter in some detail, states that "he [Bacon] praised the experimentation of the *De Magnete*, condemned the philosophy of the *De Mundo*, and used examples from both when it was convenient to do so" (1965, p. 95). The year *De magnete* was published, Gilbert was made royal physician to Queen Elizabeth and, following her death, to James I. Bacon was at that time aggressively and successfully pursuing his worldly rise in court society (he became solicitor-general in 1607), and while there is no direct evidence of any personal contact with Gilbert, one might surmise that the two men were not unknown to one another, perhaps eyeing each other now and then from afar. In any case, Bacon mentioned Gilbert many times in his own writings. Most often, this was to dismiss the latter's attempt to make "a philosophy out of the observations of a loadstone": "Gilbert . . . has not unscientifically introduced the question of magnetic force, but he has himself become a magnet; that is, he has ascribed too many things to that force, and built a ship out of a shell" (quoted in Kelley, 1965, pp. 78–79).

This was aimed more at *De mundo*, in which Gilbert sought to apply some of his ideas of magnetism to explanations of the cosmos. *De magnete*, on the other hand, if published several decades later than it actually was, would have satisfied a fair number of Bacon's demands for the "experimental philosophy" (Sarton, 1957). Gilbert not only presented a survey of existing knowledge and speculation on the magnet, but went through each theory individually and either rejected (usually) or confirmed it on the basis of his own observational and experimental evidence, ending in his famous theory that the Earth itself must be considered a magnet. As Kelley (1978) notes, the experiments are given in such detail that they could be easily duplicated today, while natural examples are equally well identified and instructions are given for the construction and use of the new instruments Gilbert employed in his work. Moreover, Gilbert indicated his own discoveries and observations by marking them with asterisks in the margin, these being graded in terms of their size with the presumed importance of the material. He thus plainly divided "old" from "new," created his own terminology, and even offered an introductory glossary. Though, again, any direct evidence is lacking,

Bacon's condemnations of Gilbert were likely intended to create some room and distance for the prophetic qualities of his own philosophy. Downplaying Gilbert in terms of his excesses was one way to maintain his own claim to patrimony of the "new science."

To call *De magnete* a complete preexisting model of Baconian method, however, would be exaggerating the point. The book is also full of speculative reasoning and a similar brand of cosmological reductionism found in *De mundo* (it was this portion of the book that drew criticism). Though Gilbert is at pains to break away from much of medieval thought, some of his ideas are animistic, Neoplatonic, and were adopted directly, for example, from Peter Peregrinus's 13th-century *Letter on the Magnet* (Zilsel, 1941). Gilbert thus had one foot squarely planted in early Renaissance soil; inasmuch as he tried to separate old from new, his work also combined the two and thus reveals itself as a (required) transitional step into the Baconian era. What, then, of Gilbert's attitude toward language? Here, the evidence is convincing regarding his "modernism." In one of his early chapters on the nature of the lodestone, he begins with the following comment:

> First we have to describe in popular language the potent and familiar properties of the stone; afterward, very many subtle properties, as yet recondite and unknown, being involved in obscurities, are to be unfolded; and the causes of all these (nature's secrets being unlocked) are in their place to be demonstrated in fitting words and with the aid of apparatus. (1893, p. 3)

The correlation between "plaine speaking," inquiry, and discovery is here neatly laid out, as is the idea that "fitting words" are a form of "apparatus" for any clear-thinking philosopher. Elsewhere, moreover, in his discussion of "magnetic coition" (Book 2), Gilbert expresses his impatience that "our generation has produced many volumes about recondite, abstruse, and occult causes and wonders," all of which have contributed, he says, toward a philosophy that "bears no fruit; for it rests simply on a few Greek or unusual terms—just as our barbers toss off a few Latin words . . . and thus win reputation—bears no fruit, because few of the philosophers themselves are investigators, or have any first-hand acquaintance with things" (p. 47). Again, the idea that language can provide a false front for knowledge seems evident—indeed, Gilbert suggests that "jargon" (in Greek or Latin, no less) acts to defeat knowledge, to keep inquiry at bay, to screen one off from direct observation. Words have the power to mislead, to falsify, and even to prevent true thought. Though such linguistic ideas are here nascent, they are obviously present, and could not have been lost on Bacon. Indeed, Gilbert's views are reflective of a general movement in the 16th century, in which various "secrete knowledges" (e.g., alchemy), till then the provinces of insider initiation and learning, were becoming suspect for these very reasons. As scholars such as Hannaway (1975) and Golinski (1990b) have pointed out, it was one of the goals of the "new philosophy" even before Bacon to overturn this tradition of secrecy by means of experiment and observation, offered in language accessible to any educated person (i.e., unjargonized Latin). Gilbert is one example of this trend—a particularly important one, in all likelihood, for his influence on Bacon.

This influence, it must be said, does not at all diminish Bacon's own significance. Inasmuch as any prophet arrives on the scene at a particular moment of transition, gathering in effective form the larger signs of historical change and sensibility, Bacon can be said to have served this role for "experimental philosophy" in England. That he had his correlatives in other nations, such as Mersenne and Descartes in France and Galileo in Italy, confirms this truth all the more.

Some scholars have viewed Bacon's program as radical, innovative, even as marking an "epistemic break." The program has been frequently interpreted as inaugurating something critical to the onset of modernity (however defined). Yet Bacon was also very much a man of his time. His emphasis on the knowability of the universe was not his own, but had come in the wake of the mentioned episode of translation from Arabic and Greek texts, an episode that helped launch the university as a new center of learning and that, by Chaucer's time, had inspired a belief in the physical order of the universe and in the ability of human beings to perceive that order and, more importantly, to record it in words. Scholasticism, moreover, was not so irrational in its aims nor bereft of investigatory powers as Bacon would have it seem. Indeed, his own attitude toward language can be found, in nascent form, in what was no doubt the central subject area for medieval learning: rhetoric. The enthronement of rhetoric as the king of all the disciplines after roughly 1300 involved granting special privilege to one of its several major areas of methodology, the *ars praedicandi*, or "science of preaching." This "science" was aimed at the writing and delivering of sermons and was therefore the most didactic of all, the most potentially influential with regard to masses of people (its rise being predicated, in part, on the spread of lay preaching). The *ars praedicandi*, as commonly taught, employed a methodology in the composition of sermons that was distinctly inductive: a passage from the Bible would be chosen for its message; broken down linguistically into its major parts (*divisio*); then further divided into *distinctio*, each of which would then be studied and used as a discrete model or pattern for paraphrase. Writing the sermon would involve assembling a virtual hierarchy of such paraphrases and their intended effects. Stylistically, moreover, unlike other forms of rhetoric, the *ars praedicandi* "held fast to the rejection of ornate style in favor of unadorned truth" (Camargo, 1983, p. 114; see also the more extensive discussion in Murphy, 1978). Such was in direct keeping with a didactic purpose for near-universal access, a purpose close to the heart of Bacon's own program. The elements of using reason incrementally to construct from a particular "object" a persuasive text, in plain and controlled prose (of a type that consciously rejects other types of eloquence), thus seem to have existed well before the 17th century.

II

That Bacon, along with Galileo and other founding members of the "new philosophy," very often relied upon traditional rhetorical forms should hardly be a surprise. Bacon himself was a royal judge, and was influenced not only by the *ars praedicandi* but by those forms of expression and fact-finding methods particular to English common law (see Armstrong, 1975). This, of course, only makes sense. Bacon's own education and learning were based on existing rhetorical approaches. In writing his manifesto—a sermon of sorts with an overt didactic purpose—it was inevitable that Bacon would employ the styles of expression invested with persuasive power in his own time—styles that involved skilled, sophisticated turns of argument, musical phraseology, complexities of rhythm, and extended metaphors.[3] No paradox exists between Bacon's stated purpose of simplicity and his own type of carefully crafted writing. Like his colleagues elsewhere, he wrote for the readers of his time. To suddenly abandon what it meant to be an Elizabethan writer would have rendered him, in effect, incomprehensible, and certainly unpersuasive. Neither a dead-end rationalist nor a cunning or inadvertent mystifier, Bacon was, among other things, a skilled author, and his prose reflects much of what was required for this role during the period. It is by no means an accident of history, or of taste, that Bacon is second in fame only to Shakespeare as a writer of this time.

It is important, then, to look at the context in which the "new experimental philosophy" was launched. Bacon wrote at a very particular moment in the history of English. The Elizabethan and Stuart eras, roughly the century after 1540, marked the final stages of a vernacular revolution. Latin was collapsing as the universal discourse of learning, and its decline was balanced by a rich flowering of English vocabulary, phraseology, and literary styles. No period before or since has seen so much linguistic invention and experimentation. As in other European countries, this efflorescence prompted a good deal of controversy over questions of "eloquence," the "right manner of speaking," and language in general as a medium of expressive and educative power.

At the center of the controversy was a debate that consciously patterned itself on a famous similar conflict in imperial Rome, one that involved a reaction against the lavish, oratorical style of Cicero, who wrote mainly for the upper classes, in favor of the more concise, everyday language of Seneca and Tacitus. The argument in England was basically against Scholastic rhetoric, with its emphasis on luxury of sound and artificial form, but it was also directed at the somewhat chaotic state of English generally in the late Elizabethan period, when spelling, punctuation, grammar, and word invention were all at the mercy or whim of individual authors, with few general standards to ensure some degree of uniformity. The demand for greater plainness was, in part, a demand for control. Whether it came from the pen of Montaigne (for the anti-Ciceronian movement had begun in France and Italy earlier than in England), from that of Robert Burton in his famed *Anatomy of Melancholy* (1621), or from Bacon himself, it declared a need less for complete asceticism than for improved directness, for a style that would carry more content than "finery," and that would therefore be available as a useful tool for a greater part of the national citizenry. To exemplify skill and command over words, one needed, as the poet Thomas Randolph put it, to be "pure, and strong, and round; / not long, but pithy; being short-breath'd, but sound / Not long and empty; lofty but not proud; / subtile but sweet, high but without a cloud, / well setled, full of nerves, in briefe 'tis such / That in a little hath comprized much" (*To Mr. Feltham on His Book of Resolves*; see Parry, 1897, p. 126). Rhetoric, then, was not to be totally abandoned, but instead honed toward a more solid and consistent delivery of "sense." What this meant, in effect, was a shift toward greater awareness of the reader, as a subject to be taught and influenced. It indicated a desire to get away from the self-absorbed, soliloquy-like concentration on sound, luxurious phrasing, and literary play for its own sake so characteristic of Elizabethan writing.

During the later half of the 17th century, the "simplicity" side in the debate triumphed. As is often pointed out, literature in general after about 1660 underwent a distinct change, becoming more simple and less fanciful than it had been only a few decades earlier, in the process moving much closer to what we recognize today as "modern" English. This change has traditionally been credited to (or blamed on) the "remarkable advance of science . . . and the dominance of the scientific attitude" (Horne, 1965, p. 188; see also Jones, 1951, 1953). But the truth is that the context for change was much larger than this. Indeed, the fact that the "scientific revolution" in 17th-century England had deep roots in the desire for *linguistic reform* must be seen as an effect of social as well as intellectual shifts.

The Baconian, or neo-Senecan position, was really a complex response to, and reflection of, many influences. One of these, perhaps the most obvious, was the rise of a Protestant middle class, with its demand not merely for greater sobriety in outward demeanor but for critique of existing traditions, and for new avenues of contact between the individual soul, the world, and God. Oxford and Cambridge Universities remained largely backward and medieval in outlook; education at this higher level continued to resist any accommodations to a present whose curiosities were fast expanding. Secondary schools, too, entered the century teaching little more than a mastery of Latin, and would have left it as such if they had their own way. Both therefore added fuel to the fire for reform, in particular for a more direct link between knowledge, discourse, and contemporary life. "The Puritan middle class," one historian of the era has written, "learned or unlearned, felt a natural impulse to replace the old, abstract, aristocratic, and 'useless' studies with the modern, concrete, popular, and useful. Their watchword was 'the public good', the union of progress and piety" (Bush, 1952, p. 21).

Added to this was an increase in prosperity and population, especially in the towns, where commerce and technology were rapidly advancing. The reading public burgeoned and periodical literature thrived: newspapers came into existence before 1630, pamphleteering bloomed a few decades later, and the literary and scientific essay became common by 1690. The older system of patronage began to fall; the new patrons of the word, the booksellers and printers, were only too ready to give the public whatever it desired. One thing that the public had little use for were books in Latin. Toward the close of Bacon's day (he died in 1626), it was no longer possible—or advisable—for an author interested in reaching a sizeable English audience to publish in anything but the vernacular. Burton evidently originally intended to compose his *Anatomy* in Latin, but changed his mind after he was told he would not find a publisher. Most of Bacon's Latin works were translated into English by midcentury. The 17th century was the first great era of translation into English, especially of classical Greek and Latin works, but also of French, Italian, and Spanish literature. The growth of lay education and public literacy went hand in hand with the increased production and availability of printed matter, but it was also tied up, inevitably, with a nationalism that saw England as a great power in the world. Indeed, the very terms "Ciceronian" and "Senecan," so avidly employed at the time, suggest the degree to which English scholars saw their homeland as a possible imperial residence of culture.

In general, the changes wrought by the Renaissance in the market for books, in the methods of expression, and in the sensibility of authorship combined to create a new type of reader. This reader was no longer either a cleric, an aristocrat, or a member of the professions. He (and now often she) was a layperson, eager to be worldly, hungry for news, educated to some degree but perhaps more desirous now of the status accruing to those who bought, read, and discussed books and periodicals. The battle over language thus spread to an audience of new readers, people who hadn't been trained in rhetoric, couldn't read Latin, and wanted accessible, understandable texts. As F. P. Wilson once remarked, with so much interest in language and its uses at the time, one might profitably ask of the century what new forms of writing it brought into being. Scholars have pointed to the beginnings of modern biography and to the vogue for the character ("a witty analysis of a social type"). The 17th century also was marked by the production of many dictionaries, grammars, spelling guides, and other attempts to impose control on the unruly beast that English was perceived to have become. But of equal or even greater importance, surely, were two forms that Bacon himself pioneered. One of these was the first true modern history, his *History of Henry VII* (1622), which broke irretrievably with the traditional chronicle form of narrative and instead offered a critical, explanatory analysis of people and events inspired by the great Roman historians Livy and Tacitus, but going well beyond them too. The other form Bacon helped introduce, of course, was the essay, which he adopted directly from Montaigne and in which he employed a chiseled, epigrammatic style that, in some ways, offers a presage to his somewhat later program for language in science.

Other changes and modifications in writing occurred. During the latter half of the century, for example, after Bacon had left the scene, the sermon became more plain in speech and more direct in appeal, seeking less complexity of expression reminiscent of Latin (Jones, 1951). Travel writing, as a popular genre, as well as the novel (whose early forms had appeared in the previous century) continued, but at a somewhat reduced level. Travelogue, in particular—which only a few years before had as its representatives the likes of Richard Hakluyt and Sir Walter Raleigh—changed its focus to less mythic and political aims in the mid-late 17th century, becoming more educational, even "scientific" in focus (Parks, 1951, Sutherland, 1969). In part, this may have reflected a more inward focus on knowledge and its worth, on nation, town, and self, also heavily indebted to the Puritan Revolution, with its years of civil strife and social upheaval. The novel, meanwhile, which had an early start in such Elizabethan writers as Thomas Nashe and Thomas Deloney, found interest among many minor writers, to be sure, but no great new progenitors until the time of Defoe. Imaginative writing did not languish; on the contrary, it expanded its readership, via tales of romance, intrigue, and gallantry, to more fully include both men and women. But a large part of its inspiration seems to have come from translations, notably from the French and Spanish (Mlle. de Scudéry, Countess de La Fayette, and Cervantes being among the most influential). The greater force and the larger amount of effort overall was clearly expended in the area of prose writing, rather than in fanciful story-telling.

The reader that emerged over the course of the 17th century, therefore, was one more likely to be interested in language as the embodiment of observation and experience. Biography, social satire, history as an analysis of character and action, the essay—all these reflected a type of authorship that both sought the reader's confidence and promised accountability. It also looked for general laws, especially of human behavior and thought. Observation of people, events, and things, and the reporting of such observation, were paramount. Projected onto nature, then, were many aspirations of the new readership, the new desire for observation and experience. This nature still comprised a kind of cryptic message, or book of messages. Yet uncovering its secrets now came to mean opening this book and copying it out for all to see. It meant a process of linguistic exposure, of making public what one had observed through a type of language that came close to observation itself, that could penetrate, reveal, repeat, and therefore—more than ever—persuade.

Given his use of the essay and his approach to the writing of history, it is not at all surprising that Bacon became the one to lay out the groundwork for such a language. He and his followers in the intellectual circle of Samuel Hartlib (active in the 1640s and 50s), and later the Royal Society itself, represent a particular movement within the larger process of vernacularization. Baconian criticisms, as well as Baconian ideas regarding a new, plain language for science, can be viewed as an attempt to seize control of this process in its final phase, to use nature as means for working out the essential conflicts between a Scholastic establishment holding the reins of higher education and a growing cadre of writers and thinkers outside this establishment who were interested in the everyday world as it actually was. Bacon's new language was not a matter of simple Protestant asceticism. In fact, it offended many Puritans by suggesting that "God's design" had to be revealed by new Scriptures of a nonbiblical sort, those of an experimental and skeptical kind. Rather, the Baconian program was an argument for using language as both a probing and recording instrument. It was not, therefore, to be either overly elaborate or overly blunted.

The *Advancement of Learning* was written in 1605, during the early stages of anti-Ciceronianism in England. Twenty years later (1623), reacting to excesses of enthusiasm in this direction, Bacon produced *De augmentis scientiarum*. Here he condemns, in no uncertain terms, "a style in which all the study is to have the words pointed, the sentences concise, and the whole composition rather twisted than allowed to flow" (Works, iv, p. 285). The new science could be plain, stoical, focused on sense. It could not, however, if it were to be effective with regard to truth, be dull or devoid of pleasure. Concision for its own sake was no less a rhetorical "trick," a "dressing up" (rather, down). It might just as easily become a formula, employed for its own sake more than for truth.

When it was chartered in 1662, the Royal Society of London, England's first true professional scientific group, proclaimed as gospel the strict anti-Ciceronian position. Several of its members did this so often, and with such importunate redundancy, that one is urged to examine the results. Before doing so, however, it is interesting to note just how far and in what type of direction this gospel propelled itself. Linguistically speaking, its most radical member was John Wilkins, who played a leading role in a short-lived movement to try and conceive a universal language, a so-called *Ars Signorum* comprised of a lexicon of neutral, unequivocal symbols that would

> remedy the difficulties and absurdities which all languages are clogged with . . . by cutting off all redundancy, rectifying all anomaly, taking away all ambiguity and aequivocation . . . giving a much more easie medium of communication than any yet known, but also to cure even Philosophy itself of the disease of Sophisms and Logomachies;

as also to provide her with more wieldy and manageable instruments
of operation, for defining, dividing, demonstrating etc. (quoted in
Jones, 1951, p. 154)

Notwithstanding Wilkins's own inability to live up to such an ideal, his
Essay towards a Real Character and a Philosophical Language (1668)
expressed the two major failures of language that the Royal Society felt
Bacon had most identified. These were the ability of a word to possess
more than one meaning and the use of—or rather the inability to avoid—
metaphor and other figures of speech. These common features of speech
were felt to prevent language from embodying the real truth of the world,
a truth that demanded that words match only the things to which they
referred. This was not exactly Bacon's vision, but it was how he had
come to be understood. The search for a universal language, a "real
character," was itself a sign of disbelief in the possibility of ever making
English (for example) a sufficient discourse for the new philosophy. Indeed, a number of different symbolic systems were tried out—hieroglyph-
ics, Arabic numerals, signs of the Zodiac, even Chinese characters—and
abandoned (Salmon, 1979). The first language planners failed, and in
more than one sense. But the sensibility that guided them, the desire to
give speech a mathematical finality, remained unscathed. By the end of
the century it had made new strides, if in milder form.

III

The second manner in which those such as Wilkins "failed" was in their
own writing. As we saw above, redundancy and metaphor were hardly
alien to their discourse. Still more, members of the Royal Society may
have looked to Bacon as their linguistic messiah in a general sense, but
when it came to their own specific styles they tended to interpret him in
a number of different ways. No such thing as a standardized form of
discourse had yet been achieved. Differences in the type of expression
used by various Society insiders, such as Robert Hooke, Robert Boyle,
and Edmund Halley, were as marked as they were between these members and outside authors. Where Halley, for example, commonly began
his articles in a fairly straightforward manner, for example, "Before I
proceed to the Theory of the Variation of the Magnetical Compass, it is
necessary to lay down the Grounds upon which I raise my Conclusions
. . . ," Boyle, in direct contrast, often wandered about in his opening

phrases as if picking flowers of remembrance out of the air: "This occasion, I have had to take notice of siphons, puts me in mind of an odd kind of siphon, that I caused to be made a pretty while ago; and which hath since, by an ingenious man of your acquaintance. . . ." (These quotes, and many other examples of scientific writing of the time, can be found in Davy [1953].)

As Roy Porter has noted, the period as a whole was marked by "unparalleled and seemingly inexhaustible battles for the mind," which led to "the coexistent flourishing of a plurality of discourses about natural knowledge" (1991, p. 7). The experimental philosopher, or scientist, was one new form of author that emerged from this turmoil. He was a writer, moreover, emerging from the soil of the past with much of it still clinging to him. Language and knowledge were still, for him and for others interested in expression, bound to ideas of conversion. This is another reason why there was not yet any such thing as a "scientific discourse" but instead a plurality of relevant discourses, each vying for the reader's attention in different ways.

But let us return, for a moment, to the "failure" of these various discourses to achieve the purported Baconian ideal. Part of the rhetorical mud that still clung to the Society, as a group of writers, and that helped cohere them as a group, came from the pedagogical aspect of their motives. They were all, in a sense, eager to teach the world the meaning and importance of the "new philosophy." It was too early in their history for them to mitigate their witness by means of the consciousness that they were preaching to the converted. This would only come later, following the success of those such as Newton, which converted many minds to science. In the meantime, however, a large amount of space and effort was given over to clearing ground, the same ground that Bacon had tried to expose. Attacking and accusing metaphorical language for its wantonness, its antipathy to truth, became one of the only real rhetorical standards observed by the Society, as a whole. As the quote given above from Wilkins might suggest, this was often done in florid fashion, with magnificent self-negation. The following lines, taken from Thomas Sprat's *History of the Royal Society* (1667), a document meant in effect to proclaim a new profession, offers one of the best-known examples:

> Who can behold, without Indignation, how many Mists and Uncertainties, these specious Tropes and Figures have brought on our Knowledge? . . . I dare say, that of all the Studies of Men, nothing may be sooner obtain'd, than this vicious Abundance of Phrase, this

Trick of Metaphors, this Volubility of Tongue . . . [We of the society] have therefore been more rigorous in putting in Execution the only Remedy, that can be found for this Extravagance; and that has been a constant Resolution, to reject all the Amplifications, Digressions, and Swellings of Style; to return back to the primitive Purity and Shortness, when Men deliver'd so many Things, almost in an equal Number of Words. [We] have exacted from all [the] members a close, naked, natural way of Speaking . . . to bring all Things as near the mathematical Plainness as they can; and preferring the Language of Artizans, Countrymen, and Merchants, before that of Wits, or Scholars.(1958, p. 5)

Sprat's book was written under the close eyes of the Society. Jones says of it that "its importance lies not only in its being the most elaborate and comprehensive defence of the Society and experimental philosophy in [the] century, but more especially in its constituting an official statement on the matter" (1961, p. 222).

Official it may well have been, but it is the aspect of its elaborateness that seems more telling vis-à-vis the question of scientific language. Leafing through Sprat's *History*—and one might note that it is a "history" in the new interpretive (Baconian) mode—or any number of other spirited attacks on Ciceronian language, we find that the uses of metaphor by Society members fall into certain consistent and revealing categories. For example, figurative language is compared to aristocratic luxury, as in such phrases as: "excessive riches," "polished phrasing," "vicious abundance," and "extravagances." It is related to rich foods; called "garnish," "starched speech," and "overly spiced." It is damned for secrecy; it is guilty of "mysteries," "occlusions of mind," and "charms and spells." Finally, it is likened to women, their sexual attractions, and their use of makeup to create false beauty; it is attacked for its "vanities," its "stirring up of passion," its power to "disfigure the face of Truth by daubing it over with the paint of language."

The enemy of science thus pictured is a discourse that embodies a combination of social evils: self-indulgent aristocrats, gluttony, practicers of deceit, and women who tempt through disguise (i.e., courtesans and whores). Scientific language, in contrast, strives to be simple and pure. It is characterized as "prudent speech," "solid," "modest," and "a marriage of words and things." Manly in aspect, it also must be the "companion of innocence and simplicity."

Whatever the general arguments for a thinglike discourse, then, the truth was that the ideal speech for science was given an animated imagery of a particular kind. To counter Ciceronian foppishness and effeminancy, science should approximate the Protestant male, the re-strained and fatherly head of the household of speech. Such, at least, was the meaning. But the metaphors used to convey it, quite obviously, were no less rich and evocative than those damned as the opposite. As with Bacon and so many others, these early members of the Royal Society wanted to record, persuade, disseminate, teach, and accrue legitimacy thereby, both individual and institutional. They sought to write for an educated public, which means they sought to write well and to use the techniques of vividness at their disposal. Their authorship, in short, was far more interesting and skillful than their message.

This is not to say, however, that their discourse was the equal to Bacon's in its richness of figuration. For many of the reasons mentioned earlier, the English language did become more sober and simple as the 17th century progressed. Well before 1750, English prose and poetry in the hands of writers such as Dryden, Defoe, Butler, and Locke, had become less prone to fancy and pure invention, more consciously aimed at a very broad audience, at something approaching "plain speaking." Whether or not one accepts Michel Foucault's diagnosis of a grand epistemic break, it is undeniable that learned discourse not just in England but in most Western nation-states had changed radically by the early 18th century, under the canopy of "reason." Looking back, happily rather than nostalgically, over the transformation that had taken place, Bishop Gilbert Gurnet had this to say:

> The English Language has wrought it self out, both of the fulsome Pedantry under which it labored long ago, and the trifling way of dark and unintelligible Wit that came after that, and out of the coarse extravagance of Canting that succeeded this. . . . When one compares the best Writers of the last Age, with [those that excel today], the difference is very discernable: even the great Sir *Francis Bacon*, that was the first that writ out Language correctly; as he is still our best Author, yet in some places has Figures so strong, that they could not pass now before a severe Judge. (*History of His Own Time*, written 1683–1704; quoted in Ure, 1956, p. xxxii)

IV

One of the most interesting things about scientific writers of the 17th century is that many of them employed more rhetorical and narrative structures than did other writers. Again it was Francis Bacon who helped

set the tone for this. In his writings on science, he made use of aphorisms, philosophical ruminations, satire, formal "letters," enumerated
ideas, fictional travelogues, and a variety of other devices. In at least
one of his nonscientific essays ("Of Travel"), he also advocates turning
other well-known forms, such as the private diary or journal, into scien-
tific-type documents, "in order to abridge [experience] with much profit."
As a whole, it would be difficult to scan the collected works of this
author and not be struck by the variety of literary forms that one encounters.

As Virgil Whitaker (1962) pointed out, Bacon's writings should be
seen as part of the encyclopedic tradition central to scholarship since
antiquity. One might say that Bacon represented a kind of end to this
tradition. His many "failures" as an author—his contradictions, his excesses, his misattributions and near-plagiarisms—need, I think, to be
viewed in the light of this tradition.[4] These problems have been revealed
time and again. Yet, as Whitaker suggests, it is crucial to understand
that Bacon helped both bring to an end and shift to a new plane the
encyclopedic inheritance: it was his good judgment to survey the conditions of knowledge rather than to try and summarize all of knowledge
itself. Bacon's "science"—whatever else one might say about it—was
enormously plural, even consciously experimental, in its expression.
Revealed in it is unending evidence of an *authorial* search, perhaps even
more than an epistemological one.

Bacon's use of a variety of literary genres was not unique. Use of
different literary forms was also common among scientists in other countries. Examples of rhetorical "tricks" and shifts abound in the works of
Descartes, Huygens, Mersenne, Kepler, and Galileo. Particularly note-
worthy in this regard is Kepler, who composed several types of dialogues,
commentaries, surveys, and astronomical tables. Kepler, too, might be
counted as an adopted son of the encyclopedic tradition, having planned
and labored for years upon a final *summa scientia* to include all knowledge of the heavens in a single vast scheme. Failing in this, Kepler decided to recast his material into a textbooklike dialogue in the classical
medieval mold (his *Epitome of Astronomy*, 1621).[5]

Galileo also wrote dialogues, philosophical tracts, journals, and
other narrative forms common to the time. In his famous pamphlet,
Sidereus nuncius (The starry messenger, 1616), where he announces his
first telescopic discoveries, he mixes his scientific observations with such
things as diarylike entries, exclamations of wonder, snippets of personal
history, and a closing meditative defense of the Copernican system. The
story of Galileo's rhetorical skill, in fact, reads as one of the most inter-

esting and important political failures in the history of modern science. This story concerns the author's famed *Dialogue Concerning the Two Chief Systems of the World* (1632), in which he attempted to use a series of sly narrative devices to fool clerical censors. The dialogue form he employed as a traditional structure (of oratorical debate), in a traditional manner (ending with an apparent lack of hard conclusions). The attempt seems to have been to avoid the appearance of opposing Catholic doctrine, merely proposing "the Copernican view" as a strawman. The work, meanwhile, is replete with subtle sarcasm, double entendres, and twists of satire, aimed directly at the presumed ignorance of the appointed censors. In his introduction, for example, titled "To the Discerning Reader," Galileo writes:

> Collecting all the reflections that properly concern the Copernican system, I shall make it known that everything [has been] brought before the attention of the Roman censorship, and that there proceed from this clime not only dogmas for the welfare of the soul, but ingenious discoveries for the delight of the mind as well.
>
> To this end I have taken the Copernican side in the discourse [that follows], proceeding as with a pure mathematical hypothesis and striving by every artifice to represent it as superior to supposing the earth motionless. (1967, pp. 5–6)

A "discerning" reader, however, would have certainly understood the bitter irony in all this. Galileo had already been summoned and censored once, and then ordered not to teach the Copernican system, which he had publicly defended on a number of occasions. In order to publish a new book on the subject, he felt forced to make use of disclaimers and disguises: "I act the part of Copernicus in [these] arguments and wear his mask. . . . I want you to be guided not by what I say when we are in the heat of acting out our play, but after I have put off the costume, for perhaps then you shall find me different from what you saw of me on the stage" (p. 131). In the dialogue itself, meanwhile, Galileo's alter ego, Salviati, responds to an early challenge by the other two participants (Sagredo and Simplicio) as follows: "It is well that you and Simplicio raise these difficulties. They are, I imagine, the same which occurred to me when I first saw this treatise, and which were removed either by discussion with the Author himself, or by turning the matter over in my mind." In short the truth is available to those who listen and think.

Unfortunately for Galileo, however, his irony was too thick and the ignorance of his censors much too thin. Whether Pope Urban VIII (a Latin poet himself) was an admirer of his or not made little difference.

The Jesuit readers of his work were among the most skilled and learned men of the time. Fully educated in the techniques of rhetoric, they perceived the ruse, understanding that they had been cast as the speakers Sagredo and, more often, Simplicio, opposed by Galileo's own Salviati. Their decision, as we know, was not a happy one for the author.

The point would seem to be this, therefore: that Galileo thought himself capable, even masterful, as a *writer*. No real distinction existed between being a literary author and being a scientific author. The field of narrative technique was equally open to both. Indeed, Galileo's hope for the *Dialogue* was something approaching a kind of ultimate literary achievement: to deceive and convince simultaneously. But to draw on existing narrative technique, as I have said, meant to draw on the past. Galileo, like others of the early 17th century, has been shown to have obeyed, even in his experimental writings, certain fundamental tenets of Scholasticism then still dominant in philosophy (Dear, 1991). Like other works of its type, the *Dialogue* is put together as a deductive exercise, moving the reader through a series of universal theorems and definitions, then general assumptions, and finally into actual experimental results, which are used, however, to illustrate the grander themes offered earlier. The scientist-author is not as immediately present as the type of direct (confessional) witness for which Bacon had argued. Galileo appears in his work mainly by proxy, through a speaker (again, Salviati), who says: "So far as experiments go they have not been neglected by the Author; and often, in his company, I have attempted . . . to assure myself that the [results] actually experienced [are those] above described."[6] The reader in a sense is thus asked to *listen* to the experiments that unfold, to hear them being related or see them performed by a teacherlike figure. He is not asked to experience them as if he himself had thought them up, step by step. The narrative convention followed is to persuade not by actual testimony but instead by demonstration. Galileo's writing, in this case, is that of the classroom.

Such aspects of Renaissance Scholasticism were common to all those who took up the dialogue form. This does not mean that Galileo gave rise to nothing new on the plane of discourse. His language, one should note, was highly simplified, without any of the elaborate rotundity typical of the time. He too wrote in the vernacular and was an anti-Ciceronian at heart, an outspoken believer in the value of offering "things" more than mere "words." Moreover, his descriptions of his experiments were sufficiently exact to allow the reader to repeat them. As is well known, he also employed mathematics in a new manner, as more of an integral part of the language of experiment. Overall it would be absurd to try and confine Galileo to the Aristotelean tradition, just as it has been mis-guided to identify in his work a complete and revolutionary leap into the modern world. With him, as with Bacon or the Royal Society, scien-tific discourse did not burst forth fully mature. Authors continued to use much of what they knew, and to use it with skill. But change did occur, and in England this took a particular form.

V

Robert Boyle has been identified by scholars as representing something of a new phenomenon on the literary scene of British science in the later 17th century (see, e.g., Shapin and Schaffer, 1985; Paradis, 1987; Golinski, 1987, 1990b). Like many of his fellow Royal Society mem-bers, Boyle shared Bacon's attitude toward language, and he seems to have been the first one to successfully take this attitude and create some-thing new out of it, something approaching the modern scientific essay form. What led him in this direction was a series of problems having to do with language. A major concern of Boyle's, for example, was the need to overturn the jargon of the day in chemistry. This meant over-coming an existing rhetorical tradition that made chemical knowledge seem arcane by shrouding it in alchemical terminology, thereby making it the intellectual monopoly of trained insiders.[7] It was Boyle's belief that "Christian charity and the welfare of the public both demanded the freest possible exchange of chemical and pharmaceutical information and were incompatible with the retention of 'secrets and receipts' for commercial gain" (quoted in Golinski, 1990b, p. 383).

To accomplish this, Boyle first turned to convention and did what Galileo, Kepler, and others had done: he donned the robes of classical skepticism and composed a dialogue, written in English for the educated layman. *The Sceptical Chymist* (1661) was where many of the author's most important chemical theories were put forward. These were expressed in a straightforward, if rather wandering, prose meant to demolish an-cient doctrines on the nature of matter and its transformation that had long dominated chemistry since the early Middle Ages (the doctrines, that is, of the "four elements" and the "three principles"). Boyle's adop-tion of the Platonic dialogue was apt. It was a device by which the "an-cients" could be defeated, so to speak, on their own terms. In it, there is much quoting of poets and philosophers (in Latin), use of frequent and extended similes or analogies, and the employment of satire, puns, and other witticisms. All of this is juxtaposed against more blunt, empirical descriptions of procedures carried out to produce various combinations of metals ("with sulphur it [mercury] will compose a blood-red and volatile cinnabar; with some saline bodies, it will ascend in form of a salt . . ."). As such, the work is a magnificent hybrid of ancient and modern, a kind of unique amalgam.

Such an amalgam, which could only have been written during this early period, was, however, limited. It was never intended, and was in fact insufficient, to deal with another problem facing Boyle (see Golinski, 1990b). This involved devising a type of writing that might bring together the details of experimental procedure with the laying out of larger concepts, what he called "the framing of a solid and comprehensive hypothesis." To Boyle's mind, science should make itself accountable to its readers; it should make itself legible. This implied offering detailed descriptions of experiments, as a kind of "historical" writing of what happened, told directly through the "I" as a personal narrator of events, such that the reader would be brought as close to these events as possible. At the same time, however, such descriptions were not enough by themselves. In order to be made "useful to philosophy," they had to lead up to a discussion of some hypothesis. Without this, they would remain anecdotal, closer to recipes than to discoveries.

Boyle termed his new creation the "experimental essay." According to some scholars, his ultimate model was Montaigne, whose "mimetic reconstitution of the self" was meant to lead the reader through a process of self-discovery (Paradis, 1987, p. 70). But one should note that the "essay," a literary form first appropriated from the French author by Bacon himself, had long become established in England before Boyle began writing his chemical treatises. Well before 1650, writers such as William Cornwallis, Henry Peacham, and the versatile Nicholas Breton had helped give the essay common currency and had done much to broaden its range of styles and subject matters.[8] Its basis was rational reflection, a didactic searching for moral/political truths, focused on the "I" as a self-instructing entity. Unlike its manifestation under Bacon, it was often ironic, informal, wandering, and indulgent; it sought to instruct and entertain not through erudition alone but also through building a sense of intimacy with the reader that often depended upon confession, the relation of personal experiences, and the like.

Boyle's use of the term "experimental essay" seems like an attempt to draw on a range of expectations already attached to the form. That is, he seems to have been consciously literary in this choice, and there-

fore seems to have hoped for a literary science. His "essay" was structured to begin with a prologue, often anecdotal in nature, in which he
gives the reasons why he was moved to perform a certain experiment;
this was followed by a step-by-step report of his procedures, sometimes
including discussion of his false starts, ruminations, minor asides, and
more; this part led to his results, which in turn (sometimes after a good
deal more of "this and that") eventually led to some hypothesis. The
style, as implied, is often migratory, even vagabond:

> But the oddness of the experiment still keeping me in suspense, it
> was not without much delight, that afterwards mentioning it to a
> very ingenious person, whom without his leave, I think not fit to
> name, well versed in chymical matters, and whom I suspected to
> have, in order to some medicines, long wrought upon rain-water, he
> readily gave me such an account of his proceedings, as seemed to
> leave little scruple about the transmutation we have been mention-
> ing . . . (quoted in Hall, 1965, p. 223)

This is the (often exasperating) style that Boyle generally uses when he is
introducing or meditating upon an experiment. When it comes to recounting an actual procedure, however, the author performs a radical
shift to a more direct, clipped style:

> Into a crucible whose sides had been purposely taken down to make
> it very shallow, was put one ounce of copper-plates; and this being
> put into our cupelling-furnace, and kept there two hours, and then
> being taken out, we weighed the copper (which had not been melted)
> having first blown off all the ashes, and we found it to weigh one
> ounce and thirty grains. (quoted in Hall, 1965, p. 265)

Such differences have been taken to indicate that Boyle was a highly self-
conscious narrator. In Shapin and Schaffer's closely argued study (1985),
for example, Boyle's strategy is said to have been aimed at creating "virtual
witnesses" out of his readers. Its method was presumably to repeat in narrative form the labored and often incidental thought processes involved in
experiments, with their surrounding events and asides, and also the more
orderly and simple manual operations performed to actually carry them
out. In this way, Boyle is described as having sought a "public demonstration" of his science *through* writing. His "literary technology" constituted
a neat design, intended to manufacture an experience of participation.

Such an interpretation of Boyle's writing is appealing, not the least
because of its own dense logic and inclusivity. Yet it may also exaggerate
the case. Boyle, it eventually appears, is rather less than a kind of ulti-

mate, authorial mastermind. He seems, often enough, a *messy* writer. His messiness, in fact, is part of the literary quality of his "science." In terms of style, his model was clearly not Montaigne or Bacon. Rather, he seems to have tried to combine two different approaches by standing on both sides of the Ciceronian fence so to speak. In his meandering descriptions, he directly recalls the prose, if not the eloquence, patterned after Latin, of such "perfumed" authors as Thomas Browne and John Milton. Boyle's language, of course, is much simpler, but his constructions are not: he hated periods. Long, broken, windy sentences follow one upon another in a seeming endless cascade or gush of fragments. From a reader's point of view, this approach to writing, however common, had its clumsy and ineffective side. Under the control of a skilled penman, such as Thomas Browne, it could approximate a chiseled, witty, epigrammatic, breathless type of monologue, at once conversational and highly literary. Without such control, it could lose its dendritic character and simply overflow its banks, drowning the reader's patience. Boyle's prose often suffered from this flow. On the other hand, however, his narration of experiments was often the very opposite, exhibiting all the terse efficiency and clarity of the best anti-Ciceronians. The prose here is sharp, precise, highly readable, and well paced. It is itself a model type of language for relating a procedure in a manner that allows visualization.

Thus, it might appear that the battle for the century's prose style was restaged and refought within Boyle's pages. Where the reader might have determined the conclusion in one instance—Boyle was criticized more than once in his day for his "personal" and "inconclusive" writing[9]—history would come to decide it in the other. Jan Golinksi has noted that Boyle's language was often "*narrative*, as opposed to exhaustively methodical; *suggestive*, rather than conclusively demonstrative; and *subjective*, in the sense that [it] included copious personal and circumstantial details of a kind that would not be thought relevant by many subsequent scientific writers" (1990b, p. 387). The "experimental essay" was thus a first step in a certain direction. It did not eliminate the individual author; indeed, it was criticized for giving him too large a place. But it did set the terms for his generalization.

VI

Within a brief time, it appears, one side of Boyle's style had become popular among Royal Society members. Isaac Newton, in particular, was

strongly influenced by the type of descriptions Boyle employed in discussing his actual experiments. Newton's writings, in fact, show little
inclination toward giving "reflections" or "opinions." It would seem
that the next important step in the history of scientific writing in English
came in the form of Newton's first published work, "A New Theory
about Light and Colours," the apparent prototype for what we think of
today as the scientific paper or article (Christianson, 1984). This was
read before the Royal Society in 1672, when its author was thirty, and
published later that same year in the *Philosophical Transactions of the
Royal Society*, England's first professional scientific journal. Given its
importance, it is worth pausing to examine the opening phrases of this
essay, with an eye toward uses of language:

> In the year 1666 (at which time I applied myself to the grinding of
> optick glasses of other figures than spherical) I procured me a trian-
> gular glass prism, to try therewith the celebrated phaenomena of
> colours. And in order thereto, having darkened my chamber, and
> made a small hole in my window-shuts, to let in a convenient quan-
> tity of the sun's light, I placed my prism at its entrance, that it might
> be thereby refracted to the opposite wall. It was at first a very pleas-
> ing divertissement, to view the vivid and intense colours produced
> thereby; but after a while applying myself to consider them more
> circumspectly, I became surprised to see them in an oblong form;
> which, according to the received laws of refraction, I expected should
> have been circular. (1958, p. 47)

Newton has clearly adopted Boyle's emphasis on providing testimony.
He even offers personal details and a touch of background. But these are
brief at best and relevant to his report. They build to the identification
of a problem, in the form of a wrong expectation, and thereafter, in the
rest of the paper, to a series of experiments described in streamlined,
unadorned (Boylean) fashion. These, in turn, are followed by a list of
enumerated conclusions ("propositions"), supported by select mention
of experimental results as evidence. Finally, the paper comes to its end in
the form of the most general conclusion of all, the principle "that the
colors of all natural bodies have no other origin than this, that they are
variously qualified to reflect one sort of light in greater plenty than another" (p. 59).

What one reads in this paper is an almost perfect reversal of Galileo's
textual method in the *Dialogue*, with its debt to medieval and Renaissance Scholasticism. There is, moreover, a nearly ideal blending of method
and philosophy—of problem, procedure, and resolution, wrapped in a

beautifully inductive format. Indeed, it so far surpasses Boyle in this regard, while being indebted to him, as to seem the product of some much later age, not the work of a contemporary. Yet so it was: from Bacon and Galileo, to Boyle, then Newton, the change was profound, as undeniable and powerful a rehearsal of the larger trends in 17th-century discourse as one could hope for.

Newton is more consistently simple and direct than Boyle. He has none of the tendency to alternate between Senecan and Ciceronian structure. In comparing these two men, one feels inevitably that the older man still has his linguistic feet firmly planted in the Renaissance; there is something courtly and epideictic about his long, sometimes fawning introductions, his piling up of phrases, his often skillful but nonetheless convoluted uses of metaphor and analogy (usually, one might note, employing scientific imagery). Newton, on the other hand, has a flatter linguistic footprint and is much easier to follow. He seems much more recognizably a part of the "modern"—one reads his language more comfortably, without the same sense of exotic distance that surrounds Boyle, Sprat, Hooke, and most other authors of the late-17th-century Royal Society. Newton is quite intractably part of the Restoration literary moment. His immediate stylistic correlative is Dryden (who was, in fact, for a time a member of the Society), the dominant figure of late-17th-century British letters, whose desire to "refine" the English language was part of his larger program to ensure "that poetry may not go backward, when all other arts and sciences are advancing" (Ker, 1900, p. 163).

The success of writers such as Dryden in effecting a greater simplification of English prose is undeniable. The new science has been traditionally held to account for this (see, e.g., McKnight, 1928; Jones, 1951, 1953). But this is too simple: Restoration writing, actually, reveals the steady maturation of the influences discussed earlier, whose overall effect was to support the Senecan position. Distaste among the new middle class for Latin-based rhetoric, mainly associated with the highly educated elite, was matched by the widespread favoritism now granted more commonplace, accessible language—the language of expanding literacy. Writers such as Newton were plainly part of this movement, intentionally or not.

Yet the comparison, in an important way, can be just as misleading as it is informative. If Boyle was the one to help found the "I" as a required ingredient for scientific writing, to give such writing the form of the essay, Newton was a follower who saw far enough from his predecessor's shoulders to leave these things in place. Moreover, if Newton is the more "modern" of the two, he is by no means entirely so. His work reveals no final abandonment of the self for the sake of a jargon-saturated truth. We are in a kind of no-man's-land between "literature" and "science." Certainly language has been placed under a kind of yoke: looking at the passage again, one can see that the overall structure is aimed to serve documentation first, expression second. There is a preparatory mention of private delight, the hint of a few moments of fascination; then, what is "pleasing" is relegated to the status of a momentary distraction, which must subside ("after a while") before the object of study can be perceived more properly. The experiment is more tied to the element of "surprise," in other words, than to "pleasure."

And yet no less is it true that we are reading Newton, and no one else—not the words of a physicist, the strict and disembodied expert, not Science the eponym for all practitioners. One makes contact here with an individual man who tells a story, gives an account for the burden of his curiosity. We hear a personal voice, a voice that remains adhesive to the "I" of Boyle, in its more general human meaning. Indeed, it is both a private and a public self—private in the sense of offering a selective confession, public through both the intended result and the accessibility of the language used. But the point remains: the latter self is rendered possible only by means of the former one. The narrative itself remains closer, as an act of writing, to a Montaignean essay "On Colors" than to a 20th-century treatise on "Photometric Albedo Analysis of Coherent Backscatter Phenomena."

It seems wrong therefore—wrong for important reasons—to say that the essay form pioneered by Boyle and honed by Newton *necessarily* involved "a literary nullification of the self" (Paradis, 1987). There is no real demand for the death of the personal at this stage, and no real evidence of such a death, nor for the elimination of elegance. One must remember (and it is interesting that scholars have made little mention of this fact to date) that these early prototypes of the research article were composed not as papers but as *speeches*, to be read first before a live audience of like-minded yet critical listeners, and only afterward to be published. No doubt this reality imposed certain qualities of personalized narrative upon scientific discourse.

In the mid- to late 17th century, therefore, even at its Senecan extreme, what we witness is an attitude toward language, not its achievement. An older belief among students of discourse was that Bacon's ideology of plainness and clarity won the day, hands down, for scientific writing, and forced the "ornament" of figurative writing into literature, thereby defining the latter as distinct from science (see Ricoeur, 1975).

By this reckoning, literature became the readerly playground of the bourgeoisie, while science, by the early 18th century, retreated into a professional speech of removal and abstraction. This interpretation, however, is wrong on every count. In countries like England, France, Italy, and Holland, the middle class remained avid consumers of scientific literature, especially in astronomy, navigation, medicine, and the budding fields of botany and zoology so greatly enhanced by the many exotic discoveries brought back from the New World. Much of the discourse of natural philosophers addressed itself to a wide audience. Moreover, it was anything but abstract. The writing of science meant the telling of stories, the building of a readerly experience. Newton's famous Fourth Law of Reasoning—"first diligently to investigate the properties of things and establish them by experiment, and then to seek hypotheses to explain them"—remained a further justification after Boyle for each experiment to be given in the form of a plot, with the scientist as the thinking, imaginative center. "To investigate" was not divorced from confessions of pleasure or surprise, nor, in Newton's case, even from mysticism.[10] Well into the 18th century, natural philosophy remained an intellectual enterprise whose goal was to comprehend "the capacities of natural bodies to act and be acted upon, and . . . their associated actions and properties" (Gabbey, 1990, p. 245). As both the generator and observer of these actions, the scientist was himself an actor in these movements of nature, and acted upon by their effects. His discourse, moreover, whatever its goals for clarity, sought in turn to act upon its readers, to convince them of a witness who was deeply involved in every action, every movement, every property evinced. One finds, in "modern" terms, that technical language was slow to give up its openness and eloquence.

VII

Between the time of Newton and that of Darwin, the writing of scientists underwent two stages of development, paralleling the general evolution of English as a whole. First, it partook of the general restraint characteristic of the Augustan Age, and the 18th century generally. But then it entered a new era of expanded eloquence and sophistication during the romantic and Victorian periods. Among scientists, a range of individual styles continued into the 1700s, expressive of the differing backgrounds, subjects, and attitudes of respective writers. Most austere and mathematical of all were the writings of British chemists, whose discourse remained largely fixed on the detailed descriptions of experiments, their results, and the hypotheses to be gleaned from them. One notes a certain hardening, in this field at least, of literary form. Yet these men still refer to themselves as "natural philosophers," to their experiments as "philosophical investigations," and to their own writings as "philosophy." Such terminology is significant, for it reveals the greater link that scientists still felt toward other areas of knowledge and discourse.

Language was simplified; Newton's influence was felt here too. Indeed, the directness and concision of his experiments, matched by his discourse (especially in the 1704 *Opticks*), cast a brilliant shadow over an entire century of British science. This was part of the larger literary milieu, characterized by the restrained and measured writing of Locke, Addison, Swift, Shaftesbury, and others, as well as by the efforts of grammarians and dictionary makers who sought to take their native tongue firmly "in hand" and give it regulated standards that might bring "the forms and constructions of the English language to uniformity and system" (McKnight, 1968, p. 399). This general concern was, in effect, a desire to project onto language (presumed to be the essential substance of the human mind) a high degree of social order, cohesion, and predictability.

And yet such values did not rule all discourse, not even the writing of scientists. Many successful researchers still wrote with an individual flair, with sufficient confidence to admit humility, humor, and personal confession into their texts, as this excerpt from Joseph Priestley demonstrates:

> I cannot, at this distance of time, recollect what it was that I had in view in making this experiment; but I know I had no expectation of the real issue of it. Having acquired a considerable degree of readiness in making experiments of this kind, a very slight and evanescent motive would be sufficient to induce me to do it. If, however, I had not happened . . . to have had a lighted candle before me, I should probably never have made the trial; and the whole train of my future experiments relating to this kind of air might have been prevented. (1775, vol. 2, p. 33)

A century after Newton's essay on colors, the "I" is still the central epistemological agent. It is still an entity that does not merely report on events but that introduces, performs, and receives the action all at once, and then reflects upon it, for the purposes of inscribing knowledge.

Priestley, in fact, is an interesting case that deserves comment. His fame as the discoverer of oxygen ("dephlogisticated air") has unfortunately overshadowed an extremely wide-ranging authorship. Besides writing on chemistry, he penned works on theology, education, government, and grammar. A political radical, his work, *Essay on the First Principles of Government* (1768) is considered to have been a critical influence on Jeremy Bentham's idea of utilitarianism. *Rudiments of English Grammar* (1761), on the other hand, argued the concept, fundamental to linguistic study today, that "correctness follows usage." Scientists, in short, were not always the professional leaders of Enlightenment sensibilities, especially with regard to language. Priestley, like other scientific writers of his day, did not break with the past for the sake of a more purely regulated discourse. Instead, in his experimental articles, he continued the tradition set by Boyle of offering a wide range of narrative subjects: opening statements on the proper philosophy of science; brief asides on his apparatus and how he obtained it; problems he encountered in setting up an experiment; doubts he may have had in his own thinking; fortuitous happenings by which he procured a chemical of exemplary purity. In attempting to embody his own political philosophy in his writing, he also offered reviews of relevant literature, sometimes reaching back as far as Boyle, always in a manner that promoted a cooperative image of scientific work, never a competitive one (Heilbron, 1975). At times, he brings his contemporaries directly into the narrative, as helpful actors, to more fully construct such an image ("Mr. Lavoisier, Mr. le Roy, and several other philosophers, who honoured me with their notice in that city [Paris]; and who, I dare say, cannot fail to recollect the circumstance . . ." [1775, vol. 2; p. 31]). He thus embeds a particular ethical and pedagogic view directly into his research papers, a view that arose immediately out of his belief that science could advance and improve the security and happiness of humankind through a communal gathering of minds that would ignore national and linguistic boundaries and prejudices.

Though not entirely emblematic, Priestley is nonetheless revealing of the breadth of expression found within technical discourse in English during his life time. This was a time, too, when scientific writings could still take a number of different narrative forms. These forms included lectures, illustrated natural histories, catalogues of species, treatises on magnetism or the planets, and long explanatory "essays" (a great many books employed this word in their titles, thus continuing the direct link with literary endeavor). Gone is the Socratic dialogue and other medieval forms: the field now resembles much more closely the new university curriculum, with its use of lectures and demonstrations conducted in the vernacular. Other forms have become important too. Much research is presented in the guise of personal memoires, diaries, and private correspondence. Benjamin Franklin's *Experiments and Observations on Electricity*, published in London in 1774, reported his important discoveries in the form of letters written to various interested individuals, scientists and nonscientists alike, in Europe and colonial America.

Many scholars have maintained that "in linguistic matters, the 18th century was largely . . . a century of authoritarianism and prescription" (Milroy and Milroy, 1991, p. 34). In the sciences, this "ideology of standardization" has been interpreted from the great nomenclatural systems of Linneaus (botany) and Lavoisier (chemistry), each of whom tried to condense an entire field around a self-referential scheme of terminology that presumably embedded all essential knowledge and the logic required to extend it further. But if viewed close up, especially in England (and, increasingly, in America), scientific discourse at this time reveals that true centralization of technical style was lacking. Samuel Johnson had no correlative in natural philosophy to his epochal *Dictionary* (published in 1755). The Royal Society, despite its early hopes, never exacted the type of obedience to prescribed standards and never set itself up as the stern constable of discourse as did the Académie des Sciences in France—and even in this case, the failure of total standards is more than apparent.[11]

⑤⑤

The effort to refine and purify English along the lines of classical form and regularity continued into the early 19th century. Indeed, it did so with a particular moral passion, reacting both to the spread of popular education and against the spread of popular mass-circulation newspapers, novels, and political pamphlets. The former greatly encouraged the use of grammar texts, while the latter introduced many "low" and "vulgar" expressions that the new grammarians felt had to be expunged from cultivated speech and writing. Ironically, much of the standardization that had taken place in prior decades had effectively brought the written language to much closer accord with the spoken one. This had been one of the results of continual simplification. It was also a reason why the burden of propriety had actually shifted to speech itself (speaking well having become, in large part, more a sign of social accomplishment than writing well). England had become a world colonial power and its sense of lofty decorum in linguistic matters matched its larger self-image on the stage of the world. Cicero had been a conscious model of the Elizabethan era, Seneca and Tacitus of the Commonwealth and Restoration periods, and in the Augustan Age of the early 18th century such imperial poets as Virgil and Horace were adopted as models. From here, however, as the self-proclaimed "New Rome," English writers turned away from such overt models and instead relied on their own sense of propriety to make their demands for control.

In response to this trend of further purification, at least in part, the romantic rebellion took place. This movement included a conscious effort to rebel against 18th-century standards for language use and styles of writing. A powerful revival of medieval and Renaissance words and expressions took place: Spenser and the Elizabethans became a model for those such as Walter Scott and Robert Southey, Chapman's Homer for Keats, Latin for Charles Lamb. Wordsworth, the leading romantic poet, proclaimed his intent to employ the daily language of the countryside. Romantic-era writers also looked to continental literature for their inspiration, as in Thomas Carlisle's use of German narrative forms, Coleridge's interest in French, Byron's in Italian. The result of all this expanded linguistic curiosity was a vast influx into English of new and recovered words, figures of speech, and modes of expression. The language thus experienced a magnificent period of ferment and re-sophistication, which before long would settle into the often high-flown yet varied elegance of Victorian prose, which fully and consciously reinstated metaphor as an essential aspect of learned argument.

This new turn to the value of eloquence also had its effect on science. Indeed, the shadow of Newton, so strong in the 1700s, now seemed to grow weak, at least in some areas. In certain cases, scientific discourse seems to have gained an elegance above and beyond anything that had gone before:

> As the first theorists possessed but a scanty acquaintance with the present economy of the animate and inanimate world, and the vicissitudes to which these are subject, we find them in the situation of novices, who attempt to read a history written in a foreign language, doubting about the meaning of the most ordinary terms; disputing, for example, whether a shell was really a shell,—whether sand and pebbles were the result of aqueous trituration,—whether stratification was the effect of successive deposition from water; and a thousand other elementary questions which now appear to us so easy and simple, that we can hardly conceive them to have once afforded matter for warm and tedious controversy. (Lyell, 1830–1833, vol. 3, p. 1)

Such a flow of language, with its labyrinth of extended yet coherent phrasing, had not been seen since the days of the early Royal Society. Here was all the complexity and prolixity of Robert Boyle, yet relaced

into a beautifully intelligible whole. In Lyell, one sees the division between "literature" and "science" vanish once more. Such writing was to
be found elsewhere, in a range of authors. A full 250 years after Galileo
and Bacon, scientist-authors have returned to the practice of quoting
classical authors or literary figures in their scientific texts. Here, for example, is Sir Roderick Murchison, in his epochal proposal for use of
"Silurian" to designate a new stratigraphic system:

> In allusion to this term I have only further to add, that it is to be
> hoped that no naturalist will, from its sound, fall into the mistake of
> an early English writer who is ridiculed by Camden for having mis-
> applied the line of Juvenal,
>
> > "Magna qui voce solebat
> > Vendere municipes fracta de merce Siluros,"
>
> supposing that the British captives were exposed to sale at Rome,
> when the poet spoke of *fishes*, and not of men! (1835, p. 52)

Darwin himself, on the first page of the *Origin of Species* (1859), includes a
reference to none other than the Greek philosopher whose works were often viewed as anathema to progress in science two centuries earlier:

> Aristotle, in his "Physicae Auscuoltationes," after remarking that
> rain does not fall in order to make the corn grow . . . applies the
> same argument to organization; and adds . . . "So what hinders the
> different parts [of the body] from having this merely accidental rela-
> tion in nature? . . ."

By this time, that is, Aristotle could be easily recouperated to a type of
technical authorship without obvious literary limits.

In many ways, in fact, the *Origin* too is easily consigned to the
realm of "literature" today, not merely by virtue of its ideas but in terms
of its language. Academic scholarship, in fact, discovered this some time
ago. Since the early 1960s, Darwin's work has lost its canonical aura of
"science" and has been subjected to many types of literary analysis, with
the effect of revealing the breadth of narrative strategies in Darwinian
prose. Some scholars, for example, have outlined the use of techniques
adopted from dramatic literature, pointing out how the "life cycles"
and "extinction" of various species are presented in ways that play upon
a reader's sentiment through evocations of tragedy or heroic epic (Hyman,
1962). Other analysts have identified the conventions of travel writing
in the text or the covert romantic influences that create a mingling of

personal and poetic response (see, e.g., recent summaries in Kohn, 1985). There has been interest, too, in the metaphors chosen and employed by Darwin, and in how these helped make his writing both attractive to a broad audience and technically legitimate to the scientific community (see Beer, 1983). Finally (but this is hardly a complete survey of the rich literature in this area), some scholars have gone so far as to see Darwin's discoveries as being "peculiarly linguistic," emerging out of the writing process itself and its drive for a "rhetoric of persuasion," stating (for example) that "Darwin discovered neither evolution nor natural selection, but how conventional language could be made to confess a truth it had suppressed systematically" (Campbell, 1990, p. 85).

Taken all together, these approaches suggest Darwin as a writer of nearly infinite resources and expertise. Yet Charles Darwin, one should note, was not the master stylist of his age. His writing, by the standards of early Victorian prose, was characterized by a certain ordinary eloquence. His literary skills, true enough, were often better than those of many, but they were also worse than others; he was by no means the consummate narrative strategist that today's analysts would often like him to be. What his writing reveals, at base, is something else. At the point in history when Darwin lived, it was still very much required for scientists to be interested in writing *as writing*. It is not simply that they could tell better "stories"; nor is it true that they were allowed greater spontaneity in their use of language, a free hand at invention. It is more that the repertoire of materials and techniques, the latitude of expressive possibility, the toolbox of authorship, in other words, was far larger than that allowed scientists today. Building a variety of different discourses and using a variety of literary tactics was part of what was demanded and expected, by readers and writers both. Scientific writing simply partook of the greater literary field. Much technical writing, after all, was still aimed at the educated public generally: both reader and author were therefore entities of greater scale than today. One might recall that Lyell and Huxley were esteemed, even granted public honors, as great literary stylists in their day and were viewed as such on the basis of their technical writings (while others, like James Hutton, gained renown as the dismal opposite). No doubt, it could be said that the language of geology, still in its infancy, was not that of other fields. Moreover, Lyell in particular is perhaps unmatched by any scientific author of any age. His uses of metaphor, his many classical allusions, his reintroduction of historical writing and travelogue, and his analogies to styles of logic in philosophy, art, and ethics—all of which, it might be said, gave geological thought a larger cultural placement and relatedness—

reveal a style as sophisticated and cosmopolitan as that of any literary figure writing in romantic or Victorian England.

Let us then turn to chemistry for comparison. By the early 19th century, roughly contemporary with Lyell, the actor-I had begun to disappear in this field. The "literary nullification of the self" had begun. In the wake of such writers as Henry Cavendish (1731–1810), a principle approaching that of maximum sense/minimum expression had come to be viewed as commensurate with exactitude and careful procedure. The job of the scientist was to uncover, carefully and methodically, a precise fragment of the "Divine plan," whose essence was reason and order. It was no longer to insert himself into this order, to join with its simplicity and "plainness." Language was now to be used as a tool. This view was widely shared: to take one example outside England, Lavoisier (1743–1794) sought to adapt the philosophy of Condillac to the question of scientific discourse, stating that "Languages . . . are also analytical methods, by means of which we advance from the known to the unknown, and to a certain degree in the manner of mathematicians" (see Paradis, 1983, p. 218). Lavoisier's own style, however, hardly measures up to a utilitarian standard; his writings, like Priestley's, are full of rhetorical confessions, admitted digressions, personal details, redundancies, and the like. If these are handled selectively, in a manner to make their author appear reasoned and logical in his thought, they are still saturated with the narrating "I," protagonist of all action.

It is really only with Cavendish in England that the passive tense enters chemical discourse in force. In his famous *Three Papers, Containing Experiments on Factitious Air* (1766), Cavendish regularly employs phrases such as: "these experiments were made . . . "; "quantities were found to be equal . . . "; "it was observed that . . . "; and "By this means, the inflammable air was made to pass through . . . " The action, moreover, is most often performed by the objects of study themselves: "the mixture continued burning . . . "; "one grain of zinc yielded 356 grain measures of air . . . "; "the bottle and cylinder became filled. . . ." It is thus that Science stands forth in this discourse, the experiment itself and its ingredients, not the scientist, who, historically speaking, has suddenly vanished. More than this, he is changed from a narrator to a mere recorder. The actions his body has taken become events identical in objectification with those of the apparatus, the chemicals, and so on. There has occurred a kind of reversal; with the retraction of overt authorship, there is a shift from witness to reportage. Cavendish, in a sense, marks a decided jump toward the more truly modern.

A half-century later this approach to chemical discourse has been established. But not entirely, and not, by any means, as a reflexive standard. At about the time of Lyell, certainly, it was common to read something of the following by a well-regarded member of the Royal Society: "The researches I had made on the decomposition of acids, and of alkaline and earthy neutral compounds, proved that the powers of electrical decomposition were proportional to the strength of the opposite electricities in the circuit, and to the conducting power and degree of concentration of the materials employed" (Humphry Davy, quoted in Leicester and Klickstein, 1952, p. 244). This, too, seems to have come a long way on the road to the present. As one of the oldest of institutional sciences (second, perhaps, only to astronomy), chemistry would seem to have abandoned literature and the self for the coming of the professionally jargoned phrase, dense with truth. Yet Davy, in his day known as a "poet scientist," and greatly praised by those such as Coleridge, deserves a closer look. On the very next page following the above excerpt, for example, after a detailed reporting of experimental procedure, comes one of the most famous passages in 19th-century chemical discourse, a passage in which one finds the author breaking forth in a very different sort of language:

> Under these circumstances a vivid action was soon observed to take place. The potash began to fuse at both its points of electrization. There was a violent effervescence at the upper surface; at the lower . . . small globules having a high metallic lustre, and being precisely similar in visible characters to quicksilver, appeared, some of which burnt with explosion and bright flame, as soon as they were formed, and others remained . . . covered by a white film which formed on their surfaces. (p. 245)

Even though delivered in the passive tense, these are words of effect more than of "information" or "data." True, they are perhaps restrained compared to what Shelley or Byron might have made of then. Yet they nonetheless leak excitement, a quality of immediate danger (which, after all, was a daily reality for chemists of the time). In the midst of the depiction, which resembles a painting in all its flourishes of color and drama, Davy's own emotional state is naked and clear. This type of writing, in fact, occurs elsewhere in his work, like a kind of punctuating spectacle. Long, extending sentences, without incidental details, yet full of trembling, almost breathless witness, return again and again: "The globules often burnt at the moment of their formation, and sometimes violently exploded and separated . . . [flying] with great velocity through the air in a state of vivid combustion, producing a beautiful effect of continued jets of fire" (p. 246).

In his day, Davy was not attacked but admired for such writing. Like geology, chemistry too, aged don among the sciences, was far from willing at this point to give up all of its traditional claims to vitality of language. The rule of mimesis, inherited from Boyle—by which science became commensurate with a recounting of personal experience—has by no means died the death of a thousand cuts. Cavendish was, indeed, ahead of his time.

But let us turn back to geology and map, at a reconnaissance level, its stylistic progress to the present. Because this was a science that was professionalized late—amateurs and classically trained "natural philosophers" dominated the discipline into the 19th century—it seems all the more interesting to observe the timing of its literary fall into final modernism. In the days when Lyell wrote his magnificent textbook, geology was still struggling to become an accepted part of the university curriculum, both in England and America. Much of its literature, like other sciences too, was still being presented in the form of oral lectures made to one or another learned society, prior to revision and to actual publication. Geologists tended to be gentleman-scholars, dabblers in other fields. As Mott Green has stated, geologic thought in general was "notable for the acrimony of its controversies" (1982, p. 192), a situation that prompted many writers to employ high levels of persuasive eloquence and to begin their work with declamatory statements of philosophy.

Fifty years later, by the 1880s, the situation had changed. With the full triumph of the professional journal, and with the resolution of most major debates, geologists were no longer writing speeches for drawing rooms, lectures for fervid converts, or books for mass audiences. They were now writing articles for other geologists. Though students and the public were still often appealed to (especially for money), geological inquiry had become a matter of "research," an established ingredient in the new and rapidly growing major universities of the United States. It was, moreover, professional in other practical ways. By the later 19th century, geologists were regular consultants to mining, construction, and railroad firms. The government had employed them during the Civil War in a number of different capacities, for example, to assay coal and iron ore, to conduct territorial surveys, to help build bridges and to explore the western United States (it was for this reason that the U.S. Geological Survey was established in 1879). By the mid-1880s, a number of states with coal and other mineral resources had set up their own interior surveys and established their own annual reports. Taken as a whole, the discourse of geology by this time was a complex mixture of "pure" and "applied" inquiry. It varied from extended, interpretive discussions of landscape and scenery, sometimes provided in diary form, and sometimes turned into books aimed at the public (as in the account by John Wesley Powell, first director of the U.S. Geological Survey, of his trip down the Colorado), to theoretical discussions of mountain building, to short, terse reports about chemical experiments performed on obscure minerals. For such a young science, this diversity seems impressive, unanticipated. Its relative suddenness, of course, had everything to do with the economic possibilities and political powers attached to geological knowledge. Between "research" and "exploration," geologists could no longer write as amateurs who didn't have to be concerned with such problems as originality or documentation. To be accepted, their writings now had to meet the demands of being a "contribution" before they could be released as "science."

In these writings, there is little of Lyell's contemplative moralizing. In its place, one tends to find a particular kind of rationalizing, often in the form of a warrant for the types of observation and methods to follow. The Newtonian standard, so visible elsewhere, has finally entered geological discourse. It is an important change, not to be denied: within this span of time, between the 1830s and the 1880s, moral–ethical language has been declared "nonscientific" and will never again find regular appearance within the professional language of scientists. What one commonly encounters now, toward the end of the 19th century, is language more like this:

> The deep gorges which so facilitate the examination of the strata and of their displacements, are themselves of interest as monuments of erosion. To account for their existence and unravel their history is to review the laws of erosion with great wealth of illustration. Results so extreme can have been produced only under conditions equally extreme; and natural laws are often best tested and exemplified by the consideration of their operation under exceptional circumstances . . . (Gilbert, 1876, p. 85)

The tone is restrained, impersonal, less wide ranging than the prose of Lyell. And yet Gilbert's writing has a quiet grandeur to it. Note three elements of style in the above passage: the conscious employment of metaphor ("monuments of erosion"); the use of established humanistic tropes ("unravel their history," "wealth of illustration"); and the utilization of dramatic effect ("results so extreme"). The author could be subtly urbane and pedagogic too, as when he discusses how one should approach the analysis of erosion:

All indurated rocks and most earths are bound together by a force of cohesion, which must be overcome before they can be divided and removed. The natural processes by which the division and removal are accomplished make up erosion. They are called disintegration and transportation. (p. 86)

In a few short lines, Gilbert offers a skillfully condensed theory. Several rhetorical devices are evident. According to a common rule of 19th-century pedagogic thinking, he presents the reader first with concrete, everyday objects ("rocks and most earths"), and then with the actions or processes involving them. Technical terminology is finally introduced only after these appeals to common experience have been made. The embedded message is not merely pedagogical but philosophical as well: science itself returns to such ordinary experiences (a Baconian idea), and derives its logic therefrom. In this passage Gilbert employs a skillful, rhythmic decrease in sentence length. A juxtaposition of the most central elements is repeated three times and expanded upon ("divided and removed; division and removal"; "distintegration and transportation"). Such repetition, which resembles a mathematical formula, was also a common aspect of 19th-century pedagogical writing, being intended as an aid to memory.

In Gilbert's day, as in Darwin's, such urbanity was still looked upon as legitimate—indeed even admirable—within the borders of the "scientific." Before another 50 years had passed, however, most of the sociocultural elements that made this attitude possible had disappeared. The period spanning the last 2 decades of the 19th century and the first 3 decades of the 20th century was a time when scientific writing underwent enormous simplification, equivalent in a general way to the change that occurred between the early and the later 1600s. The reasons involved, however, were very different. While the desire for a more "plain and mathematical" language in Boyle's day was linked to hopes of reaching a larger audience, the call to make scientific writing more "functional" and "efficient" after 1900 was a result of increasing insidership, of a general belief that scientific language was its own special realm, with an elite relationship to truth, one that could not be effective if diluted by "literature" or "philosophy." This belief was itself an outgrowth of several things. One was the rise of positivism as a general attitude toward knowledge and human society. Positivism drew deeply on the literal fact of the greatly increased pace of technical invention and discovery, and applied this toward the notion that society (i. e., Western society) was capable of rapid advancement and progress if it could but give first place to "scientific thinking"—by which was meant fact-based thinking and expression. This appealed to the rising middle classes, who viewed their prosperity as an immediate effect of "progress." And it further grew out of important elements of nationalism: the obvious power of technology in furthering national strength during an age of intense rivalry among the major European states helped ingrain the idea, as once put by T. H. Huxley (in his well-known 1880 address, "Science and Culture"), that a truly "civilized" or "educated" outlook would be

> wholly unable to admit that either nations or individuals will really advance, if their outfit draws nothing from the stores of physical science. An army without weapons of precision . . . might more hopefully enter upon a campaign on the Rhine, than a man, devoid of a knowledge of what physical science has done in the last century, upon a criticism of life. (1893, p. 144)

In these words, Huxley was responding directly to Matthew Arnold and other literary intellectuals who had tried to draw attention to the overzealous faith in scientific knowledge and to propose a more humanities-based education as a means to weld progress with "virtue." These intellectuals failed in their effort; at the level of general public and governmental support, the "Two Cultures" debate was decided in favor of science. Huxley's own eloquence, ironically enough, helped secure its own demise a few decades later on. Indeed, even before 1900, many fields in the humanities made the effort to "modernize" themselves by adopting the supposed external forms of "science"—as in scientific history, scientific philosophy, political science, and so forth.[12]

Within the technical professions, meanwhile, the positivist philosophy (or attitude) argued for a split between "science" and "literature." The unparalleled success of German research had helped set the pace in defining the "scientific" as no longer interested in the merely descriptive or speculative but instead in documentation, mathematical reasoning, and theory building. Other fields could perhaps advance by adopting scientific ways, but science itself, the feeling went, could do no better than to refine and further hone its methods of inquiry and of writing. As Huxley's quote implies, there was a strong note of nationalism here too. The full ascendancy of the research university in America and Europe both had been underwritten by government support. Its effect was to encouraged the breaking apart of scientific disciplines into an ever expanding array of academic specialties, each with its own self-defining professional jargon and set of professional journals. The scale of scientific work grew tremendously between 1880 and World War I: the total number of journals more than doubled, with more and more articles being written on a vastly greater range of topics, with many of them much more constrained in scope, smaller in scale, and finer of resolution than ever before. With more work being done, there was a need for more rapid dissemination of "results" and also for the coining of much new language for the many new phenomena being identified. Technical education changed: students were actively recruited to the sciences and no longer had to obtain a classical education first (e.g., at the high school or undergraduate level) before entering upon their apprenticeship (Montgomery, 1994).

These comments only suggest the broad outline of what occurred with regard to "science" between roughly 1870 and 1930. The great privilege granted to the "scientific" at this time was no simple phenomenon. It was, in reality, a complex and multitiered response within Western culture as a whole to a particular period in its own development, a period that came in the immediate wake of the Industrial Revolution and that propelled the societies comprising this culture very quickly into a "modern" world of technological advance, urbanization, social stratification, and intimidating uncertainty. It was during this period that science and technology moved to the center of Western capitalism, and were recognized as such.

These profound alterations in the worldview of science had their inevitable effects on language. By 1880, Baconian ideas were in total disrepute. An empirical knowledge, slowly building upon itself, hesitant toward "theory" and tending toward descriptive writing, was now seen as wasteful and useless. Any attempt at literary style in technical papers was more and more regarded as archaic and amateurish. The audience, or reader, had shrunk to other professional scientists; the realm of "experience" and "observation" was now either confined to the laboratory (thus dependent on technological apparatus) or to portions of the external world transformed into technical documents: maps, diagrams, charts, tables, and so forth. Knowledge had become both more abstract and detailed, more reliant on specialized terms that acted as a denotative shorthand and that effectively stood in the way of older, discursive manners of expression. By 1920, Darwin, Lyell, and Gilbert too were stylistic relics; they would have been hard pressed to find a publisher.

Indeed, by this time, typical writing by well-known geologists, at its most relaxed and introductory, is more in the vein of the following:

> The origin of all those rocks which have been termed *gneissose* has been long regarded as the most difficult problem of petrology. Nowhere at present do we observe the formation of any rock-masses

similar to gneisses. Thus, if we try to explain their origin as due to the actual causes, it is only possible if we assume that these have been active at depths below the surface not directly accessible to our observation. It has long been evident that the large group of rocks termed 'gneisses' must comprise rocks of extremely different origin. For some of them very probable explanations have already been given. (Sederholm, 1923, p. 97)

Truly, a different kind of author has stepped upon the stage of geologic discourse by this time, one who is apt to quicken his stride and check his watch. The matter-of-factness and functionalism of a Cavendish has now taken over; conscious literary effects and high style are gone.[13] Eloquence has become a matter mainly of functional delivery, of Bacon's "plaine speaking." One senses that it is not other geologists who are being addressed, but only those in the subfield of petrology (the study of rock origins). The "we" of rhetorical engagement—the "we" that does the observing, the explaining—does not really include the general public any longer. Reader and writer are no longer interested in "literature"; having become smaller in scale and scope, their appetite for a more efficient content has become more pronounced.

And yet, despite these obvious things, to the eye of a contemporary geologist, this passage from Sederholm still has an older feel to it, a kind of "early modernism." It possesses a certain informality that appears most of all in the frequent use of modifying phrases: "of all those," "which have been," "if we try to." These lend a kind of conversational tone to the writing, a character of invited, not hermetic, interest to the logic. The suggestion remains that the nongeologist, though not addressed directly, might nonetheless still find a way to gain entry to this landscape and have access to things of value. Insidership is not yet complete and total.

In fact, personal reflection and even philosophizing were not yet utterly banished from all scientific writing at this stage. A great number of journal articles written by prominent physicists, for example, Heisenberg, Einstein, Dirac, and Schroedinger, were concerned directly with such philosophical issues as the nature of matter or observation and its limitations. In geology, meanwhile, as late as 1935, one of the most eminent of Britain's researchers, E. B. Bailey, could still escape the tyranny of the journal and collect important and detailed work in a book—a book whose title, *Tectonic Essays*, blended technical and literary traditions going as far back as Boyle himself. Known for unbounded energy and skill in producing detailed geologic maps of vast unexplored regions (the Scottish Highlands in particular), Bailey could begin his volume as follows, without concern for editorial censure:

> Even those who have more sympathy with man's endeavor than with
> the affairs of Nature may take an interest in the Science of Tecton-
> ics. Knowledge, after all, is of human creation; and, as a rule, the
> knowledge of the structure of a mountain chain comes as the reward
> of glorious struggle, both physical and mental. (p. iv)

In the 1930s, such words could no longer be written unself-consciously
in a technical monograph. But they could still be written, and published.
They could still express acceptable commitment, more than mere eccen-
tricity or impunity. True, their type had become increasingly rare; indeed, in many fields it had nearly disappeared altogether. Bailey himself,
moreover, *was* known as an anomaly, an indomitable original who advanced the geologic knowledge of Britain like "a force of Nature." But
the point is that there was still room in his day for some expression of
this individuality. As an author, he was not required to entirely disappear inside the formation "geologist."

IX

It is really only in the postwar era that such expression dies out altogether, as part of the ordinary repertoire of scientific writing. Even in
Bailey's day it had become a historical phantom, a dissolvable residue.
By the 1950s, the solvents of instrumentalism had removed it forever
from the content of "science." This does not mean that philosophy or
literary turns of phrase vanished entirely: removed from the "scientific,"
they nonetheless continued to make an occasional (though ever decreas-
ing) appearance around its authorial fringe, in prefaces and introductions, for example, or during moments of intense debate or revolutionary change. The point, however, is clear: what began as the center of
science, its linguistic origins, had by this time been forced to the periph-
ery. The forms of elegance that had urged Bacon, Boyle, Newton, Lyell,
and Darwin into authorship were now a kind of vestigial reflex, brought
into action only when "science" was placed within brackets, viewed from
the outside.

   This occurred in conjunction with the continued simplification of
language at nearly all levels of speech and writing, a process characteristic of the 20th century and which has been fulfilled under the influence
of technical modes of expression—picked up, for example, under the
aegis of "factualism," by the whole of the mass media since at least the
turn of the century. One might note that it has been often stated, and

overstated, that the "experimental philosophy" was responsible for ushering in the Augustan Age in English literature: Sprat, Newton, et al. have been presumed to be the force that flattened English of its earlier Elizabethan and Jacobean sophistications. This (as I hope I have suggested) is a matter of poor interpretation. It is, I believe, a projection backward for what has clearly taken place much more recently. Science at that time, after all, had none of the prestige and little of the adoration it has acquired since the last quarter of the 19th century. In 1700, it was still relatively new on the scene; 200 years later, however, and it had been "proven" the sole engine of progress, the bringer of truth, power, wealth, cures from disease, knowledge of humankind and the universe, the chance for social betterment. Voices everywhere sought to adopt something of its tone and thereby partake of its status and certainty. A curt, descriptive, impersonal style was taken up by many professions, by the social sciences, by educators, by textbook writers, and eventually by most nonfiction authors, especially in journalism. To any student of language, it is evident that this has remained the case down to the present. The connections to "science" are perhaps complex and multitiered; but they are obviously there, in ways that they were not in the 17th century.

Within scientific practice itself, on the other hand, all the major developments underway by 1900 simply gathered steam and standardizing impetus thereafter. From an activity of individuals or small university groups, it became one of teams, backed by major commercial, academic, and governmental institutions. Little Science became Big Science, and Big Science, especially following World War II, became Giant Science, building upon itself with ever greater amounts of manpower, ever more complex hierarchical organizations, unparalleled federal support, and a burgeoning demand for publication. By 1950, technical labor had become critical to all kinds of corporate productivity; to military power and the cold war; to governmental expertise; to national interest in, and support of, universities; to a hundred different aspects of "postindustrial" society. The education of scientists had by this time shifted almost entirely away from the humanities. The old classical curriculum, which lingered on even unto the 1930s and which emphasized an appreciation for the uses of language, was not merely avoided or discredited in the postwar age; it became extinct. War, big business, the "atomic age," "miracle medicine," and more had taught scientists that it was their knowledge that was the key to power. This was stimulated further by the tremendous privilege granted science and "scientific thinking," from the 1920s onward. In the name of efficiency and relevance, professional scientists, during the years of their education, were now urged to avoid taking courses in English, history, foreign languages, or philosophy. The "Two Cultures" debate was fought and won on the playing fields of "training." Indeed, as one indication of this, it is at this particular time, the 1950s, when scientists finally stopped writing (and presumably reading) the important histories of their own disciplines, something they had done as a regular part of their work for the entire preceding 300 years (see Laudan, 1993).

It should also be pointed out, in this connection, that the late 1940s and 50s are the historical moment when older scientific professionals perceived a marked decline in writing ability among their younger colleagues. This was the time when writing guides ("How to Compose and Publish a Scientific Paper") first appeared on the scene. It was also when research papers began to be regularly written by more than one author (by the 1960s and 70s, authorship by teams of four or more became standard; by the 1980s, a dozen or more names could be attached to an article). Publishing articles no longer had any link whatsoever to opportunities for displaying one's high standards of learning, culture, urbanity, and so on. In the postwar era, publications became coins of the realm in the race for status, career positions, better quality students, and grant money.

Add to all this the enormous increase in terminology and one can see the reasons why scientific discourse has continued unbroken on its path to instrumentality. The general attitude among postwar scientists toward functionalism and their increasing disconnection from humanities education no doubt greatly aided the general change toward a more flat, simple style. Certainly on the whole they ceased to take an interest in the language they used, as a mark of learnedness or cultivation. Any obvious literary elements in technical prose were no longer rewarded by journal editors and peer review boards either, but instead tended to be censored. And on a higher level, this reflected something significant. For it can hardly be denied that an impersonal style of expression, with its "literary nullification of the self," perfectly matches the large-scale contemporary anonymity of technical labor, its massive character, now so often built upon research teams, institutional programs, national priorities, and the like. If it is jargon that now drives the narrative along, having replaced nearly all aspects of literary nuance; if it is taboo to include in this narrative any personal biographical details; if writing itself has become a type of exercise or chore for scientists, a necessary toil for producing "contributions"; and if it is a rare thing for these professionals to take pride and pleasure any longer in their acts of authorship—then this is no doubt partly because this narrative and its authorship have both become elements in a larger series of standardized communicational events, ingredients in a mode of production. The cold and faceless qualities of scientific writing today express the corporate realities that frame the act of writing. It is, after all, no longer merely science that is produced in quantity today, but scientists as well.

🖘🖘

It can hardly be a surprise, therefore, that even at the simplest level, say that of the textbook, technical discourse in geology has come to sound like the following:

> Geological study of rocks confirms that they have been folded and uplifted; local motions within the crust are thus established. Records of the Earth's former magnetic field from paleomagnetic studies and magnetic anomalies provide evidence that continental masses and oceanic crust have moved with respect to the poles. (Wyllie, 1971, p. 1)

What might Gilbert have said of such a passage? Darwin? Huxley? By the 1980s, moreover, a full century after the time when these writers were all active, professional geological discourse had gone still further in this same direction. One expects, that is, from the vast majority of geoscientists writing today, a type of language much more on the order of this:

> Regional patterns of present-day tectonic stress can be used to evaluate the forces acting on the lithosphere and to investigate intraplate seismicity. Most intraplate regions are characterized by a compressional stress regime; extension is limited almost entirely to thermally uplifted regions. In several plates the maximum horizontal stress is subparallel to the direction of absolute plate motion, suggesting that the forces driving the plates also dominate the stress distribution in the plate interior. (Zoback et al., 1989, p. 291)

Such a language does not breathe or undulate, it hums and clicks. It is now much closer to a mechanism, a technology, tumblers falling into place. The very first sentence acts to cast into iron who might be an insider, who not. If one does not know the meaning of "tectonic stress," "lithosphere," or "plate seismicity," one can go no further. The limits of literacy are established immediately, and without mercy.

Such is what remains after the likes of Lyell, Gilbert, Sederholm, and even Bailey have all been boiled away. To make this yet more clear, we need only "update" the passages quoted above from the works of these men, let us say Gilbert (first excerpt) and Sederholm, and examine the result. To do this is not difficult; the residue of what would qualify as "science" today is easy enough to lay bare. It would be the following:

> Deep canyons provide excellent exposures for stratigraphic and structural analysis. Furthermore, their considerable relief provides important evidence of erosive processes. (Gilbert)

> The origin of gneissic rocks is a central problem of petrology. Since formation of such rocks is not observed at the surface, it must be assumed to take place at depth. (Sederholm)

In short, all the air must be taken out to achieve a contemporary look and feel. Again, it is expressiveness—as a form of interest in using and shaping language, as something to be enjoyed as well as employed—that seems deleted. Gilbert was writing, after all, in an era when Victorian belief in learnedness and cultivation included the sciences. Sederholm came later, obviously, but not so late that his narrative sought only to transfer the barest and bleached bones of "information." Their discourse is not at all difficult to translate into the contemporary scientific idiom, simply because so much of it has since been deemed irrelevant. Linguistically, it is far easier to take out "style" than to put it in.

Thus, over the past 100 years, something striking, perhaps even spectacular, does seem to have happened to technical discourse. The final product has come to achieve something close to Sprat's own vision of "primitive purity" or Wilkins's "greatest plainnesse"—except, of course, that it is accessible only to a trained few, something that the early members of the Royal Society would have found unfortunate, scandalous, regressive, even self-defeating. At present, however, this discourse stands at the end point of a long, historical process of vaporization, to the point where it now seeks to read like a proteinous extract from the logos itself. It is, perhaps, too simple to say merely that the humanities, as an obvious influence, are what have been removed. A conscious and broad-scale relatedness to other areas of culture is also mostly gone. To a large degree, scientific discourse has become a type of speech that formalizes its own writers, that prescribes—both in terms of obedience and transgression—the contours of authorial presence.

I dwell on these things here not because they reveal a kind of "degradation" of language, but because of their clear historical significance for what it means to be a scientific author. As shown in the first chapter, one could, of course, turn in another direction altogether and take the time to analyze the rhetorical aspects of technical language today, how it (still) seeks to persuade its readers through various strategies and demands. One could also examine the faults in actual narrative logic that are common to scientific writing, the clumsinesses, the failures, and the exaggerated claims it often commits and that show it to be something less than ideal in rationality, something therefore ordinary with regard to writing itself. One might go still further and look at the place and purpose of metaphor today, the ways in which originality leaks in, the conventions at work that allow for such things as speculation or personal bias to hide behind a mask of logical reasoning. And yet, for the actual history of scientific prose, all these types of inquiry, interesting as they are, would simply end up pointing to the very changes I have been charting above. These changes have traded losses for gains; to claim anything else would simply be wrong. Writing too is part of the cognitive content of "science." The death of eloquence, or rather its functional transformation, has been the basis for a maturing content whose form says much, in turn, about the powers we attribute to knowledge today.

All of this is said from a particular point of view. Ironically, it is a viewpoint commonly overlooked, yet in some ways more basic than that of academic study. Simply said, it is this: what has it meant to be a scientific author at different periods of time in the modern era? How have some of the important norms surrounding this author changed? In more personal terms, I have tried, that is, to look briefly at the history of technical writing from the perspective of being a scientific author myself. What sorts of things could I have said a century or two ago that I can no longer say? What types of speech were still my own, or demanded of me, that have since retreated to other areas?

In the end, it is hard not to be struck by a curious echo. It is difficult, that is, not to take note of how the contemporary realities of technical writing comport so well with the famed idea, so often quoted and discussed by those in the humanities during the past 15 years, of the "death of the author," an idea put forward by no lesser lights than Roland Barthes and Michel Foucault. In using this phrase, of course, these writers were engaging in a bit of dramatics. They were, however, interested in attacking the enormous cultural prestige granted the Author, as an ultimate, heroic, yet nonetheless fictional (constructed) subjectivity, as well as the false view of the writing process borne within this image: language as passive clay, manipulated by genius, without important determinations of its own (see Chapter 1, note 6). Yet it must seem ironic that, in their desire to formalize the author away and thus make language more supreme, these thinkers, so often held up as the epitome of humanities intellectualism, inadvertently called upon characterizations that would essentially pose *every* writer as *scientific*—in the sense that we are urged by scientific language to believe in a "writer [who] can only imitate an ever anterior, never original gesture" (Barthes, 1977, p. 146). The death of the author has often been cited as a key postmodernist notion. But in scientific discourse, it might be said to form the very fundament of modernist authorship.

X

It is often noted about science that its practitioners do not much read the writings of the past. Physicists who have never read Newton vastly outnumber those who have. Biologists familiar with the writings of Buffon or Cuvier are rare. Those in the geological sciences who might have ever been required or even urged to peruse William Maclure's *Observations on the Geology of the United States* (1817) probably total less than a thousand. For scientists today, trained as they are, the past in all its plenitude of intellectual evolution is generally resigned to the category of "literature." The humanities, said to abide in past creations, seem directly opposed by science in this regard. Science, the cliché runs, has incorporated its own "great books" into the flesh of its contemporary knowledge. The classics of Western science are thus mere husks from which the nutrients have long been extracted and metabolized. They are the stuff of lore: having lived and died, their remains, if of lasting import, are now part of the technical corpus.

This attitude—actually a socialized (professional) behavior of reading—was first put into words in what is perhaps one of the most influential documents ever written about the sciences, by the author who coined the very term "scientist" itself. William Whewell's *History of the Inductive Sciences* (1837) states the notion very plainly:

> Previous doctrines may require to be made precise and definite, to have their superfluous and arbitrary portions expunged, to be expressed in new language, to be taken into the body of science by various processes; but they do not on such an account cease to be true doctrines, or to form a portion of our knowledge. (p. 10)

"Nothing which was done," in other words, "was useless or unessential, though it ceases to be conspicuous and primary." Here, then, is articulation for one major aspect of the history of what has happened to scientific discourse during the last 350 years, an aspect I have tried to broadly map out above. What has ceased, what has been given "new language" and thus, in large part, what has been erased through incorporation, are the very forms of expression that were once so useful and necessary.

The language of science today, like the knowledge it embodies, confesses little or no memory of its genealogy. Beyond archival allegiances, to nomenclature above all, it reveals little interest in the overt forms of literary structure, style, and diversity that once—and for the greater part of its development—graced it. Scientific discourse has developed in specific directions, only one of which I have tried to show here. It has not given up metaphor ("messenger RNA"); it has not abandoned rhetoric ("The foregoing can be understood to suggest . . ."); nor has it turned its back altogether on the epistemology of witness ("In our field studies, we often observed . . ."). All these things, perhaps, are inevitable phenomena, however streamlined and selective. But this does not at all diminish the truth that this discourse has become standardized on the basis of rules and codes that significantly reduce the range of human linguistic gesture, that so limit and confine the contours of author and reader that these, in effect, become nearly identical. Built upon the granite of the "I," so new and vital in Boyle's time, so intact for Darwin, scientific discourse today has reduced this "I" to a sandy ash. What can be said about a style that has become disinterested in its own past?

My charting of change in this discourse, then, forms a kind of eulogy or lament. It does not take account of the many new powers and efficiencies that scientific language has acquired, often as a direct result of losses in other areas. These have been much analyzed elsewhere, and I have tried to add my own perceptions in Chapter 1. Nor have I taken full or systematic account of the scholarship on scientific discourse that has grown so enormously during the past decade (an historical indicator, no doubt, of the cachet that has come to be attached to the power of this discourse). Much of this work has sought to break down the old "Two Cultures" distinction, to propose technical language as entirely accessible to the same types of analyses that have long been profitably used upon the writings of literature or philosophy. As recently summarized by David Locke, a sizeable part of the relevant field now seeks to dissolve the assumed borders between "nominally heterogeneous bodies of texts—texts whose usual attributions, 'literary,' 'scientific,' 'technological,' 'political,' seem finally irrelevant to their reading as documents growing out of a common mode of thought, designated often by Foucault's term 'discursive formation' " (1993, p. 20).

From my own point of view, as both scientist and writer, this has much intellectual appeal, with regard to method. But it utterly fails in a more important area. On the level of actual experience, the experience of writing, the humanities and the sciences are truly different, and no amount of methodological analysis can eliminate this disparity. The origins of modern Western science, as told through its language, are intrepidly bound up not with the suppression of experience, as is so often said, but with its elevation and refinement. As I've tried to suggest, it now seems that the origins of literary discourse and scientific discourse are intimately linked through their mutual employment, early on, of forms regularly used by intellectuals during the medieval and Renaissance periods. Bacon's adoption of Montaigne's essay structure, which, as an exposition of personal thought and the experience-based search for knowledge, came to supplant the older tradition of composing commentaries on the work of sacred texts, represents a deep connection at the root between literary and scientific authorship. To the degree that the relevant etymology—*essai*, from *essayer*, "to try," "to put to the test," "to experiment"—had its meaning in the 1600s, it retains it today in more heightened form, as one looks back. It is too simple to say that, linguistically, science emerged out of literature, or that both developed their parallel courses unbridged, confined to separate valleys of intent.

For 200 years, from the late 17th century to at least the time of Darwin and Huxley, scientists continued to use the term "essay" in reference to their own writings. They continued, as standard practice, to insert themselves, as active (observing, thinking, feeling) agents into their texts. They continued to strive for many of the same types and degrees of eloquence as did authors in general. This remained true nearly a half-century after Whewell's "scientist" first appeared, thus marking the fact that a new and separate identity had come to exist. Geologists, chemists, physicists, and botanists lost their title of "natural philosophers"; for a host of reasons, their world of research had grown more and more distinct from that of letters, from the types of speculation and the claims to truth that had, only a few years before, qualified their labor as a branch of philosophy. Yet, even so, the new scientists were slow, reluctant, even resistant to giving up classical learning as an essential ingredient of their training, and thus to abandoning sophisticated types of rhetoric and expression as sometime models for their own work. They still considered themselves, in short, "individuals of culture." And as such, they sought an inevitable deposit for such culture in their narrative style, which, in many of its literary qualities, did not at all leave the realm of *lettres* until the opening of the present century.

It is not toward this age of last eloquence, nor for the scientist as a repository of "culture," that technical writing shows its dispassion and neutrality. Rather, it is for the scientist as one who potentially enjoys and celebrates language. This active creator of knowledge, after all, once had great opportunities of command over the speech s/he used. Today, on the level of encounter and involvement, on sitting down to write, s/he is necessarily far more often a kind of required medium, restrained by conventions that force her or his individual presence to evince itself as a wholly secondary aspect. I don't mean to say that the scientist-author is absent or entirely ritualized; that would be exaggerating the matter needlessly. The point is that this presence is produced through impersonal traces of personal choices, and is most evident where some type of breach of existing standards occurs. At times, one might still find in a scientific text such things as irony, wordplay, even signs of exclamation, offense. Yet again, among the hundreds of articles that the average scientist consumes each year, and in the research papers s/he is urged to produce, such things are as flecks of foam on an otherwise epic sea of denotation.

All these, then, are reasons why the past writing of scientists has attained its exotic quality, the quality of "literature." But also of "history." For this history, as a repository of "discursive formations" from the past, is exactly what seems so out of place with regard to technical discourse today, so hopelessly, if grandly, "historical." Foucault, one should note, never wrote about the hard sciences (except biology) and stopped his own research on scientific language, abruptly it would seem, in the early 19th century, when the sciences and the humanities were still irretrievably and practically united in many of their expressive norms. To a degree, stopping at this point defined an easy way out: the true difficulties begin only afterward, when this unity broke apart. From the point of view of the present, the scientist, unlike his literary counterpart, cherishes only from afar, from the other side of a wide and rising river, the language and customs of his forebear. S/he is, so to speak, trained to a state of dispossession. It is for this entity and this condition, that, at some level, the history of scientific discourse must be said to mourn.

Notes

1. An idea of the range of writing on scientific discourse to date can be gleaned, for example, from the articles and sources given in the following: Jack Selzer, ed., *Understanding Scientific Prose* (Madison: University of Wisconsin Press, 1994); Peter Dear, ed., *The Literary Structure of Scientific Argument* (Philadelphia: University of Pennsylvania Press, 1991); Frederick Amrine, ed., *Literature and Science as Modes of Expression* (Boston: Kluwer, 1989); Bruno Latour and Steve Woolgar, *Laboratory Life: The Social Construction of Scientific Facts* (Beverly Hills, CA: Sage, 1979); Charles

Bazerman, *Shaping Written Knowledge: The Genre and Activity of the Experimental Article in Science* (Madison: University of Wisconsin Press, 1988); Greg Myers, *Writing Biology: Texts in the Social Construction of Scientific Knowledge* (Madison: University of Wisconsin Press, 1990); Stuart Peterfreund, ed., *Literature and Science: Theory and Practice* (Boston: Northeastern University Press, 1990); and Fernand Hallyn, *The Poetic Structure of the World: Copernicus and Kepler* (New York: Zone Books, 1990).

2. See, for example, Charles Haskins's early yet still valuable work on this episode, *The Renaissance of the 12th Century* (1927).

3. Two excellent books on this subject are Lisa Jardine's *Francis Bacon and the Art of Discourse* (Cambridge University Press, 1974), and Brian Vickers's, *Francis Bacon and Renaissance Prose* (Cambridge University Press, 1968). One of the very best examples of Bacon's use of extended metaphor, however—especially in view of its content—has been noted by William Armstrong in his introduction to *The Advancement of Learning*, book 1. In this example, Bacon criticizes what he feels to be the excessive ornateness of writers inspired by Ciceronian rhetoric, and he does so in an entirely, and magnificently, demonstrative way, within a single sentence (which I break up into three parts):

1. This [the overuse of words and metaphor] grew speedily to an excess; for men began to hunt more after words than matter;
2. and more after the choiceness of the phrase, and the round and clean composition of the sentence, and the sweet falling of the clauses, and the varying and illustration of their works with tropes and figures
3. than after the weight of matter, worth of subject, soundness of argument, life of invention, or depth of judgment.

As Armstrong points out, one sees here how the echoing, extending phrases of the first and second parts, obviously intended to skillfully illustrate the very sense of what is being said, are followed by a sudden shift to simplicity and brevity, based on the repetition of monosyllabic words—weight, worth, life, depth (note, too, how repetition of the word "matter" at the end of (1) and the beginning of (3) implies the complete superfluity of everything in between). As R. F. Jones (1951) once observed, Bacon and his followers were never more rhetorically complex, metaphorical, and literary than when denouncing other authors for these very qualities. This method of polemic, one should add, was entirely conscious, indeed a skill of prose argument that was itself taught in the Renaissance university course on rhetoric. Bacon was simply more skilled at it than most.

4. In particular, the degree of purposeful theft and misattribution among medieval and early Renaissance authors was notoriously high when it came to scientific matters. It is clear, for example, in early works such as the

Etymologiae by Isidore of Seville (7th century). But it is nowhere more in evidence than during the late 11th to early 13th centuries. This formed the great period of translation, mainly from Arabic (and to a lesser extent Greek and Hebrew) into Latin, when the larger portion of Greek thought was transferred to Europe (when Greece, in a sense, was first nativized to Europe in terms of its canonical authors such as Aristotle, Euclid, and so on). During this period and somewhat afterward, confusions often arose between translators as "decanters," those who acted as linguistic mediums for the new knowledge, and as "authors," experts and originators of such knowledge. Some translators, such as Constantinus Africanus (c. 1015–1087), seem to have taken full authorial credit for works they had only translated (something the earlier Arabic translators had done as well). So potent was the status of such knowledge, moreover, especially after the rise of the universities, that cultlike followings (of students, mostly) grew up around some of the translators, whose biographies were written to heroify and monumentalize them in ways that often far exceeded their actual accomplishments. Knowledge and ideas were clearly forms of public "capital" even at this early date. A good sense of the heights and depths to which these types of purposeful confusion rose and sank can be gleaned from Lynn Thorndike's magisterial *A History of Magic and Experimental Science* (New York: Columbia University Press, 1923), Vols. 1 and 2.

5. Recent interpretations, in fact, have largely overlooked the issue of literary form and focused instead on rhetoric itself. The question of those such as Kepler or Galileo taking up a range of different formats for expressing their ideas does not appear as important to most scholars as that surrounding the detailed deployment of rhetorical technique.

The latter, meanwhile, has tended to be viewed in sociologic terms: early scientists are said to have used existing linguistic resources in order to gain credibility within the learned universe of the time, to achieve legitimacy by employing (or only quietly deforming) the given order of discourse. No doubt this is true, if a bit obvious. Such could be easily said of any author with specific ambitions for general impact. There is a more important problem, however, in the focus on rhetorical technique. This comes from an often unspoken link, or rather leap of identification, that scholars make between the detailed "moves" and strategies of a particular text and authorial intentions. Focusing on processes of the narrative itself, analysts tend to insert the writer into the action of their own discourse ("What Kepler does here is to . . ."). The effect is to make it seem as if a degree of overt choice went into every last rhetorical trope, as if each text was the result of an infinitely manifest strategy, minutely planned and carried out like a 20th-century battle plan (one might note, too, that this blurring of distinctions between author and text, in a broader sense, permits the illusion that a historical quality pervades and coats everything, as if a paleontological excavation of the author's original "mind" were in progress). Likewise, the image of writing, as a method of creation, is similarly implied to be rational in the extreme. In the end, the scientist, as author, emerges from the dust as a kind of wondrous, positivistic entity, gilded in absolute and reasoned control of the "method" of his discourse.

6. Simplicio responds to this as follows: "I would like to have been present at these experiments; but feeling confidence in the care with which you performed them, and in the fidelity with which you related them, I am satisfied and accept them as true and valid."

7. The work of Owen Hannaway and Jan Golinski, exemplary in this area, has been already mentioned. Others, meanwhile, such as Charles Webster, P. M. Rattansi, and Allen G. Debus have indicated in detail how this move in chemistry, begun in the 16th century, was linked to a strong increase in the influence of the German theorist Philip von Hohenheim (1493–1541), known as Paracelsus, whose writings only became available posthumously but then went on to have enormous impact beginning in the 1550s. Paracelsus argued for a more experimental chemistry, whose results were to be used in reforming medical practice and were to be made public knowledge by use of the vernacular. His followers took up the latter idea in particular: making chemistry more "clear and open" in language—more public in accessibility—seems to have been one way in which they sought to rally support for the defeat of their opponents within the traditional alchemical and medical societies. As Golinski writes, "The exploitation of the printed medium was an obvious resource for Paracelsians trying to take over the craft practices of apothecaries. . . . It has been noted that chemical works were printed in large numbers after the Paracelsian revival and at other junctures when the movement was gathering public support, such as the period of the English Revolution in the 1640's and 50's" (1990b, p. 374). See also Allen G. Debus, *The Chemical Philosophy: Paracelsian Science and Medicine in the Sixteenth and Seventeenth Centuries*, 2 vols. (New York: Science History, 1977). Boyle, in a sense, added himself to this budding tradition of revolt.

8. It was Bacon's work, in fact, along with translations of Montaigne, especially that by John Florio (1603), that set off the flurry of similar writings in the decades immediately following. See the discussion of this episode of influence in F. O. Matthiessen's *Translation: An Elizabethan Art* (Cambridge, MA: Harvard University Press, 1931).

9. In explaining this reluctance to come up with any final determinations, Boyle, in his *Certain Physiological Essays*, says the following:

> Perhaps you will wonder . . . that in almost every one of the following essays I should speak so doubtingly, and use so often, *perhaps, it seems, it is not improbable*, and such other expressions, as argue a diffidence of the truth of the opinions I incline to, and that I should be so shy of laying down principles, and sometimes of so much as venturing at explications. But I must freely confess to you, . . . that having met with many things, of which I could give myself no one probable cause, and some things, of which several causes may be assigned so differing, as not to

agree in any thing, unless in their being all of them probable enough; I have often found such difficulties in searching into the cause and manner of things, and I am so sensible of my own disability to surmount those difficulties, that I dare speak confidently and positively of very few things, except of matters of fact.

The author here claims modesty as a reason for offering no conclusions. But he also, in effect, reiterates his definition of "fact" in the empirical, Baconian sense. These essays, it might be noted, were ostensibly intended as a textbook for the education of his nephew and were cast in the form of a monologue to one Pyrophilus ("he of fiery disposition"), presumably the boy himself. In one of the opening portions of the book, "In Defense of Experimental Essays," Boyle states that his design is to collect "experiments and observations" in order to yield "a continuation of the Lord Verulam's [Bacon's] *Sylva Sylvarum*, or natural history."

10. Newton's own mysticism and his interest in alchemy are well known (see, e.g., Richard Westfall, *Never at Rest: A Biography of Isaac Newton* [Cambridge, MA: Harvard University Press, 1980]). In the concluding section of his *Philosophiae naturalis principia mathematica* (Mathematical principles of natural philosophy)—so often held up as the grand document for a new, strictly rationalistic science—Newton seems to perform a remarkable, almost dizzying aboutface in the space of a few lines, first stating, then, in effect, contradicting a foundational principle of his logic. Regarding the cause of gravity itself, that is, he declines to draw any conclusions, since "whatever is not deduced from the phenomena is called an hypothesis; and hypotheses, whether metaphysical or physical, whether of occult qualities or mechanical, have no place in experimental philosophy." Yet the very next paragraph begins thus:

And now we might add something concerning a certain most subtle spirit which pervades and lies hid in all gross bodies; by the force and action of which spirit the particles of bodies attract one another . . . and electric bodies operate [in repulsion and attraction]; and light is emitted, reflected, refracted, inflected, and heats bodies; and all sensation is excited, and the members of animal bodies move at the command of the will, namely, by the vibrations of this spirit, mutually propagated along the solid filaments of the nerves. . . . But these are things that cannot be explained in few words, nor are we furnished with that sufficiency of experiments which is required to an accurate determination and demonstration of the laws by which this electric and elastic spirit operates.

The point one might surmise from this is that the style of "logic" for thought and for discourse, employed even by the most rigorous of experimentalists, is still an open forum of reasoning at this time, loosely defined compared with today, more open to a range of materials and types of reflection.

11. A simple example might suffice. Lavoisier, as it happened, usually followed the basic tenets for reporting chemical experiments set down by Boyle, that is, a detailed, step-by-step description of what was done, seen,

and thought, with any subsidiary details that might add to the authority of
the work performed thrown in. Yet, like Boyle too, he often adds a number
of ancillary, personal details. In one such report on the "calcination of tin"
(metallic oxidation), Lavoisier speaks of a "Mr. Chemen, inspector of coinage" who supplied a balance, and then notes, "I have reason to believe no
more perfect instrument of this sort exists." This type of thing was fairly
common at the time, a kind of expected element, especially in the introductory part of an article or speech. Lavoisier's writing then proceeds through a
series of narrated actions and tables showing precise weights (before/after
"calcination"), but just before presenting its final conclusions, interjects the
following:

> I tried to repeat with lead the experiments which I have just described,
> but, as I said, I succeeded well only once and then with such uncertain
> and extraordinary results that I am induced to postpone its publication.

There were no rules or expectations governing this type of confession. Whether
intended or not, the effect is novelistic, reminiscent of *fictional* techniques at
the time: '. . . an outcome we must postpone until the next chapter.' Thus, one
might say, an intriguing literary image of the scientific career itself.

12. This was especially true in the United States, where definitions of
the "scientific" directly affected the expressive style of scholarship during
the early Progressive era, often making it flatter, more "factual." Earlier
writers, whose style was closer to Macaulay or other such literary authors,
were called "amateurs" (see, for example, Hofstadter, 1968).

13. The truth of this seems all the more apparent when one considers
that Sederholm, a Finn by birth and upbringing, was himself a scientist trained
in the 19th-century European tradition, that is, he was grounded in Greek
and Latin, literature, history, and philosophy. He was, moreover, an enormously learned man, capable in many languages (he wrote articles in Finn-
ish, English, German, French, and Swedish), and imbued with intense romantic feelings toward nature and his homeland. None of this, however,
persuaded Sederholm to employ a more literary style in his own technical
writings. On the contrary, he was himself an ardent believer in casting science into "plain, unadorned language." Like many other scientists of his
day, he made use of his classical background not as an inspiration to eloquence but, instead, for something quite opposite: the coining of many new
jargon terms and names, such as "anatexis," "agmatite," "Katarchaean,"
"migmatite," "myrmekite," "ptygmatic," and so on.

References

Armstrong, W. A. (1975). Introduction. In *The Advancement of Learning,
Book 1* by F. Bacon (pp. 1–47). London: Athlone Press.

Auerbach, E. (1953). *Mimesis: The Representation of Reality in Western Literature.* Transl. by W. R. Trask. Princeton, NJ: Princeton University Press.

Bacon, F. (1857–1859). *Works.* 7 vols. J. Spedding, R. L. Ellis, and D. D. Heath (Eds.). London.

Bailey, E. B. (1935). *Tectonic Essays.* London: Thomas Murby.

Barthes, R. (1977). The Death of the Author. Transl. by S. Heath. In *Image-Music-Text* (pp. 142–148). New York: Hill & Wang.

Beer, G. (1983). *Darwin's Plots: Evolutionary Narrative in Darwin, George Eliot and Nineteenth-Century Fiction.* London: Routledge and Keagan Paul.

Boyle, R. (1773). *Works.* 6 vols. Thomas Birch (Ed.). London: J. & F., Rivington.

Boynton, H. (Ed.). (1948). *The Beginnings of Modern Science: Scientific Writings of the 16th, 17th and 18th Centuries.* Roslyn, NY: Walter J. Black.

Bush, D. (1952). *English Literature in the Earlier Seventeenth Century, 1600–1660.* London: Oxford University, Press.

Cadden, J. J., and P. R. Brostown (Eds.) (1964). *Science and Literature: A Reader.* Boston: D. C. Heath.

Camargo, M. (1983). Rhetoric. In D. L. Wagner (Ed.), *The Seven Liberal Arts in the Middle Ages* (pp. 96–124). Bloomington: Indiana University Press.

Campbell, J. A. (1990). Scientific Discovery and Rhetorical Invention: The Path to Darwin's *Origin.* In H. W. Simons (Ed.), *The Rhetorical Turn Invention and Persuasion in the Conduct of Inquiry* (pp. 58–71). Chicago: University of Chicago Press.

Cavendish, H. (1921). *The Scientific Papers of the Honourable Henry Cavendish, F.R.S.* 2 vols. Edward Thorpe (Ed.). Cambridge, MA: Harvard University Press.

Chaucer, G. (1987). A Treatise on the Astrolabe. In L. D. Benson (Ed.), *The Riverside Chaucer* (pp. 661–684). Boston: Houghton Mifflin, .

Christianson, G. (1984). *In the Presence of the Creator: Isaac Newton and His Times.* New York: Free Press.

Darwin, C. (1914). *The Origin of Species.* Philadelphia: David McKay. (reprint of 6th London ed.).

Davy, H. (1952). Selections. Reprinted In H. M. Leicester and H. S. Klickstein (Eds.), *A Source Book in Chemistry, 1400–1900* (pp. 243–258). Cambridge, MA: Harvard University Press.

Davy, N. (1953). *British Scientific Literature in the Seventeenth Century.* London: George G. Harrap.

Dear, P. (1985). "Totius in verba": Rhetoric and Authority in the Early Royal Society, *Isis, 76,* 145-161.

Dear, P. (1991). Narratives, Anecdotes, and Experiments: Turning Experience into Science in the Seventeenth Century. In P. Dear (Ed.), *The literary Structure of Scientific Argument* (pp. 135–163). Philadelphia: University of Pennsylvanian Press.

Feuer, L. (1992). *The Scientific Intellectual.* 2d ed. New Brunswick, NJ:Transaction Press.

Finocchiaro, M. A. (1980). *Galileo and the Art of Reasoning.* Dordrecht:Reidel.

Foucault, M. (1970). *The Order of Things.* Transl. by A. Sheridan. New York: Pantheon.

Foucault, M. (1977). What is an Author? Transl. by J. Harari. In J. Harari (Ed.), *Textual Strategies: Perspectives in Post-Structuralist Criticism* (pp. 141–160). Ithaca, NY: Cornell University Press.

Franklin, B. (1987). *Writings.* New York: Library of America.

Fuller, S. (1993). *Philosophy, Rhetoric, and the End of Knowledge.* Madison: University of Wisconsin Press.

Gabbey, A. (1990). Newton and Natural Philosophy. In R. C. Olby, G. Cantor, J. R. R. Christie, and M. J. S. Hodge (Eds.), *Companion to the History of Modern Science* (pp. 243–263). New York: Routledge.

Galileo (1967). *Dialogue Concerning the Two Chief World Systems,* 2d ed. Transl. by S. Drake. Berkeley and Los Angeles: University of California Press.

Galileo. (1989). *Sidereus Nuncius; or, The Sidereal Messenger.* Transl. by A. Van Helden. Chicago: University of Chicago Press.

Gilbert, G. K. (1876). An Analysis of Subaerial Erosion. *American Journal of Science,* 3d ser., *12,* 85–103.

Gilbert, W. (1958). *On the Magnet.* Transl. by P. F. Mottelay. New York: Appleton.

Golinski, J. (1987). Robert Boyle: Skepticism and Authority in Seventeenth-Century Chemical Discourse. In A. Benjamin, G. Cantor, and J. R. R. Christie (Eds.), *The Figural and the Literal: Problems of Language in the History of Science and Philosophy, 1630–1800* (pp. 58–82). Manchester, UK: Manchester University Press.

Golinski, J. (1990a). Language, Discourse and Science. In R. C. Olby, G. Cantor, J. R. R. Christie, and M. J. S. Hodge (Eds.), *Companion to the History of Modern Science* (pp. 110–126). New York: Routledge.

Golinski, J. (1990b). Chemistry in the Scientific Revolution: Problems of Language and Communication. In D. C. Lindberg and R. S. Westman (Eds.), *Reappraisals of the Scientific Revolution* (pp. 367–396). London: Cambridge University Press.

Green, M. T. (1982). *Geology in the Nineteenth Century.* Ithaca, NY: Cornell University Press.

Hall, M. B. (Ed.). (1965). *Robert Boyle on Natural Philosophy.* Westport, CT: Greenwood Press.

Hannaway, O. (1975). *The Chemists and the Word: The Didactic Origins of Chemistry*. Baltimore: Johns Hopkins University Press.

Haskins, C. H. (1927). *The Renaissance of the 12th Century*. Cambridge, MA: Harvard University Press.

Heilbron, J. L. (1975). Priestley, Joseph. *Dictionary of Scientific Biography*, 7, 344–349.

Hofstadter, R. (1968). *The Progressive Historians*. New York: Vintage.

Horne, C. J. (1965). Literature and Science. In B. Ford (Ed.), *From Dryden to Johnson: The Pelican Guide to English Literature, vol. 4* (pp. 188–202). Harmondsworth: Penguin.

Howell, W. (1956). *Logic and Rhetoric in England, 1500–1700*. Princeton, NJ: Princeton University Press.

Huxley, T. H. (1893). *Science and Education: Essays*. New York and London: D. Appleton.

Hyman, S. E. (1962). *The Tangled Bank: Darwin, Marx, Frazer and Freud as Imaginative Writers*. New York: Atheneum.

Jardine, L. (1974). *Francis Bacon and the Art of Discourse*. Cambridge, MA: Harvard University Press.

Jones, R. F. (1951). *The Seventeenth Century: Studies in the History of English Thought and Literature from Bacon to Pope*. Stanford, CA: Stanford University Press.

Jones, R. F. (1953). *The Triumph of the English Language*. Stanford, CA: Stanford University Press.

Jones, R. F. (1961). *Ancients and Moderns*. Berkeley, CA: University of California Press.

Kelley, S. (1965). *The "De Mundo" of William Gilbert*. Amsterdam: Menno Hertzberger.

Kelley, S. (1978). Gilbert, William. *Dictionary of Scientific Biography*, 5: 396–401. New York: Scribner's.

Kepler, J., *Epitome of Copernican Astronomy (Books IV and V)* and *Harmonies of the World (Book V)*. Transl. by C. G. Wallis. Chicago: University of Chicago Press. (Great Books Series, 16: 845–1085.)

Ker, W. P. (Ed.). (1900). *The Essays of John Dryden*. Oxford: Oxford University Press.

Kohn, D. (Ed.), (1985). *The Darwinian Heritage*. Princeton, NJ: Princeton University Press.

Laudan, R. (1993). Histories of the Sciences and Their Uses: A Review to 1913. *History of Science, 31*, 1–34.

Lavoisier, A. (1952). Memoire on the Calcination of Tin in Closed Vessels. In H. M. Leicester and H. S. Klickstein, (Eds.), *A Source Book in Chemistry, 1400–1900* (pp. 154–163) Cambridge, MA: Harvard University Press.

Locke, D. (1992). *Science as Writing*. New Haven, CT: Yale University Press.

Lyell, C. (1830–1833). *Principles of Geology, Being an Attempt to Explain the Former Changes of the Earth's Surface by Reference to Causes Now in Operation*. 5 vols. London: John Murray.

Markus, G. (1987). Why is There No Hermeneutics of Natural Sciences? Some Preliminary Theses. *Science in Context, 1*, 37–43.

McKnight, G. H. (1968). *The Evolution of the English Language: From Chaucer to the Twentieth Century*. New York: Dover.

Milroy, J. and L. Milroy. (1991). *Authority in Language*, 2d ed. London: Routledge.

Montgomery, S. L. (1994). *Minds for the Making: The Role of Science in American Education, 1750–1990*. New York: Guilford.

Montgomery, S. L. (1989). The Cult of Jargon: Reflections on Language in Science. *Science as Culture*, no. 6, 42–77.

Murchison, Sir R. I. (1835). The Silurian System. *London and Edinburgh Philosophical Magazine*, 3d ser., 7, p. 46–52.

Murphy, J. J. (Ed.). (1978). *Medieval Eloquence: Studies in the Theory and Practice of Medieval Rhetoric*. Berkeley and Los Angeles: University of California Press.

Newton, I. (1958). New Theory about Light and Colors. In I. B. Cohen (Ed.), *Isaac Newton's Papers and Letters on Natural Philosophy* (pp. 47–59). Cambridge, MA: Harvard University Press.

Newton, I. (1952). *Mathematical Principles of Natural Philosophy*. Berkeley and Los Angeles: University of California Press.

Parks, G. (1951). Travel as Education. In Jones, R. F. *The Seventeenth Century* (pp. 264–290). Stanford: Stanford University Press.

Paradis, J. (1983). Bacon, Linnaeus, and Lavoisier: Early Language Reform in the Sciences. In P. Anderson, R. J. Brockmann, and C. Miller (Eds.), *New Essays in Technical and Scientific Communication: Research, Theory, Practice* (pp. 200–224). Farmingdale, NY: Baywood.

Paradis, J. (1987). Montaigne, Boyle, and the Essay of Experience. In G. Levine (Ed.), *One Culture: Essays in Science and Literature* (pp. 59–91). Madison, WI: University of Wisconsin Press.

Parry, J. J. (Ed.), (1987). *The Poems and Amyntas of Thomas Randolph*. New Haven, CT: Yale University Press.

Porter, R. (1991). Introduction. In S. Pumfrey, P. L. Rossi, and M. Slawinski (Eds.), *Science, Culture, and Popular Belief in Renaissance Europe* (pp. 1–15). Manchester, UK: Manchester University Press.

Priestley, J. (1774, 1775, 1777). *Experiments and Observations on Different Kinds of Air*. 3 vols. London.

Ricoeur, P. (1975). *The Rule of Metaphor*. Transl. by R. Czerny. Toronto: University of Toronto Press.

Salmon, V. (1979). *The Study of Language in 17th-century England.* Amsterdam: John Benjamins.

Sarton, G. (1957). *Six Wings: Men of Science in the Renaissance.* Bloomington: Indiana University Press.

Sederholm, J. J. (1967). *Selected Works: Granites and Migmatites.* New York: John Wiley.

Shapin, S. and S. Schaffer (1985). *Leviathan and the Air-Pump: Hobbes, Boyle and the Experimental Life.* Princeton, NJ: Princeton University Press.

Slawinski, M. (1991). Rhetoric and Science/Rhetoric of Science/Rhetoric as Science. In S. Pumfrey, P. L. Rossi, and M. Slawinski (Eds.), *Science, Culture and Popular Belief in Renaissance Europe* (pp. 71–99). Manchester, UK: Manchester University Press, 1991).

Sprat, Thomas. (1958). *History of the Royal Society,* J. Cope and H. Jones (Eds.). St. Louis, MO: Washington University Press.

Sutherland, B. (1969). *English Literature in the Later 17th Century.* Oxford: Oxford University Press.

Ure, P. (Ed.), (1956). *Seventeenth-Century Prose, 1620-1700.* Middlesex, UK: Penguin Books.

Vickers, B. (1968). *Francis Bacon and Renaissance Prose.* Cambridge, MA: Harvard University Press.

Vickers, B. (1985). The Royal Society and English Prose Style: A Reassessment. In B. Vickers and N. Struever (Eds.), *Rhetoric and the Pursuit of Truth: Language Change in the Seventeenth and Eighteenth Centuries* (pp. 11–14). Los Angeles: Clark Memorial Library.

Webster, C. (1975). *The Great Instauration: Science, Medicine, and Reform, 1626–1660.* London: Duckworth.

Whewell, W. (1840). *History of the Inductive Sciences.* 3 vols.London: J. W. Parker.

Whitaker, V. (1962). *Francis Bacon's Intellectual Milieu.* Berkeley and Los Angeles: University of California Press.

Wilkins, J. (1968). *Essay Towards a Real Character and a Philosophical Language.* Menston, UK: Scholar Press.

Wilson, F. P. (1960). *Seventeenth Century Prose: Five Lectures.* Berkeley and Los Angeles: University of California Press.

Wyllie, P. J. (1971). *The Dynamic Earth: Textbook in Geosciences.* New York: John Wiley.

Zilsel, E. (1941). The Origins of William Gilbert's Scientific Method. *Journal of the History of Ideas,* II(1), 1–32.

Zoback, M. L., et al. (1989). Global Patterns of Tectonic Stress, *Nature, 341,* (6240), 291–298.

3

Illness and Image

On the Contents
of Biomedical Discourse

᛫᛫᛫᛫᛫᛫᛫᛫᛫᛫

*By the word "information" we denote all the knowl-
edge which we have of the enemy and his country;
therefore in fact the foundation of all our ideas and
actions. Let us just consider the nature of this foundation . . . and we shall soon feel what a dangerous
edifice War is, how easily it may fall to pieces and bury
us in its ruins.*
— Clausewitz, *On War, 1832*

Nietzsche observed that the metaphors of the past become the truths of
the present. He meant this, in fact, quite literally: words with a rich
analogical power; images that bear within them new types of organizing
rationality; terms or similes that seem, as if by some sudden or forgotten
magic, to defeat complexity at a single bound—all of these, Nietzsche
perceived, have at one time or another been critical to the founding and
development of knowledge.

Nietzsche's idea was that formations of truth are not always "dis-

134

covered." They are not necessarily oracular, like an amazed or unexpected reading from "the book of nature." Instead, they are built up from the concrete and historical materials of human language, imagination, and need. Truth begins in a molten state, saturated with images. It hardens only over time, as an ongoing process.

The interactions by which truth is solidified in discourse are complex, and for the most part silent. In simple terms, the evolution of metaphors into hypotheses, theories, even "facts" is something that involves a change of habituation, a loss of origins. Over time, repetition and standardized usage gradually obscures the original figurative character of an image or term. By being endlessly repeated within a restricted context of meaning, such a term soon sinks into the institutional given of a particular discipline, becoming an element of its jargon. As such, it ceases to operate, and is no longer seen, as a figure of speech. Linguistically, its analogical (metaphoric) dimension has been erased; it is now identified directly with its referent. Those within the bounds of the relevant discipline are taught to perceive it—to read, write, and speak it—as if it had no root meaning and no history behind it. One ceases to consider the birth or ancestry of jargon in other areas of language, that the vocabulary adapted for special, technical use has a life outside the discipline. One is urged, in short, to comprehend such terms as "force" or "cell," "energy" and "nucleus" as if they had no real and long-standing connections outside the realm of scientific knowledge.

Nietzsche's idea is nowhere more powerful and relevant than in relation to the natural sciences. Alongside those terms just mentioned, one could list thousands of others in contemporary technical speech that had their metaphorical beginnings: "wave," "particle," "vein," "fault," "reaction," "immunity," "equilibrium," "blood circulation," "black hole," "cold front," "dwarf star," "dark matter," and on and on—all of which have revealing stories of appropriation and use behind them and that therefore help bring into the open a distinct cultural process essential to the ways in which knowledge is created and, so to speak, solidified. The special institutional usage of such terms today, that is, is often the result of a long journey through time and lexical evolution. But it is also a result of coinages made by particular scientists at particular moments in the evolution of particular areas of knowledge.

Harvey's choice of "circulation" with regard to the movement of blood through the body, for instance, came directly from an analogy he hit upon to solve the problem of visualizing blood flow: "I began to think," he wrote, using confessional italics, "whether there might not be *a motion, as it were, in a circle*" (quoted in Boynton, 1948, p. 496).[1] But one should note, too, that this image was conceived within the context of a larger analogical understanding of the time that often proposed the body as a machine, within which the blood moved through a series of "passages," "vessels," or "doorlets," by various "emptyings" and "fillings." While Harvey may have employed the term "circulation" as a kind of useful and approximate picturing of things, it nonetheless quickly became accepted by the scientific community as a literal description of blood's passage through the body. Its success depended, in part, on helping expand the dominant image system of the day, which, in a sense, had prepared the conditions for its visualization in the first place.

Another example would be Pasteur's introduction of the term "immunity." This was justified by a range of other metaphorical conceits he himself had already established regarding what he often called "la résistance vitale de l'animal" and the body's ability to "fortifier lui-même" (to strengthen itself). An older, more vernacular sense of "immune"—to be free of anything evil or injurious—made it an appropriate term to adopt as an analogy within Pasteur's new system of discourse and ideas. It fell neatly into place as a term for describing a state of "natural resistance" or "defense."

🔠🔠

Such metaphors, in short, have come into technical language at certain points in time with a burst of descriptive and aesthetic power. In many cases they have offered new seeds of suggestive unity (or unifying suggestiveness) around which older ideas could condense; Harvey's "circulation" is an excellent example. At other times, they have provided organizing images, even image systems, whose own internal logic later became the guiding basis for inquiry. Pasteur's concepts of disease as an "invading agent" and the body as a "defender" demonstrate this point profoundly. No doubt many other examples of specific metaphoric uses could be found. I stress here the powers of novelty, but metaphors can also be clichés—in science as elsewhere—oriented toward refocusing existing nomenclatural commonplaces (an example might be the more recent substitution of "uplift terrain" for older terms such as "mountain system," "orogen," etc.). They have also served to delete an earlier, descriptive terminology, and thus in a sense to impoverish technical language by making it more efficient (e.g. "fracture," with regard to rocks, as a replacement for a variety of more specific terms, such as "fissure," "crevice," "rupture," "crack," and "joint"). Whatever the case, however, metaphor gains depth of explanatory power by being absorbed into the literal at the cost of losing its original richness of association, by becoming a kind of "positive unconscious of knowledge" (Foucault, 1970).

The exact ways in which figurative language has served the cause of technical knowledge are many, and have been touched on here and there within the existing literature on the subject (see, e.g., Leatherdale, 1974, and the large bibliography he includes). It is not my intent to engage this literature here; it is far too vast and (in my opinion) too clouded with efforts to define the final place of metaphor in science. In my opinion, no such place exists. A detailed look at the language of any particular field will reveal that an entire geography of metaphoric usage has existed over time, that such usage has always been enormously plural, and that it will probably always remain so. The problem with a great deal of the writing on this topic is that, in trying to locate some essence of the "scientific," it treats words as if they were mere property without content. Language is spoken of as if it were made up of particles of meaning with "function" but no etymology, no history, no cultural implications. To claim that science is irreducibly metaphoric says little; the claim is neither daring nor original, and certainly not correct, at least as far as the workings of this knowledge in the real world. The point is to try and understand where words come from, linguistically and culturally, and what they bring with them after their figurative character has been defeated by standardization.

In science, as elsewhere, the power of metaphor lies in its ability to create images or even whole image systems. Such image systems are critical to any analysis of the sciences and how they have developed. This does not mean that they define a way to undermine the idea of truth itself, or of its correspondence with certain aspects of material reality. (What thoughtful scientist, in any case, would today reject the notion that knowledge is always approximate, mediated through language and images?) To use an analysis of language to deny the truth power of scientific knowledge, everywhere evident in the technologies of contemporary society, would be absurd and self-defeating. On the contrary, the idea is to reveal the dependencies of this power on larger realities. I will argue that the study of the linguistic "unconscious" of science can reveal very specific and concrete instances of how this truth has been built, as an interaction between human beings and the contexts of culture into which they have been born and through which they have lived. Considering the stories of how terms and images in science have arisen can demonstrate how distinctly nontechnical, and at times highly emotional or political, concepts have come to be codified as organizing principles within the very heart of scientific thought—to the extent of furthering certain possibilities for knowing the world at the expense of others.

Without doubt, the one-time metaphors in scientific discourse are more evident in some disciplines than in others. And among these, some are more socially significant than others—that is, if one considers the long-term impact they may have had on people's conceptions and attitudes about realities which immediately affect their lives, as well as on the development of related social, economic, and political structures. In both regards, one field seems particularly germane: that of biomedicine.

Therefore I would like to attempt four things in this chapter. First, to identify and discuss two image systems that have largely come to dominate our contemporary language about the conception of illness and disease,[2] both in the laboratory and in public. Second, to analyze the discourse of what is ordinarily thought to be the primary "alternative" to biomedical thought: holistic medicine. Third, to look briefly at the moment in history when the older and more pervasive image system of establishment medicine (the more reflexively employed, one might say) came into being. And fourth, I hope to point out, albeit in a short list, a few consequences that have come about as a result of the institutionalization and standardization of this image system at many levels of cultural (not just medical) practice. As such, my attempt here is diagnostic. It follows on the work of other writers and is, itself, at most a beginning of another kind. Two points, however, should be given strong emphasis. What follows is by no means intended as a comprehensive review of the discourse of medicine in our society—I am concerned only with those image systems that tend to enter *most routinely* into discussions of disease (and therefore of health). And again, in speaking of these systems, I do not want to imply that they still operate on a (purely) figurative level. They do not; indeed, the degree to which they have become part of our collective consciousness is the sign of their literalization via technical knowledge. The process by which this literalness developed and the extent to which it might be considered "false" or "overly limited" in any objective sense are questions that I cannot tackle here. Suffice it to say, however, that the common and rather blithe identification of "metaphors" in science is usually beside the point. Truth does not function metaphorically, nor does disease. To understanding how knowledge is made and used, it is necessary at some point to accept it as knowledge.

Disease in Popular Discourse

How, then, does our culture speak of illness and disease? What kind of discourse exists to give them a conception, a reality, a presence among us? Still more, what type of presence is this?

One might suppose that with the advent of modern biomedicine, people have become less fearful of or troubled about sickness than in ages past. As with most things, we have been taught to believe in a progressive movement from the prescientific to the scientific, in this case from a mordant era of leeches and humours to a more enlightened time saturated by a productive rationality built upon the holy trinity of basic research, clinical treatment, and pharmaceutical therapy. Today, the refrain goes, we know what disease is, or at least what it does; our knowledge has become firm. Though new and frightening examples have arisen (AIDS in particular) and have thrown many new challenges at medical understanding, the health care establishment and the public at large should be assured that modern medicine will rise to the challenge. Whatever levels of fear might be involved, our discourse of disease has become, over the centuries, gentler, more rational, more "in control," expressive of our growing knowledge. Or has it?

One discovers, quite quickly as it turns out, that the very opposite is the case. The ways in which we speak of illness and disease today—technical and nontechnical alike—are no less emotional, limited, and anxiety-laden than they have ever been. In some ways they seem even more so. To show this, let me begin with an extract from a recent article on AIDS, published in *National Geographic*:

> Organic invaders enlist the full range of immune responses. . . . Many of these enemies have evolved devious methods to escape detection. The viruses that cause influenza and the common cold, for example, constantly mutate, changing their fingerprints. The AIDS virus, most insidious of all, employs a range of strategies, including hiding out in healthy cells. What makes it fatal is its ability to invade and kill helper T-cells, thereby short-circuiting the entire immune response. (Jaret, 1986, p. 709)

What one recognizes immediately in this passage is the reigning image system for all disease in Western culture, that of war.

More generally, in fact, we speak of how disease "strikes" or "attacks." We say that illness "invades" and "spreads" within us, setting in motion "the body's defenses." We speak of "struggle," of "resistance," "the fight for one's life," how certain diseases can be "detected" through various "warning signs," "killed off" or "defeated" by the best of therapies, and finally, how the "battle against disease" therefore demands the best efforts and results from "the front lines" of research and technology, from scientists engaged in the hunt for "magic bullets," in the "crusade" for new knowledge, and so on.

At the core of this way of thinking and speaking lies a rigorous, an unrelenting use of military imagery. Indeed, in the passage quoted above the relevant terms literally flow out, one after another, in an unbroken stream of baroque aggression. Within this discourse there exists a dense, closed, and unremitting logic. It is a logic that suggests nothing so much as a formulaic rationality whose only method for trying to alleviate sickness is to "fight back" through one or another "strategy" or "countermeasure" (through intervention more than prevention). While it may no longer be the only shared language our culture has with regard to illness and disease—due in no small part to the rapid expansion and successful challenge of holistic therapies during the past 15 years—the military image system nonetheless remains the dominant one. Certainly it is everywhere around us: routinely employed by doctors, medical personnel, and patients, it is also spoken by those who produce language and images at every level of the culture, by politicians, social scientists, historians, science writers, filmmakers, novelists, poets, and textbook writers. It is the imagery employed by the mass media as well, in all its contemporary forms.

This image system of "military metaphor," as some social critics have dubbed it (see, e.g., Sontag, 1989), acts to evoke disease in particular ways. It gives to disease certain powers to excite fear and anger that it did not have in the past, when sickness was viewed as a divine test, or retribution. Today, rather, a different scheme of capitulation is invoked. When taken as a whole and stripped of its everyday, naturalized character, the language of militarism portrays its users as a terrorized and occupied people. In both tone and character, it suggests a manifesto of armed resistance against an enemy of nearly infinite power and evil intention. (Any talk of upcoming "medical miracles" and "wonder cures" merely acts to heighten this general sensibility, both by its feeble attempt at inversion and, more to the point, its overt and archaic call on the divine.) Indeed, the quality of obsessive obedience to images of battle—images that commonly portray a virus or other disease "agent" in the guise of a guerrilla (leftist?) force and the body as a sovereign and legitimate state—is one that has often linked disease in its connotative charge with the greatest evils and atrocities of the 20th century: fascism,

Stalinism, even the Holocaust (the case of AIDS is notable in this regard, having often been described in terms of "genocide" either with regard to the immune system or the homosexual community). One obvious result of this, on a grand scale, is the feeling that solutions and cures must also follow the military line: disease can be "defeated" only by massive mobilizations, involving more money, more personnel, more research, more effort of every kind. This quantitative argument arises directly, even logically, from the ideology of "battle." Win or lose, knowledge is strength; and so its factorylike production becomes essential.

Such is how our own era, in comparison to the past, has come to deal with disease in public. We have, in short, traded one type of retributive paradigm for another. Doctors and other writers on medicine sometimes try to deny this truth. "In the centuries before antibiotics, guided by the unerring prejudice of folk wisdom," writes Gerald Weissmann,

> the citizenry blamed one or another minority for the outbreak and propagation of contagious disease: the beggar, the Jew, the gypsy, the *other*. Happily, in our time, the rapid advance of medical science has in good measure made this particular sort of scapegoating superfluous. . . . We neither fret over dangers posed to our communal health by the burghers of Lyme, Connecticut, nor scold the victims of toxic shock for their habits of personal hygiene. Only the gays with AIDS remain in double jeopardy of disease and ostracism. (1985, p. 69)

Such a total denial of the actual state of affairs is stunning, even difficult to imagine—until one understands that it comes from a widespread belief among physicians that disease in the West, with a few exceptions of course, is largely a "conquered" affair, or an affair of "conquering," by modern medical methods. Definitionally, then, ignorance and prejudice have also "retreated." It is clearly thanks to this aura of self-heroic optimism that Weissman advises more money be spent on AIDS, in order that we be better able to "pin down" and "attack" its cause.

Disease as destiny has never really been shifted from upper to lower case. Instead of the simple logic related to guilt and punishment, we are now at home in the terrified rationality of militarism and war, those nonbiological realities that have caused the greatest suffering and death in almost every century (but especially our own). One might then ask: How new is this way of conceptualizing and speaking about disease? How endemic to the period since 1914, when "war" came to occupy a heightened place in Western consciousness? Yet one finds it in evidence even before this. A seminal document in which to look for its trace is the well-known *Flexner Report,* published in 1910 by the Carnegie Foundation. This was a critical survey of medical education in the United States, which, in concert with growing trends in professional medicine (see Warner, 1991), proclaimed a need to ground all training of doctors in the rigors of science and experimental research. The basic premise of the report, which set the standards for medical education ever since, was simple. "Medicine," the author wrote, "is part and parcel of modern science." This is so because

> the human body belongs to the animal world. It is put together of tissues and organs, in their structure, origin, and development not essentially unlike what the biologist is otherwise familiar with . . . . It is liable to attack by hostile physical and biological agencies; now struck with a weapon, again ravaged by parasites. The normal course of bodily activity is a matter of observation and experience, the best methods of combating interference must be learned in much the same way. (Flexner, 1910, p. 53)

Thus even at this early stage, where it is to be wedded to education, biomilitary imagery appears standardized to the point where its metaphorical origins are erased. Words such as "attack," "hostile," and "combat" are employed with exactly the same degree of narrative flatness and habituation as are "tissue," "organ," and "structure." Even the immediate conjunction of literal and (once) metaphorical uses of wartime imagery—"now struck with a weapon, again ravaged by parasites"—calls forth not the slightest jag of qualification or narrative awareness (e.g., "We might say it is liable to being attacked or ravaged . . ."). The equivalence, instead, is already part of common discourse. Flexner was a classicist, an educator, a writer on teaching and learning in the United States and Europe: his use of biomilitary imagery, indeed his effective recommendation of it as a basis for medical training, reveals that it had already reached a certain level of institutional use.

Scientific Images of Disease

The implication, however, is that Flexner derived this imagery from medicine itself. Indeed, this turns out to be true. As we will see, this imagery can be traced to the early experimental work on infectious diseases by Pasteur and Koch and had already become common currency in medical thought and speech by the 1880s and 90s. It was the military image system that helped provide, at some level, the central vision of the "wrong" that disease presented, and therefore the framework of "intervention," of therapeutic possibility, for making it "right."

What of biomedical discourse today? If popular speaking about disease has remained loyal to biomilitarism, has medicine itself? What image system or systems are to be found there to give guidance to conceptualization and research? As before, it is a simple matter to let discourse speak for itself. The following extract comes from a recent article on immune system responses, published in the prominent scientific journal *Nature*:

> The T-cell response of BALB/c mice to the bacteriophage cI *repressor* protein is directed predominantly towards the epitope contained within a single peptide encompassing residues 12-26. Similar phenomena of immuno*dominance* of a particular peptide have also been observed in other protein systems. The mechanisms that have been suggested to account for the focusing of the T-cell response are partial *deletion* in the T-cell *repertoire*, biased *antigen processing*, and *competition* for *binding* to the presenting molecule, the major histocompatibility complex *encoded* class II transplantation antigen. (Ria et al., 1990, p. 381)

The italics have been added to help point up the existence of not one, but two distinct image systems in this passage.[3] The first speaks itself through such terms as "repressor," "dominance," "anti-," "competition"—that is, through a distinctly combative imagery. Such terms are part of a broad lexicon of similar words and jargon phrases found in research articles on similar subjects. A few of these would include: "killer T-cell," "natural killer cell," "cell proliferation," "ion mobilization," "target cell," "response triggering," "humoral induction," "supressors," and "interferon." Such terms immediately recall the passage from *National Geographic* quoted above. They imply that we should expect to find, here and there, in even the most technical research writing, a parallel to what we might have assumed was a type of expression exaggerated for the sake of effect. And indeed, we would not be disappointed in this expectation, as the next passage demonstrates:

> The ability of microorganisms to establish chronic infections in immunocompetent individuals is an enigma of host–parasite interactions. Such microorganisms must somehow subvert or escape the immune defenses for long periods. In some instances, such as the acquired immunodeficiency syndrome (AIDS), the . . . virus directly attacks the cells of the immune system. Alternatively, organisms may hide inside cells and thus avoid host defenses. (Pier, Small, and Warren, 1990, p. 537)

The second image system, on the other hand, differs in both style and substance from biomilitarism. In the first passage just mentioned it can be found in such terms as "deletion," "repertoire," "processing," and "encoded." This system employs images not of conflict or attack but of inscription, deciphering, above all information processing, the language of molecular genetics. Related terms here, to be found in textbooks and articles, are no less striking in their figurative consistency than those of the military type and include: "sequencing," "switching," "decoding," "expression," "template," "responder," "messenger," "signal," "transmission," "transcription," "translation," and "DNA library." A corresponding excerpt here shows the prevalence of this second kind of imagery:

> RNA editing, a novel and unexpected type of information processing, was first demonstrated in the kinetoplasts of certain protozoans. It is a remarkable phenomenon: certain species of messenger RNA have nucleotide sequences that differ greatly from the sequences of the genes from which they are presumably transcribed. (Volloch, Schweitzer, and Rits, 1990, p. 482)

The general implications of this image system seem more complex, less threatening than those of biomilitarism. In the role of patient, the human body is no longer cast as a victim or a territory, the ground space for combat and siege, perfidious invasion and heroic resistance. Rather than a battlefield between "self" and "nonself," the human body is a complex set of operations for the production, transfer, ordering, and collecting of "data." It is therefore something close to a kind of text— but not merely. The body includes not only the elements of textuality, of "writing" and "reading," but, more crucially, the processes of their use. Much of the relevant terminology, indeed, shows distinct crossover or affinity with the jargon of a particular technology—the computer:

> Switch transcription starts upstream of the switch region, runs through the switch region, and terminates 3' of the corresponding C_H gene. The transcripts are processed in a way that an exon located 5' of the switch region . . . is spliced to the C_H exons, generating a "switch transcript," also referred to as "germline transcript" . . . .(Lorenz, Jung, and Radbruch, 1995, p. 1825)

Within the bounds of such language, one seems to leave behind the deathworld materialism of war and to enter a region of control, where flesh and bone accedes to a literal form of biotechnology.

Such innocence is little more than a mirage, however. One can just as easily point out other, far more disturbing possibilities. What, for example, are the implications with regard to surveillance and intervention? What of enhanced manipulability, the transformation of the human subject into a mere piece of technology (an old, Cartesian end point of mechanistic philosophy)? What of the opportunities for merging biomilitarism and bioinformationism, by which the ill body becomes something akin to a military information system, the understanding of disease a matter of wartime cryptography?

Let us leave this topic, however, for later discussion and instead return to another question. How does the image system of bioinformation show up in everyday language? What type of presence does it have in popular discourse? This, it turns out, is a large field. Whereas the primary idea of "disease" today remains fixed on military images in public parlance, genetic knowledge has come to be represented in a number of different ways, encompassing a host of analogies and metaphors. Greg Meyers (1990), in his brief inventory of such images, implies that many of these exceed the purely mechanistic and instead come back to two conceits in particular: the book and the map. Popular discussions speak generally of genetic "flaws" or "breakdowns" in relation to disease, but in explaining these, the language frequently relies on ideas related to "errors," "recognition," "miswriting," and "genetic mapping." While these terms do exist within the technical lexicon, their often-heightened selection by the press and among science writers has its effect.[4] It tends, that is, to place more emphasis on the softer side of bioinformationism, on textuality in its relatively pure state. It favors "reading" rather than "decoding," or "processing," "messages" rather than "signal," "data," and "markers." In popular discourse, genetic "information" is treated as if it were a kind of simple writing that is fixed and nonactive, like a volume on a shelf. This allows the gene itself to become the active, guilty entity:

> With much fanfare last week, two separate research teams announced the discovery of a gene that underlies a common type of colon and rectal cancer. . . . The hunt for a gene responsible for some cases of [colorectal cancer] intensified last spring when a team [from] Johns Hopkins University School of Medicine in Baltimore reported that the gene was located on a specific stretch of DNA along chromosome 2 . . . (Fackelmann, 1993, p. 388)

A geography of "location" thus allows, even encourages, a "hunt" for "responsible" agents. This does not sound like a softening discourse, but in fact it does end up adding elements of anthropomorphism to what is otherwise a language of relentless technology. Such does not make popular discourse in this area fully humanizing; it merely shifts and, indeed, exaggerates the burden of cause. But it also disguises the larger imagery involved by making it appear more understandable within the paradigm of human-type actions.

The relation between scientific language and popular discourse, therefore, is not a simple one. Certainly, public images draw directly, if selectively, on science but not without important deformations or recastings. This is particularly evident in the case of biomilitarism, which in everyday speech commonly appears in a magnified and more graphic form, as noted in the excerpt from *National Geographic*. And yet such magnification, as we have seen, helps bring to the surface the troubled figurative system that lurks exactly within this region of scientific truth. With respect to computer imagery, on the other hand, the effect is somewhat opposite. Here, popular usage tends to lighten and displace, not to intensify, the relevant image system. Rather than making plain the ideology involved, it directs us to a more ordinary set of figuratives, which imply that this science, through its language, is more concerned with what is "human" than it actually is.

In fact, in each case, it would appear that the deeper effort of popularization is to do exactly that: to translate (or simply transfer) the image systems of technical discourse into the realm of common experience, whether this be the activities of war or of reading. This kind of translation has often been cited as the ultimate goal of *all* popularization (under the aegis of the notion that familiarity breeds content, or at least comprehension). Explaining scientific ideas in "common language," especially in the media, has long involved the decanting of such ideas into communal images. Correspondingly, this process has also long been thought (by scientists especially) to inevitably corrupt or at least to sensationalize scientific knowledge. However true the first of these perceptions may be, the second is clearly wrong, as a maxim at least, as the case of disease makes plain. Above all, our popular discourse in this area remains focused on biomilitarism, the central conceit of medical science. As yet, bioinformationism, whether in native form or in metaphors of "reading" and "study," occupies a minor role in public speaking about disease and illness. Popularization, in other words, has neither seriously distorted nor sensationalized the dominant scientific discourse to which it is related. Instead, our public language, however disturbed and filled with fear, has emerged directly from medicine itself.

The Biomedical View: A Closer Look

To summarize: the growth of technical discourse in the medical area has been in part systematically guided by military images for some 100 years, to which, in the last four decades, computer imagery has been added. In effect, both constitute something akin to "research programmes," in the terminology of Imre Lakatos (1978), in that they have helped orient the accretion of new knowledge along fairly rigorous and mappable lines of sensibility. Not only does each image system contain a particular vision of disease, it provides a mind-set for creating new and related images that have helped contour the very shape of "objectivity" in this area. Biomilitarism, in particular, has given form to the possibilities currently pursued at a general level, for both speaking about and rendering visible the body in its diseased state of existence. Bioinformationism (a clumsy term) gives this a much more highly detailed, internal specificity. To assemble a comprehensive glossary of terms and ideas relating to each image system would no doubt provide a look at just how and where the body has been both "militarized" and "informationalized," in what aspects of its reality and its interaction with the surrounding world it has been made the tattooed messenger of these two socializing conceits.

It bears repeating that these image systems are anything but neutral or dispassionate. Rather, they emerge from the page like living forms—one wants to say *sociohistorical memories* or *ghosts*—climbing out from within what is normally looked upon as an instrumental language, transparent to hard "fact." It is not that such "fact" is absent or impossible, that our knowledge is metaphorical. To say so would be, as I have suggested, both naive and (in a sense) anti-intellectual. Medical knowledge is real, but it is mediated, probable, and partial, constrained in certain directions—and part of the reason for this is the language of medical conception.

What else, then, might we say about these two image systems? Their importance and power make it imperative to discuss them in more detail. The following points, for example, seem crucial.

Combat and Codes as History

War came into medicine and the human body thanks to the full institutionalization of the germ theory of disease, with its axial concepts of microbial "invasion" and bodily "resistance." Of course, war and disease had a long history of association prior to this, and were especially well known to medieval and Renaissance Europe. Coming in the midst of the Hundred Years' War, for example, the outbreak of the "Black Death" left its mark on Western consciousness: paintings at the time and for a century thereafter often portrayed disease in the form of an attacker, armed with arrows, spears, or the like (Montgomery, 1991, p. 343). This mixture of allegorical and literal meanings, however, did not penetrate medical theory. Nor had military imagery entered medical discourse as late as the 18th century. In the whole of Daniel Defoe's *Journal of the Plague Year* (1722), for example, there occur no such reference to the illness at hand, horrible as it was. The plague is said to "infect," to "lay upon," and to "seize" its victims, but not to "attack," "strike down" or "defeat" them.

The germ theory—which first codified such terms—was established by Pasteur during the 1860s and 70s, and was solely aimed at helping explain contagious disease. It was a theory, however, with tremendous, unprecedented power to both generate and employ empirical knowledge for many types of illness. During an era when physiology, in its experimental, laboratory guise, was generally viewed as the future of medicine, yet had not yielded any new therapies, the germ theory arrived to give medical practice a new efficacy. The vaccine (as substance and concept) opened the way for the full advent of drugs; it was the first truly successful, therapeutic result to be based on laboratory study. It helped give medicine its first real claim to the "scientific." Its language of "invasion," "damage," and "combat" justified intervention at many levels and was thereafter rapidly taken up, far beyond its original limits, by the medical community and applied to *all* kinds of disease.

Pasteur was by no means the first to conceive the theory that microbes might be the cause of one or another biological change or malady. Nor was he the only one to demonstrate this. As René Dubos points out in his classic biography (1976; see especially pp. 233–266), several other chemists in France and Germany had already done so, albeit unsystematically, as early as the mid-1830s. The germ theory, far from being Pasteur's invention or "discovery," was instead the product of a dispersed yet coalescing view of the powers borne by even the most minute forms of life. Likewise, the use of war imagery was not entirely unprecedented: as I will point out below, this imagery had actually been around, in scattered form, and available for centuries, but was never placed in systematic service due to the predominance of other image systems for disease (in Pasteur's day, these included a range of rival inheritances from the previous century, portraying disease as an effect of "functional alteration," "excess of irritation," "loss or overabundance of vital force," or "contamination," to name just a few).

The language of war in Pasteur's day had a broad base. It was deeply touched with the central importance which ideas of struggle and conflict held for the 19th century, from Darwin and Marx onward. It went hand in hand with what Peter Gay has recently diagnosed as "the cultivation of hatred" in late-19th-century society, what he identifies as "beliefs, principles, rhetorical platitudes that legitimated verbal or physical militancy on religious, political, or, best of all, scientific grounds" (1993, p. 6). Military metaphor also drew its sustenance from 19th-century organicism, with its view of the nation-state as a living body (the "body politic") subject to various forms of assault from foreign powers. It seems doubtful that this "political medicine" (see Schlanger, 1971) worked unmediated upon Pasteur's mind; yet, as I will discuss below, there are some indications that the literal reality of war—specifically, France's defeat by Prussia in 1871—did have an influence on the language he chose to employ.

Meanwhile, the discourse of "information theory" had to wait for the unraveling of the DNA structure in 1953. The terminology chosen by Watson and Crick is now well known: the DNA molecule provided a "template," a "mechanism for self-duplication," and most important of all, a "copying process" by which the reproduction of genetic information took place. Previous to this, one spoke of "gene action" and "genetic maps"; it was not "information" that the gene helped transmit but instead various "characteristics" or "specificities." Like Pasteur, however, Watson and Crick took neither a sudden nor a herculean leap on the plane of terminology. In the 1950s genetic science already spoke of a gene's "expressivity," its role in "duplication," and the importance of "linkage." The new theory, however, ended up selecting these linguistic traits over others, and giving them a new, centralizing focus. Before this, the gene itself was conceived as a relatively massive entity: "Genes act at various stages during the development of the organism to produce definite characters. . . . Some genes behave so differently in different environments that the characters they produce are strikingly different" (Riley, 1948, p. 30). Such words as "act" and "behave" reveal a world of conceptual and figurative distance from their later counterparts: "decode" and "transcribe." By the early 1970s, the new language of the "gene-as-computer" had become the discourse of genetics as a whole, and thereafter spread to all research oriented toward examining disease on the level of genetic materials.

This new language was not unique to genetics. Linguistically speaking, Watson and Crick were part of a dispersed professional community beginning to center itself on various ideas related to "coding," "communication," and "transmission." During the 1940s, for example, such workers as John von Neumann (brain/computer theory), Claude Shannon (information theory) and Norbert Wiener (cybernetics) placed the idea of "information" itself at the core of human mentality and action (see, e.g., the general discussion in Gardner, 1985, especially pp. 10–27). These efforts were themselves consciously predicated in part upon Alan Turing's work in cracking the Enigma Code during World War II, his foundational ideas in the areas of computer theory and technology, and his general notion that information is the most important product of any advanced civilization (Turing, 1936, 1950). In the humanities, meanwhile, the pseudo-scientific theory of structuralism, as put forth by such writers as Claude Lévi-Strauss (1949, 1955) and Roland Barthes (1953), made its own employment of such words as "code," "message" and "receiver." Finally, linguistics itself, most notably through the work of Noam Chomsky (1957), sought a new grounding in ideas of "structure," "parsing," and the like.

Bioinformationism, in other words, like biomilitarism, grew out of a much broader sensibility. Between roughly 1940 and 1960, this sensibility converged in fields focused on communication and on the transfer of language, including mathematical language. Mostly this work involved a utilitarian view of discourse, a view that conceived language as reducible to particulate units, structures, and probabilities. It was a view, in other words, that sought new technical conceptions of language, new "sciences." And one of its primary means of doing so was to reformulate the act of communication as a type of technology.

These ideas had significant connections elsewhere, chiefly in the political domain. Images of cold war intelligence gathering, of a spy versus spy battle over scientific and military "secrets," seem more than suggested by the phrase "break the genetic code," which became the rallying cry of biological research during the 1950s, 60s, and 70s. Watson himself, throughout *The Double Helix* (1968), does an excellent job of portraying the effort to discover the DNA structure as just this sort of "race" or "struggle" between competing laboratories. The larger implication, for communication generally, remains vague; but for the new field of molecular genetics, and the images of disease it would come to yield, the link to political and military spheres of linguistic influence seems striking and beyond doubt.

Strategic Agency and "Action Theory"

Each of the two image systems I have been discussing embodies a type of purloined rationality. This rationality, in effect, returns to a kind of "action theory" for the processes of disease. Biomilitarism is dependent on the concept of disease "agents" and their counterparts within the body. Bacteria, viruses, toxins, oncogenes, genetic "flaws"—all the major "causes" of disease—fall under this category. This idea of agency, external and internal, has received long-term support not only from the germ theory itself, but also from the correlative hypothesis of specific etiology—the "one disease/one cause" idea—to which most of the medical establishment still adheres.

Biomilitarism also sets up a polarity by which disease agents must be countered by their opposite, by entities or processes whose work is aimed at victory. Both types of agents, bad and good, have the power to set things in motion, to redirect existing situations. In the broadest terms, biomilitary agents, bad or good, can be any number of things: a virus, a type of antigen or antibody, a body cell, a complex chemical substance (protein, peptide, etc.), a deformed gene or gene-combination. The important thing is that they be granted the role of both a grammatical and epistemological subject, endowed with one or more tactical intentionalities. In a word, the agent *acts*—penetrates, mobilizes, binds, proliferates, expresses, encodes, and so on—and it does so to perform some kind of task, to achieve some kind of goal. It is thereby equipped with a kind of will that has the power to institute deliberate strategies: for example, preparation, counterattack, resistance, or communication. Each agent performs as a kind of pointillist subjectivity, an embattled "self," whose interaction with other such "selves," their products and operations, defines the reality of biological activity at this level of scale. What matters, often enough, or what determines a particular outcome, is the degree to which a specific agent (or "self") is able to gain control over a given circumstance—for example, whether a virus can be stopped or jammed by the processes set in motion by a specific drug. What is implied, finally, is a type of bioaction theory by which process itself is reduced to a consequence.[5]

This type of hidden conceptual frame carries with it an embedded logic for research and therapy. In making its agents "human," by investing them with characteristics of motive and "mind," biomilitarism invokes a counterstrategy. At the broadest level, that is, one finds that medical research and therapy come back to ideas associated with the concepts of "reconnaissance" and "defense." Much basic inquiry returns to a search for critical entities (actors), the processes they initiate (acts), and the reasons (intentionalities) why these are enacted. Environmental influences are handled in the same way, again through one or another agent or agency that visits a certain effect upon the body. Therapy, on the other hand, has always been framed by two basic options: counterattack and protective defense, with a range of optional strategies for each. It is interesting to recall, in this respect, that whereas Pasteur himself spoke openly of both approaches, of the need to "combat the invading agent" and to help the body "fortify itself," clinical philosophy as a whole over the past century began with a penchant for the former and has only more recently shifted toward the latter, again under the presumably revolutionary banner of "strengthening the body's natural defense mechanisms."

With computer imagery, meanwhile, what stands free and "intentional" are actors of a generally smaller scale: DNA, RNA, various other molecules, and chemical substances. These are the carriers, senders, and receivers of various types of information. They are the nodal members of a living circuit, in a subcellular world of data production and transfer. The rules here are less tactical than they are computational, algorithmic; less driven by conflict per se than by transmission, interpretation, and appropriation. The promise for a pursuable rationality is dependent not merely on the fixity of the agents themselves and their aims with regard to one another, but also (and perhaps more importantly) on the processes of contact between them. In the world of research, this has meant "tapping in" on the precise ways which the chemical transfer of information takes place. This process has involved not merely attempts at decoding and duplicating such transfer, but also rerouting and rewiring it in controlled ways. In therapy, the possibilities also come back to manipulating or in some way entering and altering, even replacing, the relevant path and content of "information flow." Most attempts being pursued at present—those in cancer and AIDS therapy, for example—seem centered on such ideas as "boosting," "interrupting," or "jamming" one or another set of signals or pathways.

War Codes/Code Wars

Disease today is largely conceptualized in terms of an intermingling of biomilitarism and bioinformationism. Such intermingling is variable and complex, but implies a unified biomedical vision. In the passage quoted from *Nature* above, the italicized terms show the degree of intimate mixture that often exists in biomedical discourse. This is especially true for subjects like the mechanisms of viral disease, the precise workings of the immune system, and the molecular analysis of oncogenesis. In these areas, indeed, even more striking examples can often be found, for instance:

At face value, it appears that *tolerance induction* by clonal anergy could serve as an important alternative to clonal *deletion*. The thymic epithelium could render T cells anergic to epithelial-specific peptides that may be *expressed* by *peripheral* tissues. . . . The suggestion has also arisen that an anergized but *nondeleted* T cell may serve also as a specific *suppressor*, because it can *bind* to antigen and perhaps consume lymphokines while being unable itself to produce lymphokines or to respond by *proliferation*. (Ramsdell and Fowlkes, 1990, p. 1347, italics added)

Yet in other areas of medical writing, it frequently happens that this type of mixture does not occur. Instead, military and computer terminologies tend to alternate. This can take place, moreover, at almost any level of narration. Within a single journal, for example, some articles may show a preponderance of military language, others that of computer discourse. At other times, a single article may reveal a similar alternation from section to section. Finally, even single paragraphs and sentences will sometimes go back and forth from one image system to another. The reason for this I have already touched on. Each image system represents a different historical level of "visibility," and is therefore, for the most part, aimed at a different scale or concept frame of phenomena. One finds biomilitarism applied most reflexively when the subject calls for discussion at the level of cellular or subcellular responses (the smallest causative frame known in the second half of the 19th century). It is also employed with regard to viruses and genes, and to some aspects of immunity such as the maturation of T cells. These are all, however, spoken of as more or less self-contained entities. Bioinformationism, meanwhile, reigns on the more reduced (20th century) scale of the molecular level. Its precincts are most often the detailed interior workings of the cell, the virus, or the gene, the processes of protein synthesis, of DNA and RNA function, and immediately related phenomena.

This divisional scheme is admittedly general and inexact. Many examples exist that defy it (e.g., those dealing with what is called "transcriptional activation"—the switching on of the process by which genetic information is passed from DNA to messenger RNA—in which such terms as "DNA target sequence" and "suppression phenotype" are fairly common). But it helps point up something important: namely, that the greatest and most consistent intermingling occurs where several levels of scale are being considered with regard to disease, which means where bioinformationism occurs *within* the context of biomilitarism.

Such would appear to have an obvious historical logic: Pasteur's work preceded Watson and Crick's by some 70 years. But perhaps it is not so obvious. As the passages cited above show, the latter "revolution," far from obviating or replacing the earlier one and its discourse, has instead helped expand and even animate this language still further, with distinct late-20th-century overtones ("suppressor" and "subvert" being striking cases in point). The two image systems do not at all cancel each other out, figuratively speaking. Indeed, they suggest numerous points of convergence. Bioinformationism, to no small degree, provides the discourse of combat with a new, inner layer—this, because its content, focused as it is on the role and power of "information," strongly overlaps our more general visions of struggle and conflict under late-20th-century capitalism, for which knowledge has become the key resource of power. These visions include not only such things as economic and academic competition, private ambition and international espionage, but war itself in many of its most concrete objects, its high technologies and communicational networks most of all. Bluntly put, "smart bombs" and "smart therapies" are not easily separable. To the degree that the computer has penetrated the human body in analogical form, so has it become a literal reality in general society and in military operations.

⑤⑤

What, then, is the more detailed combinatory vision of disease implied by infomilitarism? How can it be portrayed? A number of ideas seem possible. Disease might be seen as: (1) a form of war fought over the possession or control of command codes or information systems; (2) a form of attack upon, or disruption of, normal modes of information processing; (3) a form of conflict between different, competing codes (all of which need to be "broken"); or (4) a form of encounter involving bioespionage, the stealing of coded information, and the like.

None of these possibilities excludes the others. They may even be seen as different aspects of a single paradigm. Health or disease would be defined as one or another state of control over the body's informational systems. Indeed, this has already been proposed. Current research seems to be involved more and more in deciphering these systems to the nth degree. Therapy, too, has turned to such ideas as "reprogramming" invading viruses or defective systems within the body. Preventing disease, in turn, would eventually come to demand some type of automated surveillance, monitoring, the use of an "alarm system" for "early warning."

The overall implications inherent in this view I have touched on briefly. I will speak of them again below. They are not reassuring. In the long run they may even prove more fear-inducing, more legitimating of the patient's helplessness, more capable of granting power to external agencies than even biomilitarism. Once the body has itself been redefined as a technology, the potential opportunities for its manipulation and control become nearly infinite: the notion that genetic fingerprinting and screening at birth could be used to restratify society in new ways can only pale in comparison to the idea that at some point in the future, medical knowledge, in league with the most sophisticated computer advances, will make it possible for one or another informational authority, whether insurance company, credit agency, or the state itself, to keep constant, perhaps even invisible, watch over certain critical neurochemical and biochemical processes that occur within every individual under its purview. Such processes, conceivably, would themselves be categorized as necessary "information" to the making of important economic/political or medical decisions.

This may smack of post-Orwellian paranoia. Yet such exaggeration helps to point up certain connections. Epistemologically, that is, biomedical truth today is deeply involved in political, economic, and military realities.

Dis-ease

All of the foregoing brings one back, eventually, to the question of objectivity. The historical role of image systems in biomedical discourse implies that "disease," at base, remains a problematic idea in any final scientific sense. No biology to date has been able to define the idea of "disease" except in medical terms—meaning, in moral terms. Disease is always portrayed as some type of negative change, some disturbance in normal "health." There is always something "wrong," and therefore some recognition of therapeutic need. Whether applied to animal or vegetable, that is, "disease" always casts a malignant shadow. On an ideological level, one might say that the epistemological "threat" invested in it lies at least partly in this: that it confronts the privilege our culture has long accorded the unity and continuity of single existences.

Disease, in short, remains a value-laden idea. A nonmedical biology, on the other hand, would tell us that, as a septic or injurious thing, no such phenomenon exists. Such a biology would begin from the premise that any comparison between the value of one life form and another has no moral meaning. No qualitative difference would separate the most ordinary bacteria or virus and the rarest animal or most brilliant human being; nor would it make any sense whatsoever to differentiate between the spoiling of wine, the eutrophication of a lake, the infection of cattle by anthrax, or the demise of a city by plague or cholera. All that would exist instead would be the multiplicity of what we call "life," with its diverse, interactive, and changeable capacities for continuity.[6]

Today disease is given existential and moral content simultaneously by biomedical discourse. A century and a half ago, at the point when the germ theory was being framed, the accusation of "metaphor," of troubled figuration, could have been leveled with some justice. At that time, alternatives still existed; the competing terminologies that gave disease a more variable and oftentimes confused presence, proposed as they were by a range of different schools and individual researchers, were only then on the verge of being rendered vestigial or marginal by the new images of biomilitarism. Some of these alternatives, no doubt, would have proven no better, perhaps even worse, than what exists today. Concepts of disease as "contamination" or as a "loss of vital force," current in the 18th and early 19th centuries, seem of this type. Others, recently resurrected within the field of alternative medicine, do not promise much improvement either, as we shall see below.

Asking what might have been makes little historical sense, in the end. Perhaps another linguistic tradition could have somehow been chosen, perhaps not. The reality is that this did not happen. The reasons it didn't, moreover, are complex and have a great deal to do not only with the state of the medical community at the time, and with sociopolitical factors, but also with the great success the germ theory *did* have in helping explain disease and in saving lives. History, so to speak, has left us with the dominant image systems of biomilitarism and bioinformationism—at least as far as mainstream institutional medicine goes. And to the degree that the mass media, science writing in general, and other organs of contemporary science consciousness remain wholly loyal to this medicine and its chosen discourse of truth, this will remain the case for our public speech as well.

Images of Society/Images of Disease

Every view of disease is also a view of the body and of society. It is a view by which certain important norms, margins, and transgressions of a particular era are brought together, given the terror of "fact."

In the Middle Ages, this view was fairly simple. It centered on the expectation of punishment and trial, on retribution sent by God for private sin and public disobedience. During the 17th and 18th centuries, with the advent of secular rationalism, the relevant determinism shifted to concepts of an existing order and its underlying mechanisms. Disease then became a vision of discord and anarchy, these being the central aspects of deviance from Nature, from the proper economy of relations that then composed the healthy body and suggested it as a utilitarian model for social organization and process. In the 19th century, the conception of disease underwent another change. Disease was envisioned as the result of some type of encounter, whether with the external environment or some internal ("vital") force. Ideas of "struggle," "incitability," "irritation," and "chemical corruption" came to the fore; one of the most well-known statements of the time, that of Bichat in France, defined life itself as simply a set of functions that resisted death (Canguilhem, 1988, p. 44). In medical terms, conflict and resistance became conditions of normalcy, as did the image of the individual pitted against de-individualizing powers. This view was tied to the debut of individualism as an ethic, and to the influence of a host of disciplines—not the least, biology—that placed the ideas of "environment" and "competition," in some form, at the very center of their epistemologies.

For medicine, this view was neither strictly punitive nor purely mechanistic, at least at first. Rather, medicine took the organic as its theme. The body and society were both within its purview. This was because, as physicians such as Virchow clearly saw, it was often the interaction between these two (e.g., poor diet, sanitation, working and living conditions) that led to the outbreak of disease, and to some indication of its causes as well. Society was therefore conceived by many medical researchers in a way that agreed both with the larger romantic view of the state, and with Darwinian-type notions of "struggle," that is, society as a literal organic "body" engaged in efforts of conflict and survival. It was from a medical theorist, Broussais, that Auguste Comte, another of the century's most influential thinkers, adopted the society-as-organism analogy, elaborating it into a powerful explanatory model for much of human behavior.

Comte and his followers lavished organic metaphors on every aspect of social reality. History, culture, art, experience, thought—all these and more were interpreted as manifestations of "growth" or "decay." It became common to speak of the "health" of a city or of an army, the "flowering" of an industry or of an era. This made it inevitable that people would also begin to identify "diseases" of the state (e.g., revolution) and "illnesses" of philosophy or of moral behavior. And in every case, there were always implications for "treatment" and "cure." Thanks to the organic analogy, disease and society were each invested with the language of the other. Thus disease became a kind of "social ill," on the one hand, while on the other hand, poverty or unemployment became correlative "diseases on the social body." To try and ameliorate either type of "sickness" large institutional forces had to be brought into play.

A century later, during the past several decades in fact, it appears that another shift in general outlook has occurred. Broad parallels with what occurred in the 19th century are striking, and a little disturbing. Today one can still find sociology and medicine echoing each other in their visions of society and of the body. Discussions of "society" today no longer employ the organic analogy but instead speak of such things as "structures of social organization," of "systems theory," or of "value/interest distribution." There has been a change away from imagery focusing on growth and evolution toward that borrowed from information theory, communications, and related models. The body too has lost much of its earlier organicism. Well beyond the borders of medicine, biology has begun to do away with the idea of the organic altogether, substituting a series of transformations that approach the mathematical in character. Indeed, as one renowned researcher puts it, one no longer speaks of "life" very often, or of "the body"; instead, one speaks like this:

> Today biology is concerned with the algorithms of the living world. . . . The organization of living systems obeys a series of principles, as much physical as biological: natural selection, minimum energy, selfregulation, construction in "stages" through successive integration of sub-sets. . . . Regulatory circuits give living systems both their unity and the means of conforming to the laws of thermodynamics. (Jacob, 1973, pp. 300–301)

In other words, one speaks of a merger between physics, chemistry, and biology, and of their unified application. One speaks of phenomena that are neither wholly organic nor wholly inorganic, neither "living" nor "dead," but instead those which more closely resemble the technological, the assembled, the constructed. Again, as Jacob expresses it:

> There is not one single organization of the living, but a series of organizations fitted into one another like nests of boxes or Russian dolls. Within each, another is hidden. Beyond each structure accessible to investigation, another structure of a higher order is revealed, integrating the first and giving it its properties. The second can only

be reached by upsetting the first, by decomposing the organism and recomposing it according to other laws. Each level of organization thus brought to light leads to a new way of considering the formation of living beings. (Jacob, 1973, p. 16)

Here, in short, is what seems an engineer's utopia. The living subject—nay, life itself—yields up its organic paradigm to become simply one more type of "project" identical in substance with the apparatus employed to study it. The flesh, examined more and more by means of computers, being then touched, coated, engulfed, and finally constituted by the language of this instrument, is therefore brought to a state of subservience to the ideology of the tool, the supremacy of technology as the most fundamental organizing principle of biological knowledge.[7]

This concept, of course, has already come to the surface in the area of artificial intelligence. Researchers in this field, in order to elevate the importance of their favorite instrument, have long claimed that the human brain is nothing but a computer or series of computers with particular abilities and handicaps, in one felicitous rendering, a "meat machine" (Marvin Minsky, see Bernstein, 1982). We also have this view confirmed in the idea of the computer "virus." This is a usage, in other words, which may offer us a peek into the future, or at least one possible future. While today still treated as a metaphor, the computer "virus" represents a kind of literal exchange with respect to contemporary biomedical language and its view of the human subject. As the body and the computer approach each other in terms of their linguistic substance, so do they begin to share the possibility of each other's specific attributes. Put differently, to the extent that the human subject is divested of its organic life by being conceptualized in terms of the computer model, so does this same model gain life for itself through a vampirism of literal association. Thus it is that the computer has a "body" and a "mind," which can fall "ill," be "infected" by viral agents, and be administered a "vaccine" by a specialist or a team of "data doctors" (all these terms are currently in use). So, too, as the ideologues of artificial intelligence (AI) would have it, may the computer one day gain final privilege as a type of "subject" all its own, a social and political entity with certain citizenship "rights" and "guarantees," with class divisions, special interests, and the like.

Thus again one comes upon the social dimensions of this vision directly. Dimensions, again, which involve a variety of potentially frightening ways for defining, observing, ranking, employing, and eliminating human beings. And among these, perhaps the most disturbing is exactly what I have been describing on the most general level, namely, the transformation of the individual into a true object, a repository, or else collision site, for various types of detectable and usable information.

It is this reality, when combined with that of biomilitarism, that holds out the greatest worry for the future. For if the "death of Man" (Foucault) was begun some time ago—by such theories as behaviorism or cybernetics, and by whole fields like economics (with its reduction of people to an input–output model)—this cannot have gone further than it has in biomedicine itself, a region of knowledge which now studies the body, as it were, from within, as if this body were itself a vast and unfathomable galaxy of techno-substances.

The Question of Holistic Discourse: How Alternative Is "Alternative"?

We come now to a seemingly new topic. Many social critics have noted and taken to task the existence of biomilitary imagery in society. More than a few see the presence of such imagery in our everyday speech as tantamount to an infection by "military metaphor." This way of speaking, it is said, has a hegemonic yet wholly avoidable morbidity. It has poisoned our discourse and our psychology, and should be abandoned. Susan Sontag remarks: "Not all metaphors applied to illnesses and their treatment are equally unsavory and distorting. The one I am most eager to see retired—more than ever since the emergence of AIDS—is the military metaphor. . . . About that metaphor . . . I would say . . . give it back to the war-makers" (1990, pp. 94–95).

Such an appeal, with its call for a new imagery, is, for all the reasons given above, problematic. Beyond simple critique, it has little to offer. This silence is itself an indicator of the reality that biomilitarism, as the language of science, is not really a metaphorical system any longer—if by metaphor one means applying to one thing a name that belongs to something else. It cannot be "given back," whether to the war makers or anyone else. Within the language of formal knowledge, as it has come to exist, disease is not like—or incorrectly likened to—such things as mobilization, invasion, and so forth; it *is* these things, in the same way that the body is composed of "organs," "tissues," and "cells," the universe of "atoms," "forces," and "energy," the Earth of "plates," "mantle," and "core." To escape biomilitarism altogether, it is necessary to travel somewhere beyond scientific medicine, or along its outer margins. A different kind of knowledge, an alternative truth, is required.

Such a truth, it is often said, already exists. What is called "alternative medicine," or more simply, holism, has in recent years often been claimed as a true competitor to establishment medicine, particularly in Britain and the United States. And indeed the domain of natural healing today has achieved impressive successes in these countries. After more than a century of marginalization, it has re-achieved what can only be described as a penetrative stability within daily society. Holistic medicine—which can best be described as a broad and loose grouping of therapies—has grown enormously during the past 2 decades, in terms not only of the number of its adherents and its common acceptance but also in terms of the plurality of its specific approaches. As Rosalind Coward writes in her book *The Whole Truth* (1989), "Who could have foreseen, fifteen years ago, the current popularity of 'alternative' therapies with all that they imply about attitudes towards health, the body and the emotions?" (p. 1). Even a simple list of current approaches to holistic healing is overwhelming: naturopathy, homeopathy, reflexology, rolfing, aromatherapy, kinesiology, iridology, macrobiotics, orthobionomy, sound therapy, color therapy, crystal therapy, meditation therapy, rebirthing therapy, herbalism, acupuncture, and so forth. Moreover, according to Coward's figures (for England), at least one out of three people have turned to such therapies at one time or another during the past decade.

The reasons for this upsurge of interest are varied. The search for an "alternative" world of help and healing in the face of illness and frailty has much to do with a search for personal meaning, stemming from dissatisfactions in the existing realms of medicine, private behavior, and individual responsibility. Such reasons, therefore, make the question of imagery all the more pressing. At first blush, the variety of therapies might seem to argue for not one but many discourses. This, even at the beginning, would appear a profound difference, a true alternative, with regard to scientific medicine. With so many alternatives to choose from, we might expect a number of new image systems with the power to help us reconstitute disease. Yet, as Coward indicates, the apparent diversity collapses when one examines in detail the language and terminology employed by holisitic practitioners. There is, as it happens, a great unity hovering here, a kind of mutual referentiality. A few fundamental ideas and metaphors tend to rule absolutely, and to allow for the type of fluid sharing of techniques that one sees. Treatments very often overlap from one field to another and are, in any case, commonly prescribed either in groups—for example, changes in diet, the use of herbal medicines, special exercise or movement therapy—or else in some type of sequential order.

A Closed System

One finds, in fact, that the discourse of holistic medicine tends to form a closed system. After reading through dozens of books, articles, and treatises in this area, one discovers that the language employed, particularly with regard to discussions of disease, erects itself upon a distinctly limited range of central terms. The following excerpts, for example, taken from a variety of sources, help bring this reality into view:

> The basis for the holistic approach to health is rooted in the understanding that we are more than our physical bodies—that we are composed of mind, body, and spirit. . . . Only when an imbalance or blockage of energy exists on one or more of these levels does disease manifest. . . . It is primarily a "disruption of consciousness," a misunderstanding of our essential nature as radiant beings of love who are here to serve one another, which is responsible for the disharmony which contributes to dis-ease. (Serinus, 1986, p. xx)

> There is no "cure" for any disease. In all cases the *body heals itself* . . . Deficiency and toxemia are the only two causes of all disease. The key to regaining health is to put enough good things in and to keep enough bad things out . . .
> Mental states' direct effect on the immune system has been well-demonstrated by experiments performed in investigating stress. . . . The effect of stress on the body is not the stress itself, but how the person perceives the stress. If the stressful problem is regarded to be insurmountable . . . then the effect will be bad. . . . Those who feel helpless and hopeless succumb. (Badgley, 1987, pp. iii, 25)

> The thymus gland is the organ of courage. It is the basis for the dynamics of vital self-affirmation as the center of our immune system. The thymus gland *is* our vital center. Therefore, it is the appropriate place to tap when we confess our sin and guilt. . . . (Lee, 1986, p. 30)

> The energy which the life force transmits is distributed through the body along a system of invisible channels or meridians. In normal circumstances . . . the life force, *chi*, flows evenly, maintaining a balance between the vigorous yang and the restraining yin elements. But if either yang or yin becomes too dominant, the body's harmony is jeopardized and illness can result. (quoted in Coward, 1989, p. 53)

All these passages employ a terminology of dualism. On the one side, there are the positives: balance, harmony, cleansing, energy, spirit, vital force, love. These qualities are what define the normal or healthful state of things. Where they exist, equilibrium exists; therefore health exists. On the other side, meanwhile, each of these positive forces has its corresponding negative, which is then posed as a reason for (not cause of) disease: imbalance, disharmony, blockage, self-pollution, materialism, stress, guilt.

This is a dualism, then, that is by no means free of conflict. On the contrary, it most often centers on a cardinal struggle between "natural" and "unnatural," on an opposition involving cosmic forces. Such dualism is ramified and specified in other directions. Much of holistic discourse in fact attempts to claim back (under slightly different terms) a quasi-religious view of human existence based on the opposition between the material and the immaterial, for example: "We are more than our physical bodies. There is a part of each and every one of us that is eternal: our Spirit, our God-Essence—our Soul." This quasi-religious language tends to blend well with more "psychological" (pseudo-Freudian) notions that consider disease as a symptom of deeper problems, reflecting a poor mental state ("attitude"), either a split within the individual's mind or "spirit" or else a drowning out of one's "true inner nature" by the "falseness" of an "artificial" and therefore corrupt "lifestyle." In the discourse of holism, one comes upon many of the standard dualities that have been central to Western culture since the Industrial Revolution: the fundamental antinomies, for example, between society and nature, the urban and the rural, the impure and the pure, the sophisticated and the simple, the rational and the intuitive, and perhaps most important, between a corrupted present and an idealized, communal past.

Such a mixture of cosmic, historical, ethical, and person-centered ideas helps explain the broad range of therapeutic approaches in holistic medicine. But, it also indicates that one is often dealing here with a system that does not really attempt to comprehend disease in any concrete, material sense—that is, as an object of evolving study and therefore of knowledge. The attempt, instead, is to *mediate* illness through terms that situate the individual at the intersection of forces both metaphysical and moral.

This might be put differently. First, on the literal, external plane, holistic discourse does not aspire to the realm of truth, to technical definitions, descriptions, and interpretations. It does not try to do what scientific discourse does: construct a growing edifice of explanatory knowledge guided by critical correction. It prefers instead to speak a static, often lyrical, in the end adjectival language that is aimed not at explaining illness but rather in accounting for it as a kind of essence. It is less concerned with such things as "evidence" or "proof" than with the proposing of unseen and unseeable agencies, and with evaluations that sermonize about what is "good" and what is "bad" for the individual: food, for example, being the most primary ethical substance, is strictly divided for the most part between beneficial medicines and toxic poisons.

The discourse of holism, then, might be described as one of symbolic *mediation*. It is more interested in language as a means of poetic (therapeutic) placement than as a medium for inquiry. Because of this, I must conclude that it has little power to replace the reigning image systems within biomedicine, and therefore those that our culture more generally applies toward illness. To make this clear beyond doubt, two final examples might be given. Holistic authors sometimes do speak directly about the mechanisms of disease, taking into account such things as viral/bacterial infection or the workings of the immune system. What happens when this occurs? Does one encounter in this domain some hint of a new terminology, a true alternative on the plains of truth? Not at all. On the contrary, what takes place instead is a direct and unquestioned adoption of biomilitarism itself:

> Both the immune system and the nervous system, particularly in the high-stressed individual, focus their entire attention on the invading virus. . . . The long period of resistance, during which the immune system is sorely overtaxed by its fight and by the stress and diet of the individual, eventually brings about the collapse of the immune system itself. (Holistic Group, 1986, pp. 96–97)

All the central vocabulary of biomilitarism appears in such writings. No alternative image system occurs. The uptake of such vocabulary, in fact, can sometimes be every bit as extreme as in the example from *National Geographic* given earlier, as the following passage, taken from a meditation exercise meant to help ARC/AIDS patients, strikingly reveals:

> Now focus on your bones. . . . Penetrate your awareness inside of your bones, imagine the marrow inside of your bones. The marrow is a factory keeping your blood nourished, fluid, balanced. . . . As though the marrow were like underground springs feeding the rivers of blood that move through the landscape of your body. . . . Your

blood is full of richness, of nutrients. Your blood is full of all kinds of cells. There are T-cells and B-cells that flow through your blood, traveling throughout the landscape of your body. They are your body's defensive guards. Imagine them. Imagine each and every T-cell and B-cell strong, armies of them flowing through your blood, on guard against any foreign agents, any foreign viruses, any foreign invaders. The T-cells are the generals . . . they instruct your B-cells to produce antibodies, and these antibodies flow into your blood and kill anything, any life that should not be there. (Adair and Johnson, 1986, pp. 184–185)

The unmitigated use of war imagery here appears all the more striking when one considers that it is being used as a direct part of the therapy itself. Indeed, the logic of hyperbole, which moves from pastoral landscape to sudden and massive war preparations, to armed vigilance and paranoiac surveillance over anything "foreign," could not be more obedient to biomilitarism.[8]

Holism Absorbed

Aside from borrowing the imagery of biomilitarism, holistic discourse at times makes use of technical nomenclature, often to a considerable degree. This can take either or both of two particular forms. The first, and by far the more prevalent, involves symbolic adoption, or the borrowing of scientific words and phrases for metaphoric purpose. Terms such as "equilibrium," "force," and "activated response" are commonly employed to help carry into holistic speech a certain seductive weight or suggestiveness of precision, specificity, even physical truth. One sees this more clearly in phrases such as "accumulated negative mental mass" or "required melting of mental blockages," in such notions as "dissolving stress" and "releasing stored personal energy." Most of these borrowings come from physics and chemistry, for these are the fields most associated with fundamental truths and physical reality. They are also the fields that deal very often with one or another form of "energy." The rationale for adopting such language is therefore understandable. Nonetheless, such borrowing obviously represents a kind of capitulation to the social status of the "scientific," a status that natural healing seems otherwise intent on opposing. It also, one might note, if taken literally, reduces the mind or body to processes involving inanimate substances.

The second form of adoption is more complex, and perhaps more significant. Sophisticated authors on holistic medicine frequently take up scientific terms and styles of language, mixing both into a discourse that nonetheless remains allegorical. The result is sometimes very interesting and remarkable, not the least for what it reveals about the relative failure of such mixing. The following is one example:

> Neuropeptides function as either neurotransmitters or hormones in the human body. In the case of Blue-Green Manna, its "essential neuropeptides" seem capable of crossing the blood–brain barrier and directly stimulating the neurotransmitters in the brain. This phenomenon appears responsible for its observed ability to aid the creative visualization process, balance moods, counteract chronic fatigue, lift many individuals out of depression, and create a greater sense of centeredness and well-being. Its neurotransmitter effect, combined with what seems to be its effect on the upper energy centers of the body, may contribute to what some have termed its potential as a tool for psychic development and spiritual unfolding. (Holistic Group, 1986, pp. 132–133)

Read closely, this passage can be seen to move through several narrative membranes. The style begins in what is clearly a technical (instrumental) mode, yet by the end of the third sentence a connotative shift or puncture has occurred. Purely scientific terms like "neuro-," "hormone," and "brain" have given way, momentarily, to words such as "creative," "balance," "centeredness," and "well-being." The paragraph starts out discussing "neurotransmitters" and body "function," yet closes with "psychic development" and "spiritual unfolding." Nor can it be hidden from us that "energy" is not used here in any scientific sense, but instead a metaphysical one. Indeed, by the end of the passage the allegorical dimension has gained the upper hand entirely, not so much by defeating the "scientific" as by making it appear suspect. And truly, the more one reads a passage of this type, the stranger and more fascinating it becomes, resembling as it does a beaker in which two immiscible liquids, shaken into momentary clarity, begin to separate from each other through a series of momentary forms.

What happens when a merge is attempted in the opposite direction? What might the result be when holistic nomenclature and ideas are included as an ingredient in a scientific writing? The following provides a clue:

> Five modalities were applied to obtain these therapeutic goals—botanical medicine, homeopathy, hydrotherapy, nutrition, psychological counseling. Massage and therapeutic touch were optional modalities. . . . Our goal here was to compare the efficacy of botanical medicine versus homeopathy in treating HIV disease. . . . Based on

this research, we conclude that using the treatment protocol tested in this one year trial, naturopathic physicians can significantly improve mortality and morbidity rates, quality of life and neuropsychological functioning of HIV-infected patients. (Guiltinan and Standish, 1991, p. D)

In this case the merge seems successful; no shifts or instabilities occur. This is because, at base, there has been no merger at all but instead an accommodation. All mediatory, descriptive, and symbolic language from holism has been deleted; all metaphors have been erased. In the places where such language might appear there are now only nominative titles, mostly of individual therapies, which have been, as it were, neatly dropped into place as expected narrative ingredients (note, for instance, the neat translation of what is normally "herbal" into the more technical "botanical"). Both on the level of discourse and, as the passage indicates, that of actual practice, holism has been entirely absorbed within the purview of the "scientific," into such categories as research and morbidity rates, into the institutional strategy of "clinical trials," indeed into biomedicine itself. Its alternative status has not been erased, but redefined, as a new branch of establishment medicine.

Personalities and Privileges

The implication therefore appears unavoidable and convincing: whatever other claims might be made by and for it, whatever its efficacy in the area of treatment, natural healing has no alternative imagery to offer us regarding the *phenomenon* of disease, what it is and what it does. It may want to account for illness in different ways from biomedicine—indeed, by tying all disease back to imbalances in the self, to such things as artificially produced "toxins" (which symbolically carry the entire load of evil associated with contemporary life), to problems of individual life-style and attitude, it can be said to effectively denaturalize disease by making it into something that arises not out of biology but rather out of modern human activity and, above all, human choice. Ideas of contamination remain central. The Pasteurian battle between "foreign" and "familiar," between "self" and "nonself," is here merely displaced, never obviated. Again, disease is equivalent to a struggle for, and loss of, "the natural"; it is a mediatory symptom, a symbolic index of wrongs brought on by this loss. And these wrongs, in their way, return to a particular ideology of the individual.

Much of holistic discourse is given over to arguing that personality, often in the form of attitude, has a direct and even a determinative influence on health. This translates, inversely, into the idea that disease issues from one or another defect of character. This idea, like that of self-healing, is an old one, with a long and problematic (occasionally stigmatizing) history within and outside of medicine itself. After a century of neglect, it has recently reentered biomedicine, on the back of holism's popularity, in different form, known as psychoneuroimmunology (or PNI). Today, many researchers are trying to define possible links between mental/emotional and physiological states, and in particular connections between depression and disease.

On this score, too, however, one should note that despite its frequent critique of allopathic medicine for failing to carry out such research in the past, holism itself has no real models of its own, epistemological or otherwise, to help direct such work. It tends to end at the level of rhetorical formula. The body is said to have become polluted by "improper attitudes," by "negative forces" in one's past life, and the like. The vision of disease is a vision of an individual who has gone astray and needs to return to the right path. There is much talk of personality or spiritual "types." Many illnesses—from cancer to arthritis—are described as the result of infantile or childhood traumas repressed over time into emotional "blockages" that require "clearing," "dissolving," "melting," or whatever. Other diseases, meanwhile, are said to reflect diagnostic character traits, such as "excessive inwardness" or "an overadherence to the past." Most common of all, disease is connected with personal choice on a broad level:

> To the extent that you learn to accept more responsibility for your own health, your health should improve. *Negative lifestyle habits* will be replaced by positive ones, building resistance to illness. To the extent that these healthy practices arrest or reverse any present problems . . . you will be less likely to become enmeshed in the gears of the medical system. (Bricklin, 1983, p. xvii)

The possibility of "accepting responsibility" therefore means that one is already answerable for whatever "negative" choices have been made. The deeper reasons for disease, therefore, can be reduced to two possible schemes: (1) the result of toxicity brought on by "negative lifestyle habits"; or (2) personality flaws, whose reform, though promised, is rarely clear. Either way, everything falls upon the individual, at times with particular severity and, worst of all, fatalism:

> The nervous system and immune system circuits related to hopelessness switch on and stay on all the time. These people are recognized

by their constant depression, fear, anxiety, and worry. They are devoid of spontaneous humor and have no well-formed life goals . . .
Their minds haven't taken charge of their immune systems. It is the
copers and "take charge" people who beat the problem, be it ARC/
AIDS or cancer. (Badgley, 1987, p. 26)

Again, the idea that disease is an outward expression of interior "problems" is an ancient one (the god Aesclepius, after all, was thought to
heal through dreams), and up until the 18th century was part of medicine itself. What one finds here is not merely this idea's rejuvenation,
however, but a revival through specific attachment to a decidedly Occidental ideology of conservative individualism.

In the passage just quoted, for example, and through much of holistic discourse, one finds a kind of disguised privilege granted the extro-
verted or entrepreneurial individual, she or he who is blessed with the
ability to "take charge" of a situation, with the eagerness to talk out or
"clear" the mind through avid contact with others. On the one hand,
holistic discourse seems to reject mass society (Be the real you!) and its
passive anonymities. But on the other hand, "accepting responsibility"
within this discourse seems to imply that one must become a kind of
middle-class go-getter manager or administrator over one's life, and that
this transformation will only come about through an ascetic, decidedly
Protestant effort of personal cleansing. The healthy person who has taken
on such "responsibility," who has applied himself or herself by an act of
will to the work of change and self-improvement, is therefore elevated
to a kind of model. He or she is not merely healthier, but full of "well-
ness," and thus a morally higher form of being. The sick person, on the
other hand, is essentially being punished for not being the right kind of
individual. The quality of evangelism here is palpable.

Never in this discourse are allotments made for such things as class,
race, or age. No social context for the personal is ever admitted. There is
nothing that approaches a true social criticism, no examination of institutional structures, whether dealing with food production and distribution or with the development of preferences, tastes, and so forth. There
are only timeless universals to consider. All essential relationships are
seemingly fixed and unalterable. Health in this case combines the glow
of religious conversion with an aura of Emersonian self-reliance. It does
not just happen; it is chosen. This is one reason why holistic medicine
has such great appeal in Western democratic societies: it provides a kind
of disguised theological grounding for revitalizing faith in the power of
the individual to control the forces enveloping his or her existence (during an age, one might add, in which such efficacy seems sorely lacking

for the great majority of people). At one and the same time, the healthy self seems to claim rescue from a poisoned civilization while upholding one of the most central and long-standing doctrines that helped build that same civilization.

There is a negative side to all this. For those who regain health, the moral cosmos has nothing but praise. By embarking toward the realm of the healed, one is elevated; a thousand supports and encouragements, at many levels of discourse, exist. But if one fails, if one becomes or remains "dis-eased," these supports quickly disappear, or worse, are turned into attacks. Inasmuch as one is guilty of "negative lifestyle habits," one is not merely less of an individual; one has effectively placed oneself in league with the forces of imbalance, blockage, disharmony, with the entire cosmology of "dis-": the heroification of health requires an even more inflated condemnation of un-health. Indeed, the healthy person as paragon requires the sick person as a representative of failure. Disease, like health, does not simply happen; it too is chosen. And choosing disease means choosing everything that is "bad" and "negative" in modern society; there is no in-between. Illness, culpability, and punishment therefore do not merely gain a direct connection within this discourse. They become identical; they are simply different terms for the same phenomenon. Behind every sickness there lies an accusation, and a sentencing.

Pasteur and the Origins of Biomilitarism

Given that biomilitarism in particular seems likely to remain for some time the fundamental paradigm regarding our public speech on illness, how and where did it begin?

In his much praised biography of Louis Pasteur, René Dubos makes the following comment:

> The germ theory was formulated at a time when many biologists and social philosophers believed that one of the fundamental laws of life was competition, a belief symbolized by phrases such as "nature red in tooth and claw" and "survival of the fittest." The ability of an organism to destroy or at least to master its enemies or competitors was then deemed an essential condition of biological success. In the light of this theory, microbes were to be destroyed, unless they could be used for some human purpose.... Aggressive warfare against microbes was particularly the battle cry of medical microbiology, and is still reflected in the language of this science. (1976, p. xxxvi)

In light of what I said above, this is a compelling and, at least initially, satisfying pocket statement. It is, however, not only difficult to verify precisely, but even a bit misleading. Terms such as "enemy," "battle," or "invasion" rarely appeared in standard medical nomenclature prior to the successes of Pasteur and Robert Koch. The intellectual events leading up to the germ theory of disease, and thereafter the hegemony of such military imagery within the language of medicine, are complex and difficult to delineate. What they tell is not a story of immediate descent by Malthusian and Darwinian influences, but instead one of developments within medical theory and practice, within Pasteur's own life, and within the specific history of the time. Space is lacking to discuss this story in detail; however, I can make a few general points here.

It might be noted again, first of all, that military imagery had been loosely and sporadically associated with illness for centuries. An early example regarding syphilis, for example, was made by the 16th-century German physician, Joseph Grumpeck:

> In the last years, I have seen new scourges, horrible diseases, and a great many infirmities attacking the human race from every corner of the world. Among them there slipped over from the western shores of Gaul a disease more cruel, grievous, and foul than anything the world had ever seen before. (Quoted in Herzlick and Pierret, 1987, p. 15)

Of his own contraction of the illness, moreover, this writer says: "The first poisoned arrow of the hideous ill hit me in the Priapic gland, which as a result of that wound became so swollen that two hands would barely have fit around it" (p. 15).

The frequent association between war and disease was noted above. In addition to bubonic plague, eruptions of epidemic syphilis, cholera, and typhoid fever frequently occurred in the wake of major battles and sieges. War laid to waste whole cities and towns, disrupting sanitation and destroying farming communities, therefore causing a prolonged weakening of the populace and enhancing its susceptibility both to existing contagions and to those introduced by advancing or retreating armies, by messengers, or by any other "carriers" who had recently come from foreign regions. Given the frequency and the devastation brought about by such diseases, there is little mystery why medieval versions of the Four Horsemen so often replaced "Conquest" with "Pestilence", (the other three—War, Famine, and Death—remaining constant).[9] Certain diseases were themselves used as biological weapons somewhat later, during the colonial era, when the Spanish in the 16th and 17th centuries traded smallpox-infected blankets to the Incas, Aztecs, and other Indian tribes in the New World.

Yet in medical language itself, and in institutional concepts of illness, the imagery of war played almost no role before the end of the 18th century, when the first "modern" coherent medical theories were postulated. These were mainly philosophical systems, aimed at proposing one or another law or principle to explain the accumulated categories that had been built from the empirical information of visible symptoms. In one of the more famous of these early theories, the military paradigm actually makes an appearance—if only briefly, and as an incidental aspect. This was the theory of the British physician John Brown, whose *Elements of Medicine* (1780) was described 50 years later by Cuvier as having

> reduced the medical art to a small number of formulas: that life is a kind of combat between the living organism and external agents; that vital force is dispensed in fixed quantities . . . [and that] life can be snuffed out by an overabundance of vital force as easily as by its exhaustion . . . ; and that diseases and medications may be classified into two groups, depending on whether they stimulate or impeded vital action. (quoted in Canguilhem, 1988, p. 42)

Brown himself did not use such terms as "combat" or "war." He did, however, speak of life as "a forced state" that was allowed to preserve itself through the aid of "foreign powers." Yet his more central ideas were focused on the nervous system as the source of all vitality, and therefore of all disease. Illness was for him less a matter of actual combat than of over- or understimulation. The language of biomilitarism, therefore, had to wait for other developments.

Importantly, however, Brown and his relative contemporaries— Bichat, Magendie, and Broussais in France; Haller in Germany; and Boerhaave in Holland—were among the first to provide theories that conceived the body as a whole. Their precinct was physiology, the functioning of the organs, both in a normal and a diseased state. Brown and his contemporaries were often very different from each other: some were materialistic, viewing the body as a machine, composed of various mechanisms; others, such as Haller and Bichat, were adherents of vitalism, who saw the nervous system as the source of "vital force." Previously conceptualized as a collection of symptoms, a matter of nosology, disease now became explainable in larger terms. Nevertheless, "The general mode of medical theorising remained within the classical scheme of *balance*; balance [of force, of stimulation, of irritability, etc.] within the organism and between the organism and its environment. . . ." These theories were what we would call today, "holistic, environmentalist and tending to a linear scale of excess or deficit" (Pickstone, 1990, p. 729). At the same time, this type of "holism" allowed for local explanations. Excess or deficit, that is, could result from abnormalities in particular parts of the body; internal causes of disease could be correlative to the external lesions long treated by surgeons. Given this reasoning, and the larger demand for reestablishing "balance" through some type of decrease or increase in "stimulation," the physician was exhorted to intervene at all costs, and never to rely solely on "the powers of nature."

The late 18th and early 19th centuries marked a transition, therefore. Medicine to that point had been a discourse where a meeting took place between philosophical contemplation and scientific observation. Thereafter, several crucial developments brought an end to this way of thinking by the introduction of experimental work into the heart of medical practice. These developments, in fact, could be defined as ideological preconditions to the germ theory. This means that by the time Pasteur began his studies on fermentation in the 1860s, links had been formed between human biology and the more materialist sciences. Such links grew out of the following:

1. The advent of laboratory methods, involving a shift away from the study of dead tissue and toward the examination of living matter under strictly controlled conditions. This reflected the tremendous privilege granted in France to chemistry as a direct result of Lavoisier (and his famous successes over contemporaries in England and Germany), who, more than anyone, established the central importance to scientific labor of the laboratory as a "mill of fact," and of the experiment as a means to uncover, test, and confirm eternal principles. Respect for experimental method in physiology was first propounded by Magendie and then by his student Claude Bernard in the 1840s, 50s, and 60s, by which time it had also been established in Germany.

2. The related founding of a new discipline, biochemistry, that began to reveal the range of chemical reactions, products, and needs of organic life. This meant the advance of inorganic chemistry to the borders of the living, for example, fermentation, which at the time was a catchall term for many different processes of change in organic solutions, including the production of alcohol, vinegar, the souring of milk, and so on. The critical idea here was that living organisms were a kind

of "laboratory" in which chemical reactions were continually taking place. Fermentation was itself believed to be the result of certain catalytic entities, or "ferments" (as they were called), whose mere presence was sufficient to induce the relevant reactions. Debate centered on whether these were inorganic or living entities; minority opinion posited them as microorganisms, yeast in particular. The entire controversy, however, made possible a kind of slippage across field boundaries, from chemistry to biology and from biology into medicine.

3. The concept of cellular (microscopic) pathology propounded by Rudolf Virchow, which proposed, first, that cells are individual organisms, self-contained organizational systems, and second, that disease must therefore begin here, at this primary level of life, in order to affect the body in any lasting or profound manner.

4. A competing, yet also overlapping, view of disease as a result of one or another type of "poisoning" or "pollution" of the body. This idea, known since the early 1800s, was used as a means to describe the reasons for internal lesions, and was also connected to the idea of "infection." Claude Bernard, in developing his famous theory of the "internal environment" (*milieu intérieur*), whose "balanced stability" he claimed as the prerequisite to health, supported this idea of "contamination," which could come from "external factors." Bernard himself actually experimented with certain poisons (carbon monoxide and curare), to demonstrate their action upon the interior workings of the body. This work laid the foundation for drug therapy and experimental pharmacology.

These four trends formed the critical background against which the work of Louis Pasteur—chemist by training—came into being. They created a framework in which "agents" and "contaminants" interacted with the body through "internal" and "external" realms. They blurred the margins between living and nonliving. Indeed, they made for a domain of rationality by which life itself could become poisoning, and could use the body as a site for its own "experiments" in damage and disaster (this was actually an idea of Bernard's, who, like many of his contemporaries, saw the nervous system as an internal cause of illness).

Pasteur's work, like Bernard's, was not theoretical, but experimental and practical. Pasteur had few "ideas" in the broad sense; one of his gifts was to precisely and tirelessly apply a rigorous logic of implication to a single concept: that the microbe could only be properly understood as the source of tremendous biochemical power. Or, in his own words, that "life is the germ in its becoming, and the germ is life" (quoted in Dubos, 1976, p. 395). All of Pasteur's major findings, including the germ theory itself, were essentially a restatement of this belief, which he formed early in his career. During his initial chemical experiments on the optical effects of paratartrate crystals he found that simple microbes (yeasts, in fact) were somehow capable of telling one optical isomer from another. This discovery appears to have deeply impressed him; somehow, his chemical training led him to be immediately caught up, fatefully, in the study of life. Indeed, it seems to have been this first feeling of his, this perception that great and mobile powers— powers even of will and intent—were encapsulated in tiny bits of living matter, that remained the guiding idea. Pasteur used his laboratory as a site to reveal and control these powers, to alter, rework, and order their reality. To be sure, as Latour (1984) has written, Pasteur was also a skillful manipulator of certain social interests, and could use his laboratory adeptly as a "theater of proof." This was crucial, later on, to establish his credentials and convert his audience. But it does not help explain what set him on the path, or what kept him headed, in the direction of disease.[10]

A link was needed, a spark. What, after all, led Pasteur to span the gap between fermentation and human illness, between studies on wine or beer and those on rabies? Conceiving the microbe as the *necessary* source of all "contamination" could not have been accomplished by mere extrapolation; too much stood in the way. For one thing, beer and wine were not viewed as living systems, even by Pasteur. Instead, the essential copula appears to have come from a different area altogether: in a sense, from language. Or more specifically, from a simple metaphorical extension that can be seen in the very titles of two early publications: *Études sur le vin, ses maladies, causes qui les provôquent* (1866), and *Études sur la bière, ses maladies, causes qui les provôquent* (1876) (Studies on wine/beer, its diseases, and the causes that initiate them). *Diseases*, then, of wine and beer. Such was the conceptual–linguistic key that immediately rendered possible the ladder of logic leading from fermentation first to maladies of the silkworm, and then from there to contagious illnesses of animals (chicken cholera, anthrax) and finally to the illness of human beings (rabies).[11]

Interestingly enough, Pasteur himself, on more than one occasion, pondered the significance of this very connection. "Diseases of wines and microorganisms!" he says in one of his notebooks (dated 1877); "What a stimulus this must have given to my imagination and intelligence; were it only through the connection between these words . . ."

(Dubos, 1976, p. 246). Ten years later, moreover, in a speech given before the French medical academy on the occasion of introducing his completed work on beer, he says this:

> It is, in truth, a possible profanation of medical terms that they be applied to a beverage which, in the end, possesses nothing of that which constitutes life. I am myself somewhat of this opinion; however, things which appear considerably distinct from each other can also hold within them hidden affinities. The study which I put before the Academy today on the ammoniacal fermentation of urine [ammoniuria—a condition of excess ammonia in the urine] provides us with an example. In this instance, we are speaking of a human disease. How is one to explain [it]? It is [says M. Pidoux], a spontaneous disease, produced "in us, of us, by us"; in effect, in the mucus secreted by the bladder . . . Yet according to the observations that I offer here . . . it is necessary to abandon this manner of conception. It can no longer be maintained that ammoniacal urines are the result of a spontaneous disease, since they are provoked by an exterior cause. This cause is not in us and it is not born of us: rather, it introduces itself into us, under a solid and perceptible form, as something which is entirely different [from us]. . . . *The entire preoccupation of medicine must be to impede the introduction from the outside to the inside of fermenting germs, or to oppose their development once they have penetrated* . . . .(*Oeuvres*, vol. 5, p. 316, italics added)

This passage has been quoted at length because it reveals a critical movement. Beginning with the mentioned link—fermentation as a form of "disease," thus "disease" as a form of fermentation (microbial action)—Pasteur then employs this as an analogy to explain human illness in a wholly literal, material sense. He moves directly from a metaphorical frame into a formulation of the germ theory itself, and at that point, into a second conceit employing images that represent the beginnings of biomilitary discourse and that provide a practical rationale for "the entire preoccupation of medicine."

The connection offers a fascinating parallel between language and ideas. By admittedly "profaning" medical language, by "penetrating" its body with a "foreign" matter of expression (again, "diseases of wine and beer"), Pasteur makes the analogical connection that both reveals and "proves" his own theory of how disease results from the introduction of "exterior" agents.

Such a link was critical in a larger sense too. At the time of Pasteur's work, diseases were mainly viewed as events characterized by great power and local diversity. For example, animal diseases were conceived to be wholly different from those that affected humans; the former were thought to represent a local concentration of diverse "outdoor" forces and influences, such as those from the soil, the clouds, the winds, water, food, even the farmer himself. Many human illnesses, on the other hand, were presumed to occur spontaneously, as a result of certain of these influences acting in combination with particular human qualities, such as mood or personality, or interior malfunctions and secretions which then resulted in some form of "biological corruption" or "failure." Above all, disease was conceptualized as an expression of formidable and often invisible powers at work. It took Pasteur to show, per his original revelation, that microbes were in fact capable of such powers, that despite their tiny dimensions they could take command of and wreak havoc in bodies many millions of times their own size.

In the passage above, one sees the new concept coming into being. The microbe is turned into an agent; it is not acted upon or merely involved in some action but instead carries out its own action according to its own intentions ("introduces itself into us, under a solid and perceptible form"). In this way it is invested with the force of invasion, penetration, provocation. The germ itself becomes an attack upon the body. Better said, it has become an entity that holds within itself a tremendous condensation of disease vitality; in coming from "outside," it contains with monstrous density all the influences previously thought to be so hopelessly dispersed throughout the environment (internal or external). The microbe has become a kind of Trojan Horse; once "inside," it allows all the damaging powers and influences from the world to attack the bodily Troy. The link that Pasteur made between fermentation and disease therefore revealed the germ as being far larger than its own life. It became, for him and for the discourse he began, the bearer of the very possibility for contagion—a life force for disease and death.

🖺🖺

It is important to note, however, that in making this link, Pasteur was following a nonmedical tradition in language that had been long established among grape growers. Throughout the 19th century, it was apparently common practice within the wine industry to use the word *maladie* (illness) in referring to spoiled or bad wine. Attributing a kind of human illness to wine grapes suggests the importance of wine to *la vie française*, perhaps, or else the sentiment that, because of its complex properties, the amount of labor involved in its making, and the positive effects and significant social uses attributed to it, this beverage was tantamount to a "living thing" (in French literature especially wine is often compared to, or else given the traits of, women). Whatever the precise reason or reasons, whether ideological or not, "diseases of wine" was a term of art within French viniculture. As Pasteur notes in the introduction to his treatise on the subject, there even existed a nosology of these diseases: they were all given separate names, arranged in a hierarchy according to how often they occurred and how severe their effects might be, and were commonly discussed in terms of their "prevention" or, if possible, "treatment."

Pasteur seems to have adopted "diseases of wine" in part out of respect for the industry he had been called upon to help (he was himself a considerable lover of wines). He apparently wanted to make his work as accessible to nonscientists in this area as possible, and to do this by employing language with which they were familiar. He may also have understood, at some level, how strategic this kind of phrase might have been for attracting outside attention to his work, for "capturing interests," as Latour says. But it is just as possible, and is perhaps more likely, that this term of art also attracted Pasteur for exactly the reasons he himself mentions, for its "stimulus . . . to the imagination." Latour and others have made a compelling case that much of Pasteur's success can be related to the ways in which he brought his methods, in the form of vaccinations of cattle, into the actual fields, thereby proving that the powers of microbes were not in fact confined to the laboratory itself. Yet this occurred later. Indeed, at perhaps the crucial turning point in Pasteur's intellectual career, the very opposite seems to have occurred: with quiet yet momentous irony, a metaphor initially adopted from science into the fields of an industrial discourse, finds its way back to science again. Thus a kind of complete cycle took place, a circulation in which the essential change from analogy to "truth" was eventually accomplished for a mixture of personal, scientific, and patriotic reasons.

⑤⑤

The detailed progress of Pasteur's work and thought has been often discussed in detail. I will not try to trace or summarize this extensive literature here, but will instead make only a few relevant observations. At the outset of his work on fermentation, Pasteur was faced with two opposing hypotheses: (1) that of the prevailing wisdom, which held the "ferment" to be an inorganic chemical, a catalyst; or (2) the vitalist view, which saw it as a living presence. In alcoholic fermentation of wine and beer, the "germ" yeast was generally known to occur as a growth deposit. It was, however, denied any central or active role in the production of alcohol itself. Instead, it was consigned to an incidental by-product, or, at most, to a parasite whose sedimentary contribution took place only as a result of its decay at death. Yeast had actually been shown to be a protagonist in this process as early as the 1830s, by several researchers working independently in Germany and France, but such results drew little attention from organic chemists since they implied vitalism, and vitalism as a doctrine had fallen into ignominy among experimental scientists, who denounced such "speculative" ideas as a denial of the importance of physicochemical processes.

Pasteur's proof of the yeast connection came in two papers, *Memoire sur la fermentation appelée lactique* (1857) and *Memoire sur la fermentation alcoolique* (1860). In these, he outlines an experimental technique by which he was able to isolate the relevant germs and use them as "agents" to induce fermentation directly. It was from this point, then, that his studies turned toward the "diseases of the alcoholic fermenting process," which occupied him for the next decade. With the hypothesis that every type of fermentation was caused by a specific microbe, Pasteur soon postulated that any unwanted products—the telltale signs of a spoiled wine—were the corresponding creations of unwanted, contaminating germs. This he proved (inasmuch as he was unable to detect that the true causative ingredients were in fact chemical substances produced *by* these germs) and laid out in great detail in his next book, *Études sur le vin* (1866). What he had accomplished, more broadly, had been quietly unexpected: rather than confirming or denying the materialist position, he had made it commensurate with vitalism. "Fermentation [a physiochemical process] is the consequence of life without free oxygen," as he described it in his work on beer (*Oeuvres*, vol. 5, p. 435). Life itself did not merely include chemical processes, or transpire through them (this had been shown by Justus von Liebig in Germany since the 1820s); it now also had real chemical power, the power to create and destroy.

It is at this juncture that the first decided signs of military language become established in Pasteur's writings. In the text on wine, terms such as "invade," "foreign," "defeat," and "overwhelm" appear. Again, some of these terms had been employed before, in various settings. They were also, so to speak, waiting in the wings of late-19th-century medicine, as suggested by the writings of some of Pasteur's contemporaries who were interested in infectious and contagious diseases. The following phrases, for example, are from a lecture delivered to the Paris Academy of Medicine in 1865 on the subject of syphilis (by Auzias-Turenne):

> Viruses derive what they need from the infected organism and often end by exhausting the latter. . . . They either destroy it or abandon it for lack of food . . .

> A virus can be regenerated in a good terrain in which it multi-
> plies, whereas it can be weakened by an unfavorable terrain . . .
> Contagion presupposes a direct contact of the virus with the
> organism. Infection does not involve a direct contact; the virus may
> be carried through a medium which is usually the atmosphere.
> (Quoted in Dubos, 1976, p. 325)

These words were not actually published in book form until 1878 (when
Pasteur first encountered them, in a work entitled *La syphilisation*). They
show a type of discourse at the verge of discovering a centralizing frame.
All the required ingredients are there—"exhaust," "destroy," "multi-
ply," "terrain"—but the final step has not been taken, the step of "attack" and "defend" that would draw everything together. It was part of
Pasteur's success to have done exactly this: to have created a systematic
and rationalized discourse, a true image system, in which cause and effect could be narratively united.

By the time he had completed his initial work on silkworm disease
(*Études sur la maladie des vers à soie*, 1870), this language had considerably expanded, taking on new and more vivid terms like "embattle,"
"rupture," "attack," and "engagement." The title of an important chapter
in this book runs thus: "Methods for Combating the Silkworm Disease
and Preventing Its Return." Summary paragraphs, meanwhile, often run
like the following: "The principle which I here invoke clearly indicates
the advance of the contagion, the slowness of the initial development of
the corpuscles, and the resistance to death put up by those worms invaded by the parasite" (*Oeuvres*, vol. 4, p. 132), or "The multiplying of
the corpuscles takes place in the intestine [of the silkworm] only with
great slowness. During the five or six days after their initial appearance,
they remain isolated . . . and it is only after some time that one encounters this mass . . . in all the other tissues of the worm, as soon as they
begin to be invaded" (p. 151).

But even this is not the whole story. As a strong patriot, and as
a professional contracted by the government to perform his study,
Pasteur writes of the silkworm growers, including their toil and trouble,
as part of his narrative. Here, too, the language evokes a certain imagery:

> The universal ignorance of these facts [that some silkworms survive
> the disease and are used to grow resistant cultures], outside the re-
> gions which the epizoa have brought to ruin, is perhaps explained
> by this circumstance, much deserving of our sympathy: that the popu-
> lations of these regions never cease to show in their misfortune a

heroic resignation which cannot but inspire calamities whose causes appear to escape all human foresight. Among forty surrounding *departements* which are engaged in sericulture, there are several for whom this industry comprises the total economy. Their inhabitants, without ever growing weary, renew each year their efforts and their expenses, and each year add to their troubles a new misery. Such perseverance in struggling against misfortune, without expressions of unjust accusation or hostile complaint, can rightly be taken as proof of progress in education and of the good sense of the inhabitants of the countryside, and also, possibly, offers tribute to the vigilant solicitude that the head of state continues to evince toward the interests and the sufferings of the people. (p. 22)

The bitter irony of the last few lines is thick enough to cut even with the dullest blade. It is clear, that is, where Pasteur's real sympathies lie. In chastising the government for having abandoned the people to their ancient ways, for having done nothing by way of "progress in education" to solve a "universal ignorance" that is the source of obvious poverty, he is essentially claiming, through a language of "ruin" and "perseverance," that the state has failed in its role as protector and guardian. Pasteur's basic frame for discourse here is military too, though in a more declamatory sense; "populations" have been forced into "heroic resignation," to "struggle" against an invading "scourge" (Pasteur used this term, *le fleau*, often in the book).

The plight of the silkworm growers, whose industry had almost been destroyed by an outbreak of pebrine and whose situation had led the government to call on Pasteur for help, inspired the latter to invoke a language of righteous indignation, a literary, speech-making language reminiscent of official declarations related to one or another military campaign. One is led to wonder, therefore, whether Pasteur's perception of "struggle" in this case may not have further encouraged the uptake of more concrete, empirical military terms in his science.

At this point in his work, in fact, Pasteur, like most French intellectuals, had a traumatic encounter. He found himself overwhelmed by the defeat of France to Prussia in the war of 1870. This demeaning capitulation, secured by the Germans in only 7 weeks, came at the point in his own career when he had just finished work on the diseases of silkworms and was beginning to think about contagious illness in general. His general feelings about the war were a mixture of outrage and shame:

We are paying the penalty of 50 years forgetfulness of science, of its conditions of development of its immense influence on the destiny of a great people . . . I can not go on, all this hurts me. I try to put away all such memories, and also the sight of our terrible distress, in which it seems that a desperate resistance is the only hope we have left, I wish that France may fight to her last man, her last fortress. I wish that the war may be prolonged until the winter, when, the elements aiding us, all these Vandals may perish of cold and distress. Every one of my future works will bear on its title page the words: "Hatred to Prussia. Revenge! Revenge!" (Quoted in Valery-Radot, 1960, p. 184)

A year later saw him traveling throughout the country, giving speeches on the ways by which France could reestablish and "fortify" her powers. In one such lecture, entitled "A Few Reflections on French Science" (*Quelques réflexions sur la science en France*), the following lines occur:

> While Germany multiplied its universities, produced a healthy competition among them . . . organized great laboratories with the best equipment, France, enfeebling itself by revolution, forever preoccupied with the sterile search for the best style of government, remained aloof to her institutions of higher learning. (*Oeuvres*, vol. 5, p. 279)

Again, therefore, the government comes in for harsh critique in terms reminiscent of military images. In seeking an explanation for the war, Pasteur seems to have drawn indirectly on his own scientific discourse, on concepts of "resistance" and "the body fortified" gained from his work on silkworms, and also on his earlier belief in the failure of education. France had failed because it had weakened itself internally, because it had turned on itself—politically, educationally—and thinned the natural defenses of its intellectual and therefore sociopolitical state. Pasteur seems to have viewed the diseased body and the "enfeebled" nation in very much the same way. While it is difficult to discern which of these two models of discourse came first in his work and writing, it seems clear that they coexisted by this time.

Thereafter, Pasteur went on to tackle chicken cholera, swine erysipelas, anthrax, and finally rabies. He developed the first true vaccines for all these diseases, thereby securing—not without enormous opposition—the final success of the germ theory and its attendant concept of specific etiology. In the texts he produced during this time, the later 1870s and 80s, one finds a still greater use of war terminology. New words enter his lexicon: "mobilize," "infiltrate," "propagation," "vital resistance." These rose mainly out of his work on chicken cholera, which brought him fully into the study of immunity, and therefore into a more detailed analysis of "bodily defenses." By 1874, this work had urged upon him the conclusion that "under a host of [normal] circumstances, life brings to an end that life which is alien to it," and therefore, that the potency of microbes for causing disease lay in their ability "to become more and more successful at self-propagation and thereby at defeating [*vaincre*] the vital resistance of the animal" (*Oeuvres*, vol 6, p. 92). By the early 1880s, such a statement as the following (published in an article of 1881 on the life cycle of the virus) had become fairly routine: "Man carries within himself, or in his intestinal canal, and without great injury, the germs of certain microbes nonetheless ready at any moment to become dangerous whenever the body enters an enfeebled state and their virulence can find itself progressively reinforced" (*Oeuvres*, vol. 6, p. 337).

Pasteur died in 1895, perhaps the most widely known scientist in the Western world. Certainly he had lived long enough to witness the triumph of his "vitalist" ideas within medicine, to see them gain the status of established theory and practice. Koch's famous, meticulous isolation of the bacilli responsible for tuberculosis (1882) and cholera (1883) led to the establishment of the so-called Koch Postulates, a series of standard procedures for the identification of disease-causing microbes. The linguistic influence of Pasteur, the uptake of the biomilitary metaphor, is already completely in evidence in Koch's writing of the late 1870s, peppered as it is with such terms as "invasion," "penetration," "combat," "conflict," and the like. In the few years remaining before the turn of the century, a spectacular range of other bacterial contagions had been isolated in the laboratory by researchers, who likewise employed the new imagery. Vaccines, as "weapons against disease," could now be prepared for a wide array of "historical killers," including typhoid fever, leprosy, diphtheria, tetanus, pneumonia, gonorrhea, and plague. In addition, Lister's own longtime work with antiseptics, heavily indebted to Pasteur's research, and Emil von Behring's innovation of serum therapy, which took the vaccine itself one step further (using chemical products alone), both extended the pragmatic and linguistic consequences of the germ theory even beyond its original purview.

These and other findings helped transform medicine into a conjunction between laboratory-based science and clinical therapy. This transformation was by no means simple and straightforward. The concept of a "scientific medicine" in the late 19th century was deeply tied to physiology, to questions of function and structure. As mentioned, this area had been most intensely invested with physiochemical explanatory schemes and had placed itself in the laboratory even before Pasteur began his work on fermentation. Experimental physiology remained the acknowledged path to a real "scientific medicine," even though it did not offer any important advances in therapy. Relevant debate among physicians as to the value of such a medicine, and the issues of status involved, have been discussed elsewhere (see, e.g., Warner, 1991). During the 1870s and 80s, metaphors used to describe the profession by its own practitioners were commonly derived from other sciences: doctors were exhorted to put their practices "into the crucible of physiology" or to conceive of themselves as "engineers," for the sake of precision and reliability (Warner, 1991, p. 458). But by the turn of the century, after the advances noted above, therapies existed to protect, treat, and even cure people of illnesses that had claimed great numbers of human lives all through history, not least in the 19th century itself. The germ theory gave medicine what physiology couldn't; it turned medical practice (in part, and selectively) into a proven savior, a heroic science. Its methods, in a sense, succeeded in injecting into every patient the strategies of a new and eminent rationality, a logic of unprecedented "conquest" or "victory" over the forces of death. Dubos may exaggerate when he says there took place a "complete sacrifice of the physiological to the bacteriological point of view" (1976, p. 314). But it is clear that a profound movement did take place in this direction. By 1900, physicians no longer characterized themselves as chemists or engineers; they were instead dedicated to combating the "unseen enemies of mankind."

The uptake of such language meant that the view of disease had changed too. Ideas related to the influences of heredity, nutrition, climate, exercise, and individual physiology, ideas that had previously played an important role in medical discourse, and which Pasteur himself had often emphasized, now declined as subjects of inquiry and interest. The main effort became one of searching out "causative agents." Some 15 years after Pasteur's death, when Abraham Flexner wrote that all medicine was "an attempt to fight the battle against disease most advantageously to the patient" (1910, p. 5), he was speaking a language that had instead become public, institutional. By that time, in other words, the change in Western biomedical discourse had taken place. From then on, the Occident was literally at war with disease. And so it remains today.

Conclusions and Questions

One can safely assume that any newly defined diseases will fall within the paradigms I have sketched above—military and informational imagery. Certainly, this is well shown in the case of AIDS, all the more so due to the presumed viral cause of this disease, its contagious nature, and its specific power to interrupt the immunological system of "communication" within the body. New or old illnesses attributed to such causes will fall more, perhaps, within the realm of military imagery. At the same time, any disease given a genetic origin will have its corresponding discourse shifted more toward informationism. This is already clear in such cases as cerebral palsy, muscular dystrophy, rheumatoid arthritis, and most recently, diabetes and Alzeimer's disease, where there has been a literal "hunt" for a specific gene (or gene combination), presumed to "switch on" each malady and whose cure or prevention is conceived as a matter of "reprogramming" and "information repair."[12]

This leads us to another problem, about which little can be said here. The growing success of bioinformationism, the change of the human subject into a form of technology, seems to be having other effects of a disturbing sort. It has created an environment of invitation, wherein biodeterministic ideas can flourish. If, as is so often proclaimed today (by James Watson, no less), the human genome contains our destiny and defines our identity as human beings, if "genetic information" is the very core of our being and existence, then it must be the gene that is also the source of our problems and difficulties. Coded into current biomedical discourse is this suggestion of fixed and deterministic origins. Issues of responsibility and control therefore collapse within this frame, where one is finally brilliant, murderous, gay, alcoholic, or rich because of genetic fate. The type of power given back to the physician—as the one who "detects" and "fixes" whatever might be "wrong"—is no more reassuring. This, indeed, may be the true loss involved in bioinformationism: in defining life as something based on codes, it removes the complexities of the "human" altogether.

On the other side, meanwhile, it should be stressed that, at most points of popular discourse, biomilitarism continues to rule almost to the degree that it has for the past 100 years. This means that it rules, for the most part, at the level of social practice, that it is still the language employed by political, economic, and other institutional discourses with regard to disease. Thus it is the language that precedes and surrounds each individual, that in part determines his or her cultural idea and experience of disease as a distinctly *medical* phenomenon.

One might well ask, in the wake of recent literary theory, whether or not such a language can be challenged by the tools of insight provided by recent methods in discourse analysis. One of the powers often assigned to deconstructive readings, for example, is an ability to reveal flaws and frailties in logic, unity, and, ultimately therefore, in the efficacy of language to order the world in any final sense. It has become common to speak of "the inevitable failure of language as an instrument of representation," of how discourse always engages in "self-subversion" (Krieger, 1989, p. 59). Without going much further than this, it can be said that while such ideas may have vitality for certain areas of writing—say fiction or poetry, where metaphor still functions as metaphor—they can only appear limited, even effete, when applied to most areas of science. Here, as I have tried to make clear, language is no "game" or "logocentric" exercise. Its real-world influence in daily human life approaches the incalculable. No amount of ingenious "misreading" or methodological playfulness can hope to diminish the concrete emotions and social rituals surrounding the psycholinguistic state of being a "patient," of being "under attack" by cancer or the AIDS virus. If literary readings of biomedical discourse do possess suggestive possibilities, these would have to be confined to a theoretical, abstract level. They might, indeed, include the capacity to point up other, "softer" image systems within medical language (see note 3). But to propose these as true alternatives would be, at this point, an exaggeration. A vast amount of contemporary medical reality would have to be rendered tentative, called into question or even overturned, before this could happen. The advent of holistic therapy has not done this, despite common assumptions. On the contrary, in some ways it has accomplished the very opposite, strengthening and even deepening (via individual blame), the total reach of the biomilitary paradigm.

Such might be viewed as cause for fatalism. If we are "trapped" by our conceptions of disease, by the traditions of speech, thought, and fact on which they are based, where might our analyses take us, if not in circles? Awareness can, I think, grant a degree of freedom and empowerment. Knowing the conventions of imagery involved, realizing their inherited and therefore "imposed" nature, can make one less of a victim, willing or unwilling, in the face of disease. To this effect, then, one might well extend the type of inquiry presented here in a number of other directions. Some of these are indicated by the following types of questions.

- How has the discourse of biomilitarism helped redemonize the ill body? In what ways does the association of disease and war terrorize each category with the other?

- What might be the iconic status of the doctor today? Is s/he a stable image of the crusader in white, or has s/he become an emblem of disease and death, a totem for her/his veiled everyday presence among us?

- How does biomilitarism predispose the language of "disease" toward being used metaphorically in *any* situation of conflict, with respect to any unwanted phenomenon or opponent? What qualities does this impose on such a phenomenon; what specific recipes of action are demanded?

- When we ourselves suffer from disease, how does biomilitarism redefine, first, our own private sense of self, and second, our more public identity? In what ways, exactly, does it help transform us into the living presence of our disease, into a form of "affective contagion"?

- What are the politics involved in medically attaching the term "disease" to some new condition or behavior (e.g., homosexuality, alcoholism, drug addiction, drunk driving, eating disorders)? How does this affect how people view such a condition? What institutional frame or frames are called into action? More, what exactly is the process in each case that leads to the use and acceptance of this term (obviously, this process does not begin and end within medicine itself)? What does this imply about the presumed ability to distinguish "real" from "false" diseases?

- Given the visions of therapy implied by biomilitarism, and their focus on "resistance," on "counterwar," how are these expressed institutionally, in terms of social structures, buildings and architecture, dress and instrumentation, procedures and testing, or the role of technology? In what ways, for example, can we conceive the hospital itself as a kind of wartime factory or fortress? How does this correspond itself to the biomilitary view of the body?

Epilogue

Biomilitarism, from its beginnings, provided a frame in which nearly every perceptible agent or process gained an immediate role and significance. More, it offered a new, tightly unified scheme for public discourse about illness. Its images were at once magnificent, ravishing, and terrifying. And it is exactly these qualities which, one might say, have given this discourse its larger "therapeutic" function for Western culture, that of imposing a final meaning. One can only surmise, in retrospect, how this must have suited a Western world preparing for war on so many other fronts at the same time. Perhaps one might say, then, that this final meaning is itself a kind of historical holdover or fossil from a less-enlightened era. This, however, would inevitably ignore the continued centrality of war, whether hot or cold, up to the present day.

It must be said that Pasteur was not uniquely responsible either for the germ theory or for the spread and final institutionalization of biomilitarism. His work was neither that of a "lonely genius" nor that of a "heroic general" in the history of science. His work, and its founding ideas, were instead the site of a collision of influences: personal, political, cultural, and intellectual. The expansion of biomilitary discourse, in particular, has obviously been a product of many researchers working in many different countries, making choices according to a variety of shared influences, in part handed down from Pasteur himself, in part derived from other work within medical research, in part urged forward by powerful political–economic demands and supports. Such does not at all diminish the importance or stature of Pasteur's work; certainly it avoids the crude error of blaming him for the narrow sovereignty his ideas and language later came to have. Viewing his work as an intersection, rather, returns to it a measure of the historical responsibility that truly belongs to it, a responsibility of occurrence as well as of creation.

A century after Pasteur, medical visibility has gone yet another level beyond the visible. It has gone deep within, and beyond, the biomilitary realm. It has found a new universe, a new complexity of agents, processes, and structures. It has made all these knowable in terms of a new analogy, far more sophisticated and complicated in its implications than its 19th-century military predecessor. Yet it is an analogy, too, that neatly retains, nay, even updates, this predecessor, allowing it to appear in more varied and contemporary form.

One therefore wants to ask: Why is it that our scientific and public discourses seem so doomed to counter our own needs for a liberating conception in the face of disease? Will we never surpass the languages of terror and fatalism? In the end, any action to counter or alter the effects of biomilitarism must begin with awareness. If this image system continues to rule, one must know exactly how, where, and when it does so. One must also know that this sovereignty, such as it is, does not remain solely within the medical/technical context. Medicine is both the language of truth and the multiform aspects of culture that speak this truth in its myriad details; it is as much in and of us as around us. If it has the power to render any "alternatives" nonmedical—that is, suspect and fragile, forced to adopt symbolic languages—then we need to know just how this works and what social forces are at stake. It may be, as Gramsci once noted, that only by becoming cognizant of the invisible rationalities that surround and govern one's individual helplessness can the movement toward change, toward a reverse empowerment, be successfully begun.

Notes

1. Here, and in what follows, I use the two terms "metaphor" and "analogy" in a roughly interchangeable way. This usage is loose, inexact, but it helps emphasize the fact that scientists themselves have rarely made any kind of distinction in the manner by which they seek new explanations through imagistic language. A vast literature exists on the presumed, final difference between these two terms (an overview can be found in Leatherdale, 1974). In reality, the discussion, with its connections to analytic philosophy, has proved a distraction from more important issues: the content, origin, and long-term implications of the imagery of science. Moreover, it seems to miss the important point that "metaphor" does not remain as such within technical knowledge. No doubt the process by which images become literalized within science (a process I only touch on here) would be more worth pursuing in terms of any deeper understanding of scientific knowledge than would another round of negotiations about what type of figurative language is most "appropriate."

2. The distinction between "illness" and "disease" is a matter of endless debate. Institutional medicine views the former as a generic condition. According to *Dorland's Medical Dictionary,* illness is "any condition showing pronounced deviation from the normal healthy state"; disease is termed a specific cause for this condition, i.e., "a morbid process having a characteristic train of symptoms." This scheme, however simple in appearance, is nonetheless held in place by the intricate cultural scaffolding associated with the term "normal." A somewhat different, yet related, notion speaks of disease as a malfunctioning of biological and/or psychological process; illness as the psychosocial experience and meaning of perceived disease (Kleinman, 1980). Marxists have proposed illness as a biographical and historical phenomenon, disease as a type of commodity definition by scientists that helps mediate biology to capital formation (Stark, 1982).

The problem with all these attempts is that they assume, at some level, an essentialist reality regarding these two terms. Both, that is, are tied to some underlying correspondence theory that seeks to define categorical constants in human experience. This, it seems to me, is misleading. Illness and disease, in a sense, are utilitarian labels that allow for certain types of description and study, and therefore have had their specific meanings change over time in response to historical conditions and demands. This chapter uses "disease" on the broadest level, that is, as referring to institutionally appointed conditions of "unwellness" (codified in specific titles such as cancer, AIDS, diabetes, etc.).

3. These two conceits, as I say, are not the only ones to be found in the relevant research literature. Some others include that of the "family" (biofamilialism), evident in such terms as lineage, education, residence, immature/mature; and the "individual" (bioindividualism), revealed by phrases and words employing "self" or "non-self" (e.g., "self-reactive," "self-promotion," "self-selection"). Despite the interesting character of these conceits, however, they occur less often and with far smaller and dispersed vocabularies than those of biomilitarism and bioinformationism, and can be said to exist within the broader concepts of disease provided by these.

4. The metaphor of the map has a long-standing history within the science of genetics and is very much in use today. An excellent review of its recent employment can be found in Hall (1992, pp. 174–192). No doubt an analysis of when this image came into biological discourse would prove interesting. The concept of the body as a New World of sorts, a geography for exploration, and so on, was likely a very fruitful one for research, both then and now.

5. "Action theory" was a concept proposed by the renowned sociologist Talcott Parsons in the 1930s and refined in the 40s and 50s, mainly under the sway of cybernetic ideas. Disregarding its influence within sociology, one should note its profound relevance to a scientific image system like biomilitarism, which is based on the same type of primacy given the "act." Parsons should be allowed to speak for himself here:

> An "act" involves logically the following: (1) It implies an agent, an "actor." (2) For purposes of definition the act must have an "end," a future state of affairs toward which the process of action is oriented. (3) It must be initiated in a "situation" of which the trends of development differ in one or more important respects from the state of affairs to which the action is oriented. . . . This situation is in turn analyzable into two elements: those over which the actor has no control, that is, which he cannot alter, or prevent from being altered . . . and those over which he has such control. . . . Finally, (4) there is inherent in the conception of this unit . . . a certain mode of relationship between these elements. That is . . . within the area of control of the actor, the means employed cannot, in general, be conceived either as chosen at random or as dependent exclusively on the conditions of action. . . . What is essential to the concept of action is that there should be a normative orientation, [whose discrimination] is one of the most important questions . . . (Hamilton, 1985, pp. 73–74).

Parsonian theory had its heyday in the 1950s, 60s, and early 70s. The above suggests, however, that it could be put to new uses *within* certain fields of science.

6. This same point has been previously made by several authors, for example, Peter Sedgwick (1973).

7. Similar ideas have been pointed up by Stephen Rose in his book *The Chemistry of Life* (Baltimore: Penguin, 1970).

8. The point of this is not to impugn the effectiveness of holistic therapy (indeed, such therapy has been many times proven to have profound benefit for many people). It is merely to indicate that holism has no real alternative to offer when it comes to truth images for disease. It is unable, given its current language, to adequately explain, without recourse to symbolism and metaphysics, the effectiveness it does have.

9. Durer's famous woodblock print (*The Four Horsemen of the Apocalypse*) is an exception, for it was based on the actual wording of the Bible.

10. Latour's treatment of Pasteur and his success makes for fascinating reading, but I find his interpretation only partly convincing. Among other things, Latour is interested in dismantling the traditional view of Pasteur's "genius" and replacing it with another, focusing on his skill in "capturing" a variety of normally competing interests and converting them to his ideas, something he achieved by essentially transporting his laboratory to the farmer's field and there showing that microbes create their effects in nature as well as in the test tube.

Interestingly enough, Latour's own major conceit in this work is military, as indicated by his title *Les microbes: Guerre et paix* (Microbes: War and Peace). His is a type of sociomilitarism; he is intent on portraying the landscape of known reality as a war for influence, of "actors" and "agents" struggling for dominance. "Knowledge is the condition of this battlefront," he says. "It extends no further" (p. 231). In speaking of Pasteur himself, early on, he even asks: "Why should we still treat Pasteur's genius in a way that we no longer do Napoleon's?" (p. 21). Other than the fact (and a fact it is) that Pasteur ended up producing a body of work that saved millions of lives while Napoleon was more in the business of destroying them, one might question a view that throws such figures together with such happy utility.

The larger problem with Latour's antitheory, however, is that it qualifies as theory after all. As "actors," as victims of their own momentary teleologies, scientists like Pasteur become microbial. Latour wants to use narrative as his own laboratory, and so he ends up, at some level, seeking to provide one more essentialist explanation of scientific labor and thought. Amid the rant and strife of battle, a sly determinism waits: the laws of action and behavior have been shifted, given a more semiotic cast, but they remain laws just the same.

11. It might be noted that whereas in English a number of basic terms exist to denote a state of poor health—"illness," "sickness," "malady," "ail-

ment," and "disease," for example, with each bearing different nuances of generality and gravity—the French commonly use *maladie* for all of these, although occasionally they also use the term *affection*.

12. A contemporary definition of "gene" runs as follows: "a unit of hereditary information; the portion of a DNA molecule which contains, coded in its nucleotide sequence, the information required to determine the amino acid sequence of a single polypeptide chain" (from Vander, Sherman, and Luciano, 1986, p. 58).

References

Adair, M. (1986). Conscious Recovery. In J. Serinus (Ed.), *Psychoimmunity and the Healing Process* (pp. 168–179). Berkeley, CA: Celestial Arts

Adiar, M. and L. Johnson. (1986). Applied Meditations for Healing. In J. Serinus (Ed.), *Psychoimmunity and the Healing Process* (pp. 179–210). Berkeley, CA: Celestial Arts.

Badgley, L. E. (1987). *Healing AIDS Naturally*. Foster City, CA: Human Energy Press.

Barthes, R. (1953). *Le Degré Zéro de l'Écriture*. Paris: Seuil.

Bernstein, J. (1982). *Science Observed*. New York: Basic Books.

Boynton, H. (Ed.). (1948). *The Beginnings of Modern Science*. Roslyn, NY: Walter J. Black.

Brandt, A. (1988). AIDS and Metaphor: Toward the Social Meaning of Epidemic Disease. *Social Research*, *55*(3), 413–433.

Bricklin, M. (1983). *The Practical Encyclopedia of Natural Healing*. Emmaus, PA: Rodale Press.

Canguilhem, G. (1988). *Ideology and Rationality in the History of the Life Sciences*. Transl. by A. Goldhammer. Cambridge, MA: MIT Press.

Canguilhem, G. (1989). *The Normal and the Pathological*. Transl. by C. Fawcett and R. Cohen. New York: Zone Books.

Cannon, W. B. (1932). *The Wisdom of the Body*. New York: Norton.

Chomsky, N. (1957). *Syntactic Structures*. The Hague: Mouton.

Coward, R. (1989). *The Whole Truth: The Myth of Alternative Health*. London: Faber and Faber.

Cross, S. L., and W. R. Albury. (1987). Walter B. Cannon, L. J. Henderson, and the Organic Analogy. *Osiris*, 2d Ser., *3*, 165–192.

de Certeau, M. (1986). *Heterologies*. Minneapolis: University of Minnesota Press.

Dubos, R. (1976). *Louis Pasteur: Freelance of Science*, rev. ed. New York: Scribner's.

Dubos, R. (1959). *The Mirage of Health*. New York: Harper & Row.

Fackelmann, K. A. (1993, December 11). Nabbing a Gene for Colorectal Cancer. *Science News*, 144, p. 388.

Farley, J. (1978). The Social, Political, and Religious Background to the Work of Louis Pasteur. *Annual Review of Microbiology*, 32, 133–154.

Flexner, A. (1910). *The Flexner Report on Medical Education in the United States and Canada*. Report to the Carnegie Foundation for the Advancement of Teaching, Bulletin 4. New York: Carnegie Foundation.

Foucault, M. (1963). *The Birth of the Clinic: An Archeology of Medical Perception*. New York: Pantheon.

Foucault, M. (1970). *The Order of Things: An Archeology of the Human Sciences*. Transl. by A. Sheridan. New York: Pantheon.

Gardner, H. (1985). *The Mind's New Science*. New York: Basic Books.

Gary, P. (1993). *The Cultivation of Hatred*. New York: W. W. Norton.

Guiltinan, J., and L. Standish. (1991). AIDS/ARC Research Project Completed. *Foot Traffic* (local newspaper, Seattle, Wash.), p. D.

Hall, S. S. (1992). *Mapping the Next Millennium*. New York: Random House.

Hamilton, P. (Ed.). (1985). *Readings from Talcott Parsons*. Chichester UK: Ellis Horwood/Tavistock.

Herzlich, C. and J. Pierret. (1987). *Illness and Self in Society*. Transl. by E. Forster. Baltimore, MD: Johns Hopkins University Press.

Holistic Group. (1986). A Holistic Inquiry Into the Prevention and Healing of Acquired Immune Deficiency. In J. Serinus (Ed.), *Psychoimmunity and the Healing Process* (pp. 69–160). Berkeley, CA: Celestial Arts.

Illich, I. (1976). *Medical Nemesis: The Expropriation of Health*. New York: Pantheon.

Jacob, F. (1973). *The Logic of Life*. Transl. by B. Spellman. New York: Pantheon.

Jaret, P. (1986, June). The Wars Within. *National Geographic*, pp. 702–734.

Kleinman, A. (1980). *Patients and Healers in the Context of Culture*. Berkeley and Los Angeles: University of California Press.

Krieger, M. (1989). *A Reopening of Closure: Organicism Against Itself*. New York: Columbia University Press.

Lakatos, I. (1978). *The Methodology of Scientific Research Programmes*. Cambridge: Cambridge University Press.

Latour, B. (1983). Give Me a Laboratory and I Will Raise the World. In K. Knorr-Cetina, and M. Mulkay (Eds.), *Science Observed: Perspectives on the Social Study of Science* (pp. 141–171). London: Sage.

Latour, B. (1984). *Les Microbes: Guerre et Paix*. Paris: A. M. Metailie.

Lee, P. (1986). The Vital Roots of Immunity. In J. Serinus (Ed.), *Psychoimmunity and the Healing Process* (pp. 30–36). Berkeley, CA: Celestial Arts.

Lenoir, T. (1982). *The Strategy of Life: Teleology and Mechanics in 19th Century German Biology.* Dordrecht: Reidel.

Lévi-Strauss, C. (1949). *Les Structures Élémentaires de la Parenté.* Paris: Presses Universitaires de France.

Lévi-Strauss, C. (1955). *Tristes Tropiques.* Paris: Plon.

Lorenz, M. Jung, S. and A. Radbruch. (1995). Switch Transcripts in Immunglobulin Class Switching. *Science,* 267, 1825–1828.

Meyers, G. (1990). The Double Helix as Icon. *Science as Culture,* 9, 49–73.

Montgomery, S. L. (1991). Codes and Combat in Biomedical Discourse. *Science as Culture,* 12, 341–390.

Parker, H. P. (1948). *Genetics and Cytogenetics.* New York: Wiley.

Pasteur, L. (1923–1929). *Oeuvres.* 7 vols. Paris: Libraires de l'Académie de Médecine.

Phillips, D. C. (1970). Organicism in the 19th and Early 20th Centuries. *Journal of the History of Ideas,* 31, 413–432.

Pickstone, J. V. (1990). Physiology and Experimental Medicine. In R. C. Olby, G. N. Cantor, J. R. R. Christie, and M. J. S. Hodge (Eds.), *Companion to the History of Modern Science* (pp. 728–742). London: Routledge.

Pier, G. B., G. J. Small, and H. B. Warren. (1990). Protection Against Mucoid *Pseudomonas Aeruginosa* in Rodent Models of Endobronchial Infections. *Science,* 249, 537–541.

Ramsdell, F., and B. J. Fowlkes. (1990). Clonal Deletion Versus Clonal Anergy: The Role of the Thymus in Inducing Self-Tolerance. *Science,* 248, 1342–1347.

Ria, F., et al. (1990). Immunological Activity of Covalently Linked T-cell Epitopes. *Nature, 343,* 381–383.

Riley, H. P. (1948). *Introduction to Genetics and Cytogenetics.* New York: John Wiley.

Schlanger, J. (1971). *Les Metaphores de l'Organisme.* Paris: Librairie Philosophique J. Vrin.

Schwartz, R. (1990). A Cell Culture Model for T Lymphocyte Clonal Anergy. *Science,* 248, 1349–1356.

Sedgwick, P. (1973). Illness, Mental and Otherwise. *Hastings Center Studies, 1*(3) 19–40.

Serinus, J. (Ed.). (1986). *Psychoimmunity and the Healing Process.* Berkeley, CA: Celestial Arts.

Sontag, S. (1989). *AIDS and Its Metaphors*. New York: Farrar, Straus, & Giroux.

Stark, E. (1982). What is Medicine? *Radical Science Journal, 12*, 73–91.

Starr, P. (1982). *The Social Transformation of American Medicine*. New York: Basic Books.

Tsao, J., et al. (1991). The Three-Dimensional Structure of Canine Parvovirus and its Functional Implications. *Science, 251*, 1456–1464.

Turing, A. M. (1936). On Computable Numbers, with Application to the Entscheidungsproblem. *Proceedings of the London Mathematical Society*, ser. 2, 42, 230–265.

Turing, A. M. (1963). Computing Machinery and Intelligence. In E. A. Feigenbaum and J. Feldman (Eds.), *Computers and Thought*. New York: McGraw-Hill.

Vallery-Radot, R. (1960). *The Life of Pasteur*. New York: Dover.

Vander, A. J., J. H. Sherman, and D. S. Luciano. (1986). *Human Physiology*. New York: McGraw-Hill.

Volloch, V. B. Schweitzer, and S. Rits. (1990). Uncoupling of the Synthesis of Edited and Unedited COIII RNA in *Trypanosoma Brucei*. *Nature, 343*, 482–485.

Warner, J. H. (1991). Ideals of Science and Their Discontents in Late Nineteenth-Century American Medicine. *Isis, 82*, 452–478.

Watson, J. D. (1968). *The Double Helix*. New York: Atheneum.

Watson, J. D., and F. H. C. Crick. (1953). Molecular Structure of Nucleic Acids: A Structure for Deoxyribose Nucleic Acid. *Nature, 171*, 737–739.

Weissman, G. (1985). *The Woods Hole Cantata: Essays on Science and Society*. Boston: Houghton Mifflin.

4

Expanding the Earth

Seeing and Naming the Skies—
The Case of the Moon

♫♫♫♫♫♫♫♫♫♫

*Since man, fragment of the universe, is governed by the
same laws that preside over the heavens, it is by no
means absurd to search there above for the themes of
our lives, for those frigid sympathies that participate in
our achievements as well as our blunderings.*

—M. YOURCENAR, *Memoirs of Hadrien, 1958*

If it be one of the great powers of science to give the universe a voice,
then the act of naming must be counted among its most central efforts.
Naming is the means by which human beings have always given an identity to things, to themselves, to the world and everything in it. Through
names, people have reached out to seize, order, and command the cosmos, long before the advent of modern science. Since that advent, the
act of naming has taken on new force, having become critical to the
transformation of material reality into words, and from there, into de-

196

livery for other forms of control. To name a thing is to do more than merely give it birth within the realm of literary perception, oral or written. It is to create an object for study.

Names probably comprise as much as 50% or more of the total technical vocabulary of today, the voice of the scientific. One thinks of species, chemicals, minerals, particles, forces, stars, the elements of life, and so on. The materials and substances of science—both physical and conceptual—all exist through their titles. Such names, moreover, are often anything but static, mere labels or tags. They are dense with active and intended meaning. The great nomenclatural systems of Lavoisier and Linnaeus, for example, to which chemistry and biology are still largely loyal, were specifically devised such that each single name would be capable of containing the essential scientific truth attributed to its referent, whether this meaning involved a type of chemical action or a kind of biological uniqueness. Lavoisier created the name *oxy-gene,* "bitter-maker," to denote the gas with which other elements combined to form acids; Linnaeus used the title *Triandria monogynia* as a shorthand way of describing hermaphroditic plants with three stamens (*tri,* "three" + *andro,* "male") and one pistil (*mono,* "one" + *gyn,* "female"). Both naming systems were meant to embody "observation." Their importance has only grown more pressing: today, no less than a century ago, it is common to encounter eminent researchers declaring, "When you ask a biologist: What is in a name? The answer is: Almost everything" (Younes, 1991).

The problem of naming cannot be divorced from problems of knowledge. But the act of naming also cannot be separated from questions of culture. Names are arbitrary: they are not innate in any object but are decided upon, by *some* person or group at *some* moment in time. Every name represents a distinct choice. Moreover, it is an embodiment of the many facets of human reality that go into making such a choice, for example, the influences—historical and contemporary—and the ambitions and loyalties—private and institutional—that have brought a particular discipline and a particular individual to that particular moment when a coinage must be made. Names are linguistic entities that implant culture, in the larger sense, into the depths of science, in the restricted sense. This is all the more true when the giving of a name is tantamount to claiming and possessing a new region for investigation.

We can take this last phrase at face value. Astronomy, for example, has long been in the business of naming and mapping the universe, of filling it up with "place." When it comes to the planets, especially, this idea of "place" is very concrete. Though planetary titles were inherited by Europeans via Latin culture, this did not make inevitable the names that eventually came to be used for geographical features on the surfaces of these alternate worlds. That the Moon, for example—the first to be named (at a time when it was still known by the Latin "Luna") and the one to set the pattern thereafter—should become a domain over which the ghosts of famous scientists, philosophers, and saints would one day wander, was an outcome that had to be established over claims to priority by other, very different naming systems put forward at exactly the same historical moment. Each scheme reflected something different yet essential to its time, something of the nature of science but of the nature of its larger context too. The victory of one system over the others, then, is similarly reflective, yet also resulted in a particular inheritance, by which one speaks and studies today a "Moon" created some 350 years ago.

Before this could happen, moreover, before this entity, the Moon, could be named and mapped, it first had to be transformed into a particular kind of image. The story of this transformation, by which the Moon went from being a perfect celestial (Aristotelean) sphere to an object of observation, and then to a surface of true geographic imagining, is very much part of the overall history of giving "place" to the lunar orb. As on Earth, one finds that mapping and naming were entirely covalent. Nor is it a coincidence that the lunar surface was first named during the same historical era when "new worlds" were being "discovered" on Earth as well. Indeed, the mapping of the Moon, which began even before Galileo turned his telescope heavenward in 1610, sought to expand something specific on the Earth—something of urban, 17th-century Europe—upward into the heavens. No less than the New World of North and South America, the Moon proved to be contested territory: the naming systems applied to it by astronomers, like those of colonial explorers, reflect different acts of attempted possession, hopes and ambitions that melded together religious, political, scientific, and personal canonizations. They bring to the eye of the present an example of how the objects of scientific study are stamped deeply with cultural meanings, and finally how such meanings, once frozen, become intrinsic to "science" thereafter.

How the Moon Began

In this chapter, I want to provide a brief survey of these early attempts at drawing, mapping, and naming the Moon. The full story, the human tale of the individuals involved, is invoked, but, despite its fascinating and tragic elements, will have to await full discussion in another writing. My goal instead is more modest: to sketch the outlines of how the Moon came to be seen, pictured, and possessed by Europe at a particular point in its history.

To begin such a sketch, one needs to first ask how the Moon itself began. What views of it existed prior to the era of the telescope? The lunar disc, after all, is the only celestial surface some of whose features are readily visible to the naked eye. It seems surprising, therefore, that no drawings or maps were ever made of it until the Renaissance. A full 1,300 years earlier, the Greek mathematician Ptolemy had apparently drawn maps of the known world, devised a system of coordinates, and carried on an already venerable tradition of charting the stars and building celestial globes to display them. But none of this extraordinary effort at representing the Earth and the heavens ever included the Moon.

In ancient Greece and Rome, several concepts, often confused, dictated people's views of the lunar orb. As outlined by Whitaker, these included seeing it as: "(a) a mirror, reflecting the terrestrial oceans and continents; (b) a polished, translucent crystalline sphere; (c) a body of condensed fire, etc.; and (d) a terrestrial type of spherical body with seas, mountains, valleys, plains, etc." (1989, p. 122). Arguments for one or another view revolved around questions of whether there existed lunar and sublunar substances and, more often, what type of light the Moon produced. Such questions, also turned on whether or not the Moon was a "corrupted" or more "purified" realm, a place where souls were born, where they retired to, or where they were exiled for eternity. None of these theories, however, conceived the Moon as a mappable surface. The only theory that might possibly have done so, the terrestrial view— perhaps the oldest idea of all, subscribed to by Pythagoras, Plato (in the *Phaedo*), Heraclitus, and Plutarch—held that the lunar globe was a kind of "higher" Earth whose features could only be imagined, a reality to be discussed textually, not described pictorially. This is best exemplified in one of the most extended discussions given the subject in classical literature, Plutarch's dialogue *On the Face which Appears in the Orb of the Moon*, where an attempt is made to disprove the mirror theory by means of optical and geometrical logic (Mathematics in general was the medium by which ancient astronomers discussed the stars and planets). One of the speakers, Lucius, states that "the moon is very uneven and rugged, with the result that the rays . . . [are] coming to us, as it were, from many mirrors" (Cherniss and Helmbold, 1957, p. 111). Later in the dialogue, the narrator (presumably Plutarch himself) goes still further than this, noting that during an eclipse changes of color occur along the length of the advancing shadow, leading one to surmise that

> It is likely . . . that the moon has not a single plane surface like the
> sea but closely resembles in constitution the earth. . . . It is in fact
> not incredible or wonderful that the moon, if she has nothing cor-
> rupted or slimy in her but garners pure light from heaven . . . has
> got open regions of marvellous beauty and mountains flaming bright
> and has zones of royal purple with gold and silver . . . bursting forth
> in abundance on the plains or openly visible on the smooth heights.
> (p. 141)

The description is vivid, the logic, stemming from actual observation, impeccable. Yet the whole remains conjectural, a matter of imagination, of words. At this stage in the history of the Moon (in the West), text does not call upon image. Drawings that portrayed the planets, such as those of Ptolemy or Aristotle, depicted them as mathematical points or concentric spheres. The purpose, like that of poets or philosophers, was "to tell the stars," as Aratus says in his *Phaenomena* (another work of great influence). While Plutarch's vivid description, complete with the logic behind it, would return almost in its exact form a full 1,600 years later, in a striking reincarnation by none other than Galileo who finally brought text and image together, this span of time indicates all the more, the gulf of vision that had to be crossed.

Plutarch is our historical marker that the lunar face would one day be literally seen as a rugged, earthly landscape, that its dark and light regions would become great oceans and land masses. The tradition he represents, however, though hallowed in his own day, did not remain so in the Latin Middle Ages. It was forced underground, especially by the ascent of Aristotelean physics and cosmology, which became intellectual doctrine following the rise of university learning in the 13th century. For reasons having to do with certain compatibilities regarding the Christian view of the universe, the interpretations of Aristotle provided by the Muslim Spanish astronomer Ibn Rushid (Averroes) gained authority over all other systems of thought and formed the basis for the Scholastic tradition in medieval science rehearsed up through the Renaissance. Aristotle's cosmology was based on the idea of the planets as a series of concentric crystalline spheres. Regarding the Moon itself, Averroes simplified Aristotle's view by interpreting him as saying that, because the Moon (like the other planets) was perfect, no real parallel or kinship existed between it and the Earth: the dark and light "spots" (*maculae*) either mirrored those of the terrestrial surface or represented different condensations of some primal lunar matter (Ariew, 1984). In the words of one influential medieval Aristotelean, the 14th-century philosopher Albert of Saxony, "The moon is simple in substance, [which] would not prevent it from exhibiting differences in density and rarity between its various parts" (Ariew, 1984, p. 221). In short, the lunar surface is homogeneous, smooth, and of no real interest.

This theory of the Moon, which was dominant up to roughly the 15th century, suggests why, during all this time, no one seems to have attempted to draw, sketch, or paint a version of the Moon's surface, as an object of visibility. The Moon, of course, was an important entity to medieval society, having a role in time keeping, weather predictions, fertility rituals, agriculture, and so on. Though often represented, it was typically figured either allegorically, as a face or a human figure (e.g., the goddess Luna, scepter in hand), or else mathematically, as a solid circle, a series of phases, or the like. Such images were the rule in antiquity; early Christian science inherited this tradition. Episodes of intellectual bloom, such as the Carolingian and 12th-century "renaissances," both of which brought "strikingly before the eye manuscripts in the natural sciences with [their] philosophical–astronomical images of the world, the constellations, and the planets" (von Euw, 1993, p. 266), made absolutely no difference in this tradition. In fact, by adding new aesthetic depth to older forms of visual conception, they set new standards for their continuance. As late as 1400, no images of the Moon itself, as it presents itself to the eye, existed in Western culture.[1]

Again, this truth seems less a matter of stunted cognition, of seeing per se, than of drawing as a cultural motive. Naturalistic portrayal of the Moon was not absent because ability went lacking (indeed, what would it have required in terms of technique to draw the Moon with spots on its surface?). It went undeveloped because the medieval world was full of developed meanings that argued against such a portrayal, that demanded the prestige of "inner" over "outer," spirit over flesh, symbol over (or rather *as*) form. The Moon was not drawn or painted up to a certain time because there was no *reason* to draw it or to paint it. The types of meaning embodied in it, whether religious or cosmologic, were not of a type to demand such accurate observation. Indeed, what existed with regard to the heavens, even up to the time of the early Renaissance, was "a qualitative science expounding a hierarchy of essences," as Lynn White once described it (1947, p. 422). White's discussion of growing naturalism in the Gothic period, and the more extended treatment given this topic by Goetz (1937), supports the idea that any interest in naturalistic representation was devoted to such things as flowers and herbs, the human face and figure and clothing, animals, buildings, weaponry, items of daily usage, but not to the heavens, or, for that matter, to such things as mountains, rivers, and rocks. It seems to have been the things of proximal, immediate experience—an experience imbued with feudal (agricultural) realities—that drew the eye and hand of artists. The rest, the inorganic universe, remained distant, massive with Scholastic essence. The Moon was one of the celestial spheres, an undifferentiated level within the divine hierarchy of the heavens.

Earliest Drawings of the Moon in Western Culture

This changed, very suddenly, in the early 15th century. Indeed, so suddenly did it change that the achievement which brought it remained unique and unexploited for a century thereafter.

The lunar face, and a great part of the Earth's surface as well, finally became part of Western representation thanks to a single artist working in Flanders between 1400 and 1440. He was Jan Van Eyck (1385?–1441), the early Flemish master who was instrumental in perfecting the use of oil paints to convey the subtle effects of natural appearance and whose command of realistic technique, so striking today, so penetrating in absolute detail, was no less legendary in his own time. Van Eyck's significance in the history of art remains a source of endless controversy (see, e.g., Harbison, 1989), but his importance with regard to "observation"—in particular, recording the Moon as an object of naturalistic perception—can hardly be debated.

Yet, this aspect of Van Eyck's work has gone unnoticed until now. Scholars in the history of astronomy have attributed the first naked-eye drawings of the Moon to Leonardo da Vinci, specifically to three sketches that appear in his notebooks (Reaves and Pedretti, 1987).[2] My own study, however, shows that this priority should be overturned (Montgomery, 1994). In addition, Leonardo's drawings are all full-moon images, two of which seem quite rough and preliminary (the third, however, is more well observed). Van Eyck, by contrast, carefully painted three different phases of the lunar cycle—all in daylight—thereby indicating his distinct interest in its visible forms *and* changes. The works in which the Moon occurs are the *Crucifixion* (c. 1420–1425), *St. Barbara* (1437), and the *Knights of Christ* panel (lower left) of the famous *Ghent Altarpiece* (1426–1432). The first of these, actually a diptych including as its right panel *The Last Judgment*, displays the Moon most clearly of all, and it is this image that I will concentrate on here (Figure 1). In *St. Barbara*, the planet appears as a thin crescent in the upper right corner (only a 1- or 2-day-old Moon), yet with the full orb plainly shown in

FIGURE 1. The *Crucifixion* (1420–1425) by Jan Van Eyck, showing the first naturalistic drawing of the Moon in Western culture. Reprinted by permission of the Metropolitan Museum of Art, Fletcher Funds, 1933. (33.92a)

outline, the dark portion being faintly lit. Done in grisaille on wood, the work is monochromatic and resembles a drawing more than an actual painting, but it is highly significant from a scientific standpoint, for it provides the first naturalistic rendering in Western culture of earthshine.[3] In the *Ghent Altarpiece*, meanwhile, the full Moon is placed immediately below a precisely executed cloud formation, showing several surface features that are a bit obscured by the general haziness of the sky.

It is in the *Crucifixion* that the lunar surface hangs most strikingly before our eyes. Van Eyck placed it in the upper right-hand corner, just below an arm of the cross to which one of the two thieves is bound (Figure 2). It is a gibbous (three-quarter) Moon, chalky in color, and on it the major lunar maria are all shown as they would appear in the morning hours (the biblical account has Christ being crucified prior to "the ninth hour"). It is small, the same size as the heads of the people; it rests by itself, alone in one portion of the sky, as if on a slide, just above a group of snowcapped mountains and below a fluffy bank of cumulus cloud (in fact, the sky is populated with all manner of precisely rendered cloud formations). The mountains (showing signs of glaciation) lead forward to a broad, meandering river, then a castle atop a craggy knob (probably meant to be Jerusalem), rendered with Van Eyck's legendary detailing, then to the foreground scene, inhabited by all manner of onlookers, each one dressed in an entirely different set of apparel and holding a separate pose. In the lower foreground weeps the Madonna, comforted by a small group, and beneath their feet is an outcropping of rock, whose solid texture and solution pits make it plain to the trained observer that the rock type is limestone.

The Moon itself, therefore, is a small part of a much larger canvas of figures and objects, assembled almost like a museum of collected observations. It is a single result among a multitude of careful, direct visual studies, which include many aspects of the inorganic world. Indeed, in Van Eyck, one finds the full extension of gothic naturalism into this broader world, whose details are offered with a naked realism that far surpasses anything achieved by the whole of the Italian Renaissance (for whom such things as rocks, rivers, mountains, etc., remained props). One true measure of this naturalism, is the very high degree to which Van Eyck's world can be investigated by contemporary scientific perception and methods. His strata and landforms, that is, permit geologic analysis; his clouds, meteorologic discussion; his topography, geomorphic dissection; his Moon, astronomical inquiry. The irony here, that Van Eyck superseded the intellectual burdens of his own time, that his eye saw and his hand recorded things overlooked by his contemporaries, raises the question of his motives.

FIGURE 2. Detail from Figure 1, showing close-up of the Moon. The total width of the above detail is roughly three inches. Reprinted by permission of the Metropolitan Museum of Art, Fletcher Fund, 1933. (33.92a) .

These, however, must unfortunately remain the subject of specu-lation. Scant biographical information and the lack of any personal writings or notebooks have long ensured this must be the case. Even Van Eyck's own apparent motto, with which he signed a number of his works, *Als Ich Can* (As I Can) remains enigmatic, a phrase variously interpreted as both humble and arrogant. It is possible that the painter was practicing his art as a special type of religious concentration, based on a belief in the material world as a kind of second Bible, a scroll of God's truth written in the details of Creation, with every surface worthy of attention for having attracted the care and effort of the Creator. It is possible too, that, as a court painter, living in the wealthiest trading center in all of Europe, a place filled with exotic new things, that Van Eyck painted the visible universe as a kind of territory of the eye, a realm of unlimited property and human ownership. By this time, after all, the immediacy of agricultural–feudal experience, so central to gothic art, had long departed from the lives of ambitious artists. Urbanism, travel, war, and the rapid rise of the new principalities and nation-states had done much to impose a different materiality on the concept of "land" in European culture—mountains and rivers, for example, were now critical boundaries denoting possession—all of whose components now became, more than ever before, the signs (not symbols) of earthly power.

In the end, the latter interpretative frame appears more rich with possibilities. Viewed through the lens of "observation," Van Eyck's achievement appears as a kind of historical talisman. He reveals, that is, two crucial things: first, that the world of inorganic objects was finally coming into its own in Western representation; and second, that the pictorial image in Western culture was about to evolve from a narrative role (telling stories, visually manifesting the written word) to an epistemological one, where it became a central means of inscribing and conveying knowledge. Erwin Panofsky's well-known comment that "Jan Van Eyck's eye operates as a microscope and as a telescope at the same time" (1953, p. 182) should be reconsidered in this light. Indeed, it can hardly be a coincidence that this eye and these instruments issued from the same locale of Europe. It is as if the painter had the power to offer us, *ante facto*, a vision of the transformational capabilities that human sight, documented and guided by an accurate hand, would one day come to have. With regard to the Moon in particular, Van Eyck's objectification of it was not repeated until a century later, when Leonardo da Vinci produced his own versions. Though his images were never published in his lifetime, Leonardo took a step beyond Van Eyck by placing the lunar face on the sketching table, thus showing that it could be abstracted from the sky, taken down and changed into "evidence." This was a long distance indeed from the textual argument still raging at the time, regarding the pure or impure, luminous or reflective "substance" of the Moon. Observation had leapt well ahead of conception, and the Moon of Scholastic astronomy had begun to set.

First Map of the Moon:
William Gilbert and the Politics of Naming

The next great step was no less momentous. Sometime during the 1590s, the English physician and student of magnetism William Gilbert (1540–1603) assembled a number of his cosmological ideas into written form, and included with them what can only be called the first true map of the Moon (see Figure 3). This image, like Leonardo's, never appeared in its author's lifetime. Gilbert was known in his day as the discoverer of terrestrial magnetism, which he set forth in *De magnete*, published in 1600. When he died soon thereafter, his papers were collected by his younger half-brother under the title *De mundo nostro sublunari philosophia nova* (New philosophy of our sublunary earth), presented to Prince Henry, and thereafter circulated among other scientists, such as Francis Bacon and Thomas Harriot (Kelley, 1965; the *De mundo* was later published in 1651, in Amsterdam). Scholars are generally divided with regard to calling Gilbert's image a "map" (cf., e.g, Whitaker, 1989, and Strobell and Masursky, 1990, with Kopal and Carder, 1974). But a map it surely is, and a very interesting one.[4] Not satisfied to merely draw the dark regions or "spots" of the Moon, Gilbert superimposed a grid over the entire image. This was probably done for the sake of correctly placing the various regions, but it reveals that Gilbert both drew and plotted these areas at one and the same time (the grid resembles latitude and longitude lines). Then Gilbert did something even more utterly unprecedented: he named the darker regions—indeed, he even named portions of them—using earthly geographic forms.

Gilbert's actual concept of the planets was complicated. Suzanne Kelly has described it as follows: "The planets were solid, light-reflecting globes, closer to the Earth than the stars, not connected to any [Aristotelean] sphere, surrounded by their effluvia, and moving from the 'impulse' given them at the time of their creation" (1965, pp. 36–37). Each planet had its own unique quality: the Moon, for example, lay within the "orb of virtue" (a term more or less equivalent to magnetic field) of the Earth, acted as its "companion," and could be described as a miniature version of the latter. Siding with those such as Plutarch, Gilbert saw a distribution of land and ocean on the Moon's surface as the explanation for the variations in brightness, rather than the then-current notion of opaque and transparent areas. He viewed the dark areas, however, as land and the bright areas as oceans (this made sense within the paradigm that the Moon shone by reflected light only, water being known as more highly reflecting). It is thus with Gilbert that the ancient belief in a mimesis between Moon and Earth finally becomes formalized at the level of pictorial display and nomenclature. Indeed, Gilbert himself was at least partly aware of this. He actually bemoaned the fact that no one before him had done this, especially in antiquity, since the lack of any older maps prevented him from discerning any changes that might have occurred in the Moon's face since classical times.

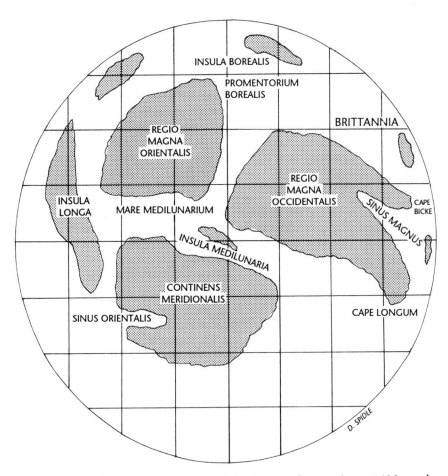

FIGURE 3. William Gilbert's map of the Moon, drawn about 1600 on the basis of naked-eye observations. This is the first such image known. The original manuscript version is unsuitable for reproduction. Grid lines exist on the original. Adapted from Whitaker (1989, p. 121).

The names Gilbert chose for the Moon's features were mainly descriptive. Each title was meant to contain or reveal the location of its referent, and thus the rationality for its choice. "Regio Magna Occidentalis" indicated the Great Eastern Continent, one of whose salient portions was "Cape Longum," another "Sinus Magnus" (Great Bay). At the upper edge of the map, there occurs "Insula Borealis" (Northern Island), and between the three continents shown, the "Mare Medilunarium" (Middle Lunar Sea). This approach to naming appears simple enough, indeed, even "primitive." Yet one should take note of a certain effect stemming from its style. By giving each feature a descriptive title, this approach creates "place" as a textual phenomenon. Geography is here an inscription of phrases, a code of placement more than of observation per se.

In so doing, Gilbert was mainly following modern tradition in European mapmaking for the Earth. This tradition had begun just over a century earlier with the first illustrated editions of Ptolemy's *Geographia*.[5] It is on these early maps that terms such as "sinus," "regio," "mare," "insula," "borealis," and so on were established as generic titles, and it is here as well that different portions of large areas were given locational descriptors: "India intra/India extra," "Java major/Java minor," "Magnus Sinus," and so on. By Gilbert's day, of course, cartography had developed enormously beyond the rudimentary maps created by Ptolemy to illustrate his text. Yet such descriptors were still very much in use for various portions of the globe, especially more distant ones (meaning "distant" from Europe).

The era of colonialism had elevated the craft of mapmaking to a very high level of importance, and the status and distribution of maps, as objects of documentation, of study, and of private possession, had spread to every major capital of Europe. In Elizabethan England, for example, maps were highly prized as curiosities, room decor, works of art, and sources of knowledge. The map's increasing status was in part the result of a profusion of travel literature, which bloomed from the 1540s onward, culminating in England with Richard Hakluyt's eloquent and widely praised *Principall Navigations, Voyages, and Discoveries of the English Nation* (first issued in 1589, later enlarged into three volumes in 1598–1600), known in its day as the "epic of the English nation," and one of the most widely read and influential texts during the English Renaissance (Shakespeare drew some of his knowledge and scenery from it). Hakluyt is credited with introducing a great deal of up-to-date late Renaissance geography into England, in part through his role as a scholar during his 16 years at Oxford, but more significantly through his writings and his propaganda for English settlement in the

New World. One of his major purposes in writing the *Principall Navigations* was to make known the "facts" that the English had always been a great seafaring people, that the voyages of the Anglo-Saxons centuries before had been among the great accomplishments of the age, and that exploration and adventure were "in the bloode" of the English people. Hakluyt was attempting to create a public memory of past heroisms that could be reclaimed in the present. His work was part of an important movement in late 16th century letters, sometimes called "antiquarian," but more accurately described as an effort to define, celebrate, and revitalize the endemic virtues of the English, in the face of growing hostility from Catholic Europe. Spenser and Shakespeare were both part of this movement, though in different ways. Hakluyt, meanwhile, along with other influential writers on geography such as John Speed (see Livingstone, 1992), saw England's survival and future greatness as both being tied to increasing its population and expanding its territory, especially in North America, where Spain had not yet gotten a firm hold. Spanish settlement, meanwhile, had proven the riches to be culled from exploration; such wealth was making Spain enormously powerful and increasing its ability to impose its will upon other nations. Spain was England's enemy, the great barrier to England fulfilling its "true destiny." Indeed, the truth of this destiny was soon proved in 1588, in the great battle with the Armada. English victory had been magnificent, breath-taking, foretold; but it also had to be secured if Spain was not to rise again to its former glory. Hakluyt was instrumental in persuading Sir Walter Raleigh to devote his energy and resources to founding the first British colony in Virginia in 1584–1586. Often noted is that the name "Virginia" (probably coined by Raleigh himself) was applied to the whole of the new territory and had a political–moral dimension, implying the superior virtue of England's chaste queen. The same, of course, was true of "James Towne," founded a few decades later (1607), though without the suggestive link between a new and a "virgin territory."

All of this was at work behind a single designation Gilbert chose for his map: "Brittania." William Gilbert, I might note, was royal physician first to the British navy, then to Queen Elizabeth herself, and finally to James I. He no doubt admired Hakluyt as a writer and a scientist, and may well have known him personally. In the heady days of the 1580s, as tensions gathered against Spain, it would have been difficult, if not impossible, to locate anyone close to the Crown who was unroused by the avid patriotism, vision, and ambition of this geographer magnus of the British nation. Moreover, it is evident that Gilbert himself had frequent contact with maps; his interest in magnetism led him directly into concerns with navigation, geography, astronomy, and cosmology. He possessed a large library and, a collection of celestial and earth globes (these, along with other possessions, were left to the Royal College of Physicians' library, tragically destroyed by the Great Fire of 1666). It can hardly be doubted that Gilbert was familiar with the geographic knowledge—and the geographic politics—of his time.

"Brittania" was an ancient Roman title for England, given to a land of distant and difficult conquest. Before the 16th century, the name "Brittania" was used in an historical sense only, but under Henry VIII and Edward VI, when efforts were made to annex Scotland, it came back into political usage as a kind of emblem for English manifest destiny. William Camden's popular work *Britannia, A Chorographicall Description of the Most Flourishing Kingdomes of England, Scotland, and Ireland, and the Ilands Adioyning, Out of the Depth of Antiquitie* appeared in 1586, in the same turbulent decade as Hakluyt's book, and also promoted the concept of geography-as-destiny.[6] It was this merger of political, military, geographical, and literary trends that Gilbert codified in using "Brittania" for one of the island portions of the Moon. England had beaten Spain; it would soon absorb the lands surrounding itself and expand deeply into the New World—did it not, therefore, have a "deserved" place on a new world of another kind, a world otherwise uncolonized, untouched, still in its "virgin" descriptive state? "Brittania" planted a claim on the Moon and implied, historically, that England, the new naval power of Europe, might one day send ships of a different kind to these distant seas and lands.

One other name coined by Gilbert deserves mention. The name "Cape Bicke," whose referent I have been unable to determine, nonetheless seems coined after a person or place of importance to the author. Like "Virginia," it signifies directly the power of the colonial "discoverer" to canonize those in his favor in the form of a textual space that then becomes transformed into universal, geographic space. This power, of course, had a still older vintage in European geography, beginning at least 300 years earlier with the maritime trade empires of the Italian city-states and with Portuguese exploration of Africa. In total, Gilbert's onomastic scheme for the Moon indicates that a blending of traditions had come to be attached to the idea of a "new world," traditions that included the classical, the contemporary political, and the colonial. All of these, in the end, had the capability of coming together as an expression of the cultural sensibilities of the time.

Finally, Gilbert's reason for drawing the Moon was different from that of Van Eyck or Leonardo. By his day, it had become common to include in scientific books many naturalistic drawings, artistic decorations, visual aids, and accents of all kinds, both for the instruction and the entertainment of the reader. Indeed, Gilbert himself is exemplary in this regard: his *De magnete* contains a large number of excellent illustrations included to demonstrate his points and experiments. By this time, textuality was no longer sufficient; images now carried a weight of demonstration and evidence. In a certain meaning, his map was itself a type of visual experiment, an attempt to "demonstrate," through inscription, his conclusion that the bright areas of the Moon's surface were water, the dark areas land, and the whole a true territory that might one day be England's.

Harriot's Drawings: The Mathematical Tradition

Gilbert's map reveals that an interest combining Van Eyck's observation with Leonardo's isolating "study" had come into being by this time. The Moon was now in the hands of science, but also in those of geography, and therefore of political traditions. Gilbert had viewed it primarily in a chorographic manner: his names are correlative with the earliest phase of colonial exploration.

Another tradition of image making, one that Gilbert's map did not draw upon, was related to the mathematical–computistic view of the heavens. This tradition represented the planets either as solid spheres or as points in fixed spherical orbits. The relevant drawings were in the form of diagrams, not pictures, and commonly included tables designating positions, times, eclipses, and the like. These images could be highly abstract, purely geometrical, or else mixed with allegorical figures (see von Euw, 1993), but they were never naturalistic. As a result of the 12th-century era of translation, when many Greek (including Euclid) and Arabic works of science were translated into Latin, these diagrams began to employ systems of description that we recognize today as mathematical, for example, the use of letters or numbers to designate various points in space.

An attempt to marry this tradition of astronomical diagrams with naturalistic portrayals of the Moon was made by the English mathematician, cartographer, and astronomer Thomas Harriot (1560–1621), considered by many to be the foremost scientist of Elizabethan England. Like other brilliant men of his day, Harriot put his hand to many areas of intellectual endeavor. He is well known as the close friend and technical adviser of Sir Walter Raleigh, who, under Hakluyt's inspiration, sent the young Harriot to Virginia on an early colonial voyage in 1585. Together with John White, artist and mapmaker, Harriot planned to produce an extensive, highly detailed account of everything he saw on this voyage, a kind of experience-based encyclopedia of the New World. Ultimately, all that resulted was a 30-page pamphlet, the *Briefe and True Report of the New Found Land of Virginia* (1588), with Harriot's text and a scattering of White's remarkable drawings and maps, the latter of which depict in striking detail the Native Americans they met, as well as flowers, fish, crabs, turtles, insects, and a host of other flora and fauna.[7] Harriot apparently did no drawings himself, but wrote the literary account and worked on the navigational aspects of the journey.

Harriot appears to have been a reserved man, a "perfectionist," hesitant to publish his work, which was very extensive in several branches of mathematics (see, e.g., Lohne, 1973; Pepper, 1974; Shirley, 1983). The *Briefe and True Report*, in fact, was the only text of his ever to appear in published form. In this text, Harriot is generally descriptive, concise, as trim in language as is White in image, though without the latter's detail. Harriot took care to learn something of the Algonquin language and to apply Native American names to everything that was foreign to him—indeed, a fair portion of the work takes the form of a sort of botanical glossary. He was conscious, that is, of several geographies of the New World he was exploring and he set these down in textual form. The *Briefe Report* went through many editions in only a few years and was translated into Latin, German, and French; later it was reprinted in volume three of Hakluyt's *Principall Navigations* (1598–1600). Its popularity reveals the public interest in the "new" geography and the exotic. Indeed, the account of his visit begins in a prophetic way with regard to his later "voyage" to a different type of New World: "I have therefore thought it good, being one that have been in the discovery . . . to impart so much unto you [that] by the view hereof you may learn what the country is, and thereupon consider how your dealing therein may return you profit and gain," (Rowse, 1986, pp. 108–109). It is interesting to think of what might have resulted had Harriot, then in his 20s, viewed the Moon through his telescope at this point in his career. Rich with visual experience, with the conjunction between maps and actual territories, indeed with the power of imagery and words to capture the "distant" and "foreign," what might he (and possibly White) have made of the Moon, seen in magnified form? We will never know, of course. Following this expedition, Harriot retreated into more theoreti-

cal work on mathematical navigation. In the 1590s, we find him working directly with Hakluyt on the technical construction of maps, on new
methods for determining and plotting position via techniques he himself
had devised (see Pepper, 1974). From this point on, Harriot acted as a
technical adviser only, and the world of his visual use was mainly restricted to flat mathematical diagrams.

It was a decade later when Harriot became one of the first learned
men in England to get hold of a new invention, the telescope, which had
emerged from Holland probably sometime in 1608–1609 and was circu-
lating among country fairs and carnivals in several parts of Europe where
it was being shown to a spellbound public (Van Helden, 1989a). Harriot's
own interest in the telescope must have been instantaneous, since he had
been working on various problems in optics for some years. On July 26,
1609, he became the first person to produce an image of the Moon on
the basis of telescopic observations, preceding Galileo by only 4 months.
Harriot apparently used a 6-power telescope, and he drew a 5-day-old
Moon (Whitaker, 1989). As Figure 4 shows, this image is a rather crude
rendering of the crescent phase, with an irregular terminator (shadow
line) and several of the maria roughly darkened in (the Mare Crisium,
the Mare Fecunditatis, the eastern portion of the Mare Tranquilitatis,
the Mare Nectaris). It is more a sketch than a drawing and is completely
without names or designations of any kind. Moreover, it lacks any commentary or written observations. It would appear that Harriot did not
consider this drawing to be very important: he did not spend much time
on it, did not expand it textually, and did not draw any other such images in the days, weeks, or months immediately following. The instrument may well have been of poor quality, but even if it were as good as
or better than the one used by Galileo, it seems unlikely that Harriot
would have produced a more naturalistic version. It seems clear, in other
words, that the image—as image—did not interest him very much.

Harriot's drawing is extremely significant, however, and not simply for its historical priority. It shows that the classical world of textual
imagination has been overcome, for now the impulse to visually record
what the Moon looks like was far stronger than any desire to describe it
in words. The problem for Harriot was that he simply was not equal to
the task. Here was the chance for Van Eyck's eye to prove its legacy; yet,
in this case, the talent was lacking. It had to wait several months for
another embodiment. A year later, after having received one of the first
copies in England of Galileo's *Sidereus nuncius* (Messenger from the stars,
published in March 1610), complete with its drawings of the lunar surface, Harriot was inspired to produce new images himself, this time with

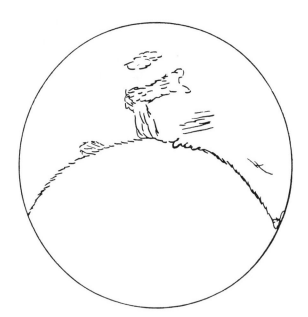

FIGURE 4. Thomas Harriot's earliest drawing of the Moon, dated July 26, 1609, as seen through a six-power telescope. This is the first such drawing known, preceding Galileo's by a mere four months. Adapted from Whitaker (1989, p. 122).

actual topographic features, obviously in direct imitation of his Italian colleague (Bloom, 1978).[8] It is evident that Harriot's interest was re-kindled by Galileo, for in 1610 he made no fewer than four drawings of the Moon in July, seven in August, and nine in September (Shirley, 1978). Those drawings made in July, like that of the year before, include no commentary. But in August, observations finally begin to appear: on August 21, "nothing notable"; on the 22d, "inaequalities scarce sensible . . . a little ragged with a peninsula"; on the 26th, "some partes having greater eminences & some valleyes with shadowes"; September 11, "The appearance was notable, ragged in many places . . . with some ilands and pronomontoryes . . . "; and by October 23, "it shewed mountenous . . . that also was montaynous with an opening in the middest & some black passage from it . . ." (quoted in Shirley, 1978, p. 303).

What one sees here are words and images catching up with a perception in search of a discourse. The progress of this commentary reveals that Harriot was discovering an ability to see the Moon directly, sensorily, in terms of established languages of textual and visual inscription, both of which had been supplied by Galileo. Indeed, this discovery is something he seems to achieve not all at once, but in stages, beginning mainly with oceanic forms (peninsula, islands, etc.) and finally progressing to land proper. At first Harriot seems to be adrift, but then he gradually gains the ability to see "place" on a planetary scale, not as a cartographic reality, but as a matter of actual visible things—islands, seas, valleys, and mountains, things that jostled each other in their dense proximity, and that called directly on their earthly titles. This seems all the more true in that Harriot felt the urge to name only one feature, "the Caspian," which apparently corresponded to the Mare Crisium, the most isolated, "landlocked," of the lunar maria (Shirley, 1978, p. 303). His literal transfer here was very different from that of Gilbert, who projected only a descriptive earthly geography onto the Moon's surface. Harriot, in fact, may have been one of the few men who had seen Gilbert's *De mundo* in its earliest form, possibly even before 1609 (see Shirley, 1983, p. 387). Perhaps because of this, he had been ready at some level to go further and to "see" a literalized version of terrestrial forms on the Moon. This was no mean feat, however, and Harriot's apparent struggle to do so, after much time, study, and even outside example, can be taken as a sign that comprehending the Moon in this manner meant a local change in the capacity of human perception.

We have other evidence of this. Letters written to Harriot by his one-time student and later scientific colleague, Sir William Lower, reveal a fascinating before-and-after account of "seeing" the Moon in the historical umbra of Galileo. Harriot sent his first telescope to Lower late in 1609 or early 1610, advising him to pursue his own observations. In a letter dated February 6, 1610, Lower writes back to thank Harriot for "the perspective cylinder":

> According as you wished I have observed the moone in all his changes. . . . [Near] the brimme of the gibbous part towards the upper corner appeare luminous parts like starres, much brighter than the rest, and the whole brimme along lookes like unto the description of coasts, in the dutch bookes of voyages. In the full she appeares like a tarte that my cooke made me the last weeke. Here a vaine of bright stuff, and there of darke, and so confusedlie al over. (quoted in Whitaker, 1989, p. 120)

Lower is groping here, seriously and playfully, to find an apt description. Words fail him; Harriot's own drawing (which he had also sent) provides little help. Stars, coasts, a "tarte," a confusion of light and dark: Lower is trying to make sense of what he sees and can only produce a surplus of images, a narrative "confusedlie al over." He descends from heaven to Earth to the breakfast table, and finally confesses that he was overwhelmed. The key lies within the very images he has invoked, the most ancient of all—why doesn't he "see" it? The Moon as a territory of land and sea rises here, like a phoenix, yet remains invisible. Moreover, Holland, the very origin of the European telescope, native country to Van Eyck and the tradition of exact artistic description, which by this time had come to include the drawing of maps and the profiling of coastlines, is called upon for a moment, but then collapses into a humorous dismissal. No discourse yet exists for what is seen, and like Harriot, Lower is not the one to discover it.

Yet in late spring or early summer of that same year (1610), Harriot wrote to tell Lower of Galileo's discoveries and drawings regarding the Moon, Jupiter's satellites, and previously unseen stars. On June 11, Lower wrote back as follows:

> Me thinks my diligent Galileus hath done more in his threefold discoverie than Magellane in openinge the streights to the South Sea or the dutchmen that were eaten by the beares in Nova Zembla. I am sure with more ease and safetie to him selfe & more pleasure to mee. I am so affected with his newes as I wish sommer were past that I mighte observe the phenomenes also. In the moone I had formerlie observed a strange spottednesse al over, but had no conceite that anie parte thereof mighte be shadowes. (quoted in Whitaker, 1989, pp. 120–121)

A grasp of shadows, a few guiding words, and the Moon for Lower is transformed into a rugged territory of Earth-like substance. The geographic metaphor returns: Galileo's eye has sailed as far as Magellan's ships and those of the Dutch. The discoverer of one planet's geography is placed beside those who had circumnavigated another. Calling Galileo a type of Magellan both expresses Lower's new "seeing" of the Moon as another Earth, and also helps him preserve and stabilize this perception by giving it a meaning entirely contemporary. Lower is saying, once again, that the Moon is a landscape which, at some time, may be crossed, charted, and colonized.

The year following this exchange saw Harriot produce yet another image, the first drawing of the full Moon as seen telescopically (see Figure 5). This image combines an attempt to actually outline the dark regions of the lunar surface with the use of letters and numbers whose only purpose is to help correctly position these regions, not to designate actual features. It is thus only partly a map, being part diagram as well.

Because it shows none of the topographic features (e.g., craters) that Harriot had drawn the year before, after seeing Galileo's book, it seems a kind of regression, pictorially speaking. The Moon is returned to being a massive object, divided only into light and dark as it had been for thousands of years before. Harriot's failure seems one of comfort. Perhaps conscious of his imitative efforts after Galileo, he tried for something different, a full-moon image drawn in the manner he best knew how: as a mathematical diagram. Indeed, even his pictorial images of the year before, one notes, are surrounded and encased in a textual sea of few words but many computations, formulas, and calculations of vari-

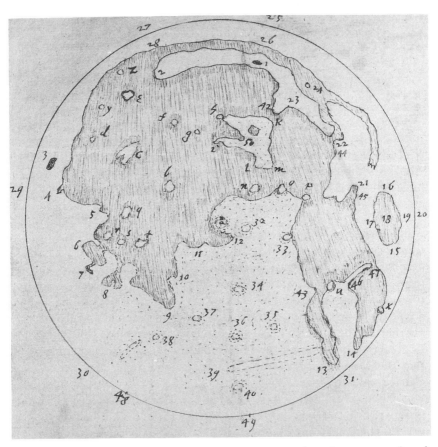

FIGURE 5. The first drawing of the full Moon as seen by telescope, produced by Thomas Harriot in 1611. Courtesy of Ewen Whitaker and the Earl of Egremont and Leconfield.

ous types.[9] Harriot was most happy with the Moon as a bearer of mathematical opportunities and visibilities. Galileo, with his own magnificent perception, helped him see the lunar surface geographically, but this is not what had the most important meaning for him (and besides, it had already been claimed!).

Galileo: Maps without Names

This perceptual change of Galileo's, by which the Moon finally became a fulfillment of Plutarch's prophecy, could only occur when the lunar surface was "seen" to have earthly features—when it was literally perceived as another Earth. The Moon, in other words, did not require features that *looked like* those of Earth: it had to have *actual* mountains, valleys, islands, peninsulas, seas, bays, and plains. At an early point, the borders between analogy and visual reality either broke down, dissolved, or, as suggested by Gilbert, never existed. No new concepts or titles were invented for the features seen on the Moon's surface. No new realities, different from terrestrial ones, were seen or conceived. These features were simply called by earthly labels, and thus, in J. H. Elliot's phrase, the "old world in the new" was inevitably secured both on the plains of inscription and perception.[10]

The story of Galileo's telescopic observations has been told in detail too many times to repeat here (see, e.g., Drake, 1976; Whitaker, 1978; Westfall, 1985; Van Helden, 1989a, b). If I have spoken more of Gilbert and Harriot, it is because they have always been assigned minor niches in the scholarly temple and therefore their true importance, as major indicators of change in Western perceptions of the heavens has been overlooked. If, meanwhile, art actually did precede science in this area, it does seem clear that the latter fully caught up in Galileo's drawings, which deserve all the attention they have received (and more).

Galileo built his own telescopes, eventually producing an instrument of 20-power magnification (much more powerful than Harriot's), which he turned toward the 4-day-old Moon on November 30, 1609. In his work *Sidereus nuncius*, he tells us exactly how he came to know of the telescope, briefly, how he fashioned one himself, and then what he did with it. He thus leads us through his experience and his thought, laying out the voyage of his discoveries. Galileo's little book offers nothing if not a traveler's account of what he did and saw, a log combining text and image in a manner much like a documentary. "In this short treatise I propose great things for inspection and contemplation by every explorer of Nature," the book begins (Galilei, 1989, p. 35). A bit further on, we are told that "It would be entirely superfluous to enumerate . . . the advantages of [the telescope] on land and at sea. But having dismissed earthly things, I applied myself to explorations of the heavens. And first I looked at the Moon from so close that it was scarcely two terrestrial diameters distant" (1989, p. 38). Galileo, the explorer, announces he is abandoning the Earth for the heavens; but note too that geography is the first thing that his narrative evokes, the first thing that comes to mind with regard to the telescope ("land and sea"), and it is the Earth that measures and defines the very closeness of the Moon. When he arrives at the Moon's surface, moreover, Galileo mentions the "darkish and rather large spots," noting that "every age has seen them," but then points out "other spots, smaller in size and occurring with such frequency that they besprinkle the entire lunar surface," such features having been "observed by no one before us" (p. 40). The author is therefore entirely conscious of his position as an explorer, someone who sees for the first time what other eyes have not seen. In naming himself the "messenger from the stars," he is being coy to a degree—humbly claiming to be only the receiver and deliverer of this message—yet he is also very much its sender as well, positing himself as one who has recently returned from a distant place, ready (as he states on the title page) to "[unfold] great and very wonderful sights . . . to the gaze of everyone" (p. 26).

These sights, though unknown, are not entirely unfamiliar, however:

> It is most beautiful and pleasing to the eye to look upon the lunar body . . . from so near. . . . Anyone will then understand with the certainty of the senses that the Moon is by no means endowed with a smooth and polished surface, but is rough and uneven and, just as the face of the Earth itself, crowded everywhere with vast prominences, deep chasms, and convolutions. (p. 36)

It is, therefore, the Earth that Galileo discovers. Wrong in one sense, those who viewed the Moon as a mirror were right in another. Indeed, this becomes far more striking later on in the work, as we will see in a moment.

Between November 30, 1609, and January 19, 1610, Galileo produced a large number of drawings (11 of which have been preserved) and a series of vivid descriptions, which together solidify the Moon, textually and visually, into a formidable geographic entity of magnificent detail. His textual descriptions are focused on the progress of sunrise; the four drawings he chose to be engraved for *Sidereus nuncius* show quarter- and half-moon phases only (no full-moon images), in which this progress is shown. It is evident that he recognized the "essence" of the Moon in the terminator shadow—this "essence" being the sight of landforms coming into view across the dawning surface, as the smaller "spots" revealed their true character (see Figure 6):

> [All] these small spots just mentioned always agree in this, that they have a dark part on the side toward the Sun while on the side opposite the Sun they are crowned with brighter borders like shining ridges. And we have an almost entirely similar sight on Earth, around sunrise, when the valleys are not yet bathed in light but the surrounding mountains facing the Sun are already seen shining with light. And just as the shadows of the earthly valleys are diminished as the Sun climbs higher, so those lunar spots lose their darkness as the luminous part grows.
>
> Not only are the boundaries between light and dark on the Moon perceived to be uneven and sinuous, but, what causes even greater wonder, is that very many bright points appear within the dark part of the Moon, entirely separated and removed from the illuminated region and located no small distance from it. Gradually, after a small period of time, these are increased in size and brightness. . . . Now, on Earth, before sunrise, aren't the peaks of the highest mountains illuminated by the Sun's rays while shadows still cover the plain? Doesn't light grow, after a little while, until the middle and larger parts of the same mountains are illuminated, and finally, when the Sun has risen, aren't the illuminations of plains and hills joined together? (pp. 41–42)

Here, finally, are the "mountains flaming bright" that Plutarch had written of so long before. Here is the Moon brought within the orbit of terrestrial discovery and exegesis. One can only imagine the overwhelming admiration such discovery must have elicited from Thomas Harriot and the rest of the European astronomical community, including one Johannes Kepler.

Galileo had sent a copy of his book across the Alps to Prague to this royal mathematician of the Hapsburg Empire, seeking Kepler's direct endorsement. He had done this via the Tuscan ambassador, who, after arriving in Prague, was leaving again in 11 days; during this brief period, Kepler composed a reply, in the form of an open letter entitled *Dissertatio cum nuncio siderio* (A conversation with the messenger from the stars). In this work, Kepler states his faith in the veracity of all

Galileo's findings and his commitment to do battle on the latter's behalf, as his "shield bearer," against any and all "reactionaries, who reject everything that is unknown . . . and regard everything that departs from the beaten track of Aristotle as a desecration" (quoted in Koestler, 1959, p. 372). Kepler had no telescope as yet, and so he could not verify the discoveries himself. Yet the Imperial Mathematicus (whose own works had been sent repeatedly to Galileo, with only minor response) also felt it necessary to put a truthful frame around some of his colleague's claims.

Kepler, in fact, reminded Galileo that the ancients had long provided evidence for mountains on the Moon, and noted that the relevant ideas had been recorded in Plutarch (Galileo, however, no doubt knew this quite well). Kepler says that he himself, while still a student, had used the methods of the ancients to estimate the height of some features on the lunar surface by measuring the length of shadows. He informed Galileo that he has broadened these early efforts into "a complete geography of the moon." Moreover, he writes:

> There will certainly be no lack of human pioneers when we have mastered the art of flight. Who would have thought that navigation across the vast ocean is less dangerous and quieter than in the narrow, threatening gulfs of the Adriatic, or the Baltic, or the British straits? Let us create vessels and sails adjusted to the heavenly ether, and there will be plenty of people unafraid of the empty wastes. In the meantime, we shall prepare, for the brave sky-travellers, maps of the celestial bodies—I shall do it for the moon, you Galileo, for Jupiter. (quoted in Koestler, 1959, pp. 372–373)

This passage is striking: here is Kepler, trying to plant his own claim on the Moon in the midst of declaring it, in no uncertain terms, the colonial territory of the future. In his enthusiasm, he is the most bold of spokespeople for his age. To map the Moon is to create its geography, to "prepare" it for conquest. Columbus has already become an icon; he is the reference point for the vessels and voyages to the stars. Strikingly, Kepler here foresees—in an instant of mingled hope and greed—that spaceships with names like *Viking, Mariner, Pioneer,* and *Voyager* will one day be sent out to explore the solar system, bringing with them the entire tradition of colonial scientific enthusiasms, the demands for knowledge, the desire for riches, for new territory, for fame.

But Kepler, like Harriot, was unequal to his vision in one major respect. He gave it no images, no pictorial language. He produced no maps of the Moon, of any kind. Indeed, his "complete geography" was no geography at all, merely a collection of notes accumulated over the years for an account of a fictional journey thereto—a fantasy, therefore, that was published only after his death, with the title *Somnium* (The dream). This literary effort was perhaps fitting to the futuristic quality of his vision, but it served not at all the cause of his claim. Kepler was not the man for the Moon; even as a literary effort, the *Somnium* lapsed. Its author never sought to improve on Galileo, or to stake a true claim to precedence. He never used the telescope to enter the fray of lunar image making. Geography and the map—these powers of the eye and hand he left to others.

What of Galileo's engravings? They are impressive, to be sure. Together with the text, they create a new double language for the Moon, at once pictorial and descriptive. Galileo's book was the creation of a "textual community" (to adopt a phrase from Brian Stock, 1983), a new means for reading, writing, seeing, and interpreting the lunar face. But there are limits to this discourse, especially to its pictorial side. One sees clearly in Galileo's images the irregular terminator, with its "horns" and "islands" protruding into the shadowed portion of the surface. One sees the maria complexly outlined, as well as a number of the craters (which Galileo termed "cavities") beautifully sketched, their darkened portions on the sunlit side, just as described. One does *not* see mountains and valleys, however, as described. Whereas Galileo exaggerated one crater enormously (presumably Albategnius, toward the lower center), no doubt to enforce his point about the Moon's unevenness, he did not reveal mountain ranges or single topographic peaks, exaggerated or otherwise. This he surely could have done, for example, across the northern or southern limits of a thin crescent phase where the ruggedness would have been especially visible. Such would surely have been possible, even given the limits of resolution of the engraving. Perhaps, as Whitaker says (1989, p. 124), the author had to portray things as convincingly as possible "if he was to bury two millennia of misconceptions," and so he felt compelled to limit his observations to the orb as a whole. But there still remains an interesting discrepancy between the descriptive detail of the narrative and the more massive scale of the images.

It has been pointed out that, as a Florentine, Galileo had occasion to imbibe much of the artistic sophistication that 16th-century Italy offered (Edgerton, 1991). That he was trained in drawing, and knowledgeable in artistic technique and method, is proven by the ink-wash images he gave to the engraver for printing. These show that he could create an excellent, naturalistic-type portrait of the Moon with some aspects of three-dimensionality. Winkler and Van Helden (1992, p. 207), who remark that "Galileo's verbal portrait of the moon is . . . much more compelling than the accompanying pictures," ascribe this to the author's belief that words were more important to conveying his message than images. In short, according to the conventions of the day, textual witness was more forceful than visual witness. Given the lack of naturalistic representations in astronomical writing up to that point, this seems a reasonable conclusion. However, I would like point to something else.

As an inheritor of Renaissance art, Galileo, the scientist, would also have been heir to Renaissance cartography, the plotting and drawing of features on maps of the Earth and its various parts. Indeed, no man of science in the 16th and 17th centuries could ignore the map as a vast new canvas on which enormous amounts of previously unknown information merged with artistic, textual, architectural, and mathematical aspects. Maps—not paintings—were the new and great documents of the age, by which European civilization inscribed itself upon the world. One needs to reconsider, in these terms, the idea, so often repeated, that Galileo perceived the resemblance of the lunar surface to the Earth. Normally, this is mentioned as a simple fact: part of Galileo's genius was that he was the first to actually *see* this resemblance, whose truth Harriot (lacking a similar exposure to Renaissance art and ideas of perspective) had missed. Yet, there are two problems here. First, as any glance through a telescope will show, the Moon hardly looks like the Earth at all. Indeed, its bleached or blackened surface, pockmarked with circular craters and corrugations, mottled throughout, resembles no terrestrial phenomena that Galileo would ever have seen.[11] Furthermore, no telescopic views of the Earth existed as a standard for comparison. The only thing that approached such a view was a map or globe, both of which offered artificial projections. What Galileo had to "perceive" therefore was an idea of the Moon visualized. His sight was conceptual as much as perceptual, but in a particular direction. In placing the Moon on a surface, he had to invert the geometry of witness—from looking up, into the sky, through a narrow "cannon" (as he called the telescope, thereby implying its power to fire the eye outward into space), to a downward gaze, at a projected image on a printed page. The transfer involved had to occur according to an idea that could make the Moon visually comprehensible—as suggested by Harriot and Lower's confusions, a photographic type of realism could not have done this—that could transport it back to Earth in the form of a legible two-dimensional surface. One notes, as well, that Galileo does not speak in his text about whether or not the lunar surface is rocky, watery, or whatever; he is not concerned with its composition, only with its gross geographic forms, the very same forms of knowledge that had come to be placed on maps.

That Galileo exaggerated one of the craters in his drawings has often been mentioned. I have not seen it discussed in the relevant literature, however, that he distorted other features too, most notably the irregularity of the terminator to which he gave an excessively scalloped appearance (see Figure 6), or the overly smoothed look to lighter areas within the western maria. Moreover, the prominent "explosive" appearance of several craters (Tycho, Copernicus, Kepler) is missing entirely, but would have been plainly visible, given the number and resolution of other craters Galileo drew (see the comparison of images given in Whitaker [1989], particularly his Figure 8.5). Galileo, that is, comprehending the Moon in terms of terrestrial features, chose at some level to make its surface look more Earth-like than it was, removing the most alien features and contouring others in accord with certain conventions of geographic representation for maps. I suggest this was done not nec-

FIGURE 6. One of Galileo's engraved drawings of the lunar surface, which appeared in the first edition of *Sidereus nuncius*. The image shows exaggerated irregularity of the terminator (the shadow line) and a greatly enlarged crater (Albategnius) below the center. This drawing of the last quarter phase was made on December 18, 1609. Courtesy of Ewen Whitaker and the Earl of Egremont and Leconfield.

essarily at a conscious level, but as a consequence of his essential ideas about the Moon. Galileo, that is, did not invent the forms and modes of presentation that appear on his images. The visual literacy he gave the Moon, the shadings, shapes, contrasts, and so forth can all be found on maps of the time. The naturalism of his drawings, meanwhile, is meant to be viewed as total, or near absolute; each image stands alone, naked, uninvaded by text of any sort, a self-contained embodiment of "observation" (again, this reminds us of Van Eyck). This distinguishes them radically from Harriot's images, with their armor of computations and formulas.

For Galileo, the image must convey its own language, apart from words (though complementary, these two media define two different sorts of witness, both essential, as in Gilbert, to the new "experimental philosophy"). If drawn using some of the techniques of linear perspective developed in Italian art, it is also intended to be more immediate, more "northern" in its realism. The point was to try and portray what was seen in the lens, not how it might look, ideally, to a staging of visual events through the medium of perspective drawing.[12] And yet it is also clear that Galileo's "realism" is being employed toward a certain type of propaganda (in the name of truth, of course). Galileo, who dedicated so much intellectual energy, to the overthrow of Aristotelean notions, seems intent in these images to do the same, and with actual visual proof. Looking at them, we are shifted across the boundary between "observation" and "artifice," between the Moon as seen via the telescope and the Moon as transported to Earth via art and the map. These images are far more than "visual aids." They are attempted fixatives of sensibility, of perception, and of belief.

I am suggesting that Galileo drew the Moon according to certain conventions of pictorial rhetoric in late Renaissance mapmaking that governed the delineation of such things as coastlines, islands, peninsulas, headlands, basins, and so forth. These conventions were guides that helped him sort through the mass of complex visual impressions (it is possible too, that the engraver had a certain role in this, being perhaps experienced in the production of maps). Comparison with actual lunar photographs strongly suggests that in most of his images lands and seas were made out of a far more ambiguous, undecipherable surface of features. Doubtless it was impossible to strictly reproduce the actual view in the telescope, which was visually confusing and likely untransferable to a flat page without some type of model or assisting schema. Several conventions in particular appear to have been helpful. One of these was the tendency to draw known coastlines—those that had actually been

charted by sight—in a heavily scalloped fashion, often with armlike protrusions reaching out around bays or inlets and with watery areas shown
in darker shading (see Figure 7). Unexplored or poorly known coasts
and islands, meanwhile, were commonly given smoother edges to designate their uncertain status. Edges of coastlines were more darkly shaded
than the open oceans in order to give the appearance of added relief and
contrast (Figure 7). In the most famous atlases of the day, such as those
created by Ortelius (1570s and 80s), Mercator (1580s and 90s), and
Hondius (before 1610), lakes were also darkened around their edges,
giving them a subsided, craterlike appearance. Mountains, too, had been
added by the mid to late 16th century, but were depicted crudely, as long
rows of bumps or bouldery masses, in all cases having an emblematic
function rather than a descriptive or naturalistic one. Shading here, to
give some character of three-dimensionality, was done either on the eastern or western side, to suggest shadowing at sunrise or sunset (the eastern or sunset side being somewhat preferred). At times, valleylike areas
were discernible by this shading within a particular mountain chain or
grouping. These were among the kinds of guides that appear to have
aided Galileo in giving the Moon a pictorial rhetoric.

There can be little doubt that, as a mathematics professor and an
astronomer, Galileo had been exposed to these conventions many times.
The atlases I noted were among the most widely bought and sold items
of knowledge of the age (Ortelius's work, for example, was one of the
great literary–scientific events of the era, going through dozens of editions before 1600, not counting pirated, plagiarized, and other unoffi-
cial versions). Maps were among the necessary possessions of any educated man in the early colonial period—especially scientists. Galileo's
own links to such fields as navigation and magnetism (he was an avid
reader of Gilbert's work), his long-term interest in the tides, and his
many years spent at the University of Padua, then an intellectual nexus
for scientists of every type, all must have brought him in contact with
maps many times. Indeed, while a young man living in Florence he had
been called to the local academy to speak on the geography of Dante's
Inferno (see Drake, 1970). While he may not have been trained as a
cartographer, his evident skill as a draftsman/artist would surely have
sensitized him to conventions of pictorial representation such as those
described above. Intentionally or not, Galileo seems to have employed
these conventions toward a realization of his vision that the Moon was
Earth-like, that Aristoteleanism should be abandoned. If he did not include mountains on his images, as I have noted, it was because there
existed no way at the time to draw these in a convincing fashion: had

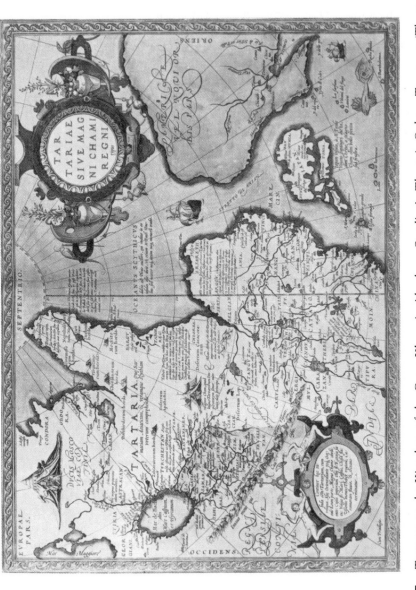

FIGURE 7. Tartary, or the Kingdom of the Great Khan, in Abraham Ortelius's *Theatris Orbum Terrarum* (Theatre of the World, 1570), showing many of the cartographic conventions of the late 16th and early 17th century. Some of these may have been used by Galileo in his drawings of the Moon. Courtesy of the National Maritime Museum, Britain.

they been attempted, the result would have cast severe doubt on the authenticity of his naturalism.

This, too, may be why he chose to impose no names whatsoever on the lunar surface. In one sense, this may seem strange—given his many drawings and his sense of being the Moon's first "explorer," wouldn't the impulse to name this surface have naturally occurred to him? Didn't he give titles to the four Jovian moons he discovered, satellites he baptized the *Medicea Sidera* ("the Medici stars") in honor of the patron whose support his book was intended to procure? Why did he name four moons, on which he could see nothing, but leave blank and naked so crucial a body as the Moon itself, on which he discovered a fabulous plurality of features? The only explanation for this is the one already given. Galileo, that is, wanted his drawing to be perceived—indeed, he probably perceived it as such himself—as a recorded sight, transcribed to paper. To act as proof, to fulfill the requirements of "discovery," it had to be convincing as an "observation," seemingly peeled from the eye, stolen and saved from the act of perception,[13] yet legible by the requirements for visual literacy of the era in which it was produced. Galileo achieved all this, providing a map, an artwork, and an observation all at once. The Jovian moons were the sign of his dependency on the existing system of patronage; to name them thus was both an act of political obedience to this system and, at the same time, no great thing. The Moon, however, Galileo left in its pre-Gilbertian, native state. It was as if, in opposition to Gilbert, he also felt a need to leave this surface of discovery free from certain terrestrial claims. And in leaving it this way, it was he who was to become astronomy's patron.

⬚⬚

For more than a decade and a half after the publication of *Sidereus nuncius*, no progress was made in creating pictures and maps of the Moon. The first book to follow Galileo, G. C. La Galla's *De phenomenis in orbe Lunae* (1612) simply republished the images from *Sidereus* (both books, in fact, were printed on the same press and probably used the same plates; see Van der Vyver, 1971a). In the images produced subsequently, up till the 1630s and 40s, one finds a kind of stylization of Galileo's pictorial language, often rewedded to the mathematical tradition by the addition of designating letters. Versions by P. Christoph Scheiner (drawn in 1614), Charles Malapert (1619), Giuseppe Biancani (1620), and Christopher Borri (1627) actually show a progressive decay or regression from Galileo's cartographic naturalism. This tended either to turn the Moon back into a diagram, with surface features coming again to look like spots and simple circles, or else into a type of baroque decoration, without topographic character altogether. Borri's image, for example, shows the lunar maria as crudely outlined blotches, each labeled with a letter and bordered by a rising line of bubblelike craters. Scheiner's drawing employs a different pattern of stippling for each of the lunar maria, with no craters visible, the whole resembling an ornate wall motif. Slightly later, images drawn by Francesco Fontana in 1629–1630 added new features, such as the explosion-craters Copernicus/Kepler and Tycho, and sketched out the maria in more detailed shape. But these, too, are fanciful and decorative, scattering little cell-like craters randomly across the surface.

For the most part, these images were made either by mathematicians (Scheiner, Malapert) or by amateurs (Borri, Fontana). Following Galileo's lead, they put north to the top—even though this convention had not yet been fully established on terrestrial maps—and generally provided more drawings of half- and quarter-moon phases than those of the full Moon. On the other hand, their visual relationship to terrestrial map conventions is far more tentative, and it is largely for this reason that they are so crude and schematic in comparison to Galileo's images.

Return of the Text: Literary Exploration of the Moon's Geography

By the 1620s, the Moon was being drawn in terrestrial terms for a new purpose. One of the great problems of colonial navigation, which had application to land surveying and mapmaking as well, had to do with determining longitude. Unlike latitude, which could be estimated relative to the height of a fixed point in space (e.g., the North Star), east-west position was complicated by the Earth's rotation. When the compass proved unusable for the purpose, most scientists realized that longitude had to be calculated on the basis of time.[14] The problem was to find some way in which time could be determined at exactly the same instant yet at different locations. Galileo's observations of the lunar surface seemed to offer such a technique, which would involve noting the exact times when certain features—mountains, for example—went into shadow or reappeared in sunlight during lunar eclipses. For this, however, a highly accurate map of the Moon's surface was essential. And it was just for this purpose that the next phase of lunar mapping and naming was undertaken, by several different teams simultaneously.

Thus the problem of longitude was very much tied to European politics. It was linked, in particular, to the ongoing competition for trade routes, colonies, and resources; to naval conflicts at sea; and, no less, to the internal struggles of various governments to fully command the territory within their borders. This last should not be underestimated. One tends to think of the colonial era as a time in which the new nation-states of Europe were engaged primarily in overseas expansion. This process, however, was complemented by these same states' efforts to gain control over the ordering of space within and along their borders. Indeed, as Jacques Revel has said of this time period, "Knowledge of the territory became inseparable from the exercise of sovereignty" (1991, p. 137), meaning that accurate cartography, the precise mastery of space in plotted form, had become coterminal with the waging and winning of wars, the establishment of a centralized administration, the ability to exploit and mobilize resources of all kinds—the very power, in other words, to create and solidify a nation-state. The determining of longitude was a gap, a noted weakness, in this growing power of spatial command. Thus it should be no surprise that King Philip III of Spain would offer 6,000 ducats to anyone who could solve the longitude problem (Whitaker, 1989). But the historical rewards must have seemed far greater than any monetary gain: whoever succeeded would become a national and an international hero, during a period when the mixture of rivalry and cooperation among scientists sometimes matched that between nations. The Moon, as a type of new celestial compass, was drawn into this larger circumstance, and it was through this that it finally gained its nomenclature.

This, however, did not happen immediately. Even while it was being mapped for this purpose, the Moon was also being given a politicized textual reality of another type: it was being "mapped" in literary fashion. These efforts, which involved both fanciful accounts of voyages to the lunar surface and books of popular science, began at exactly the same historical moment as the new mapping and nomenclatural efforts: interest taken in the Moon by scientists and sovereigns had its correlative in the new educated public's own curiosity for tales and speculations. Even more than for Gilbert, the Moon became a surface for the collision of historical movements.

The cosmic voyage, however, was by no means a new literary genre. Whether by dream, by whirlwind, or by flying chariot, the journey to heaven was a genre whose history stretched back to classical antiquity. Early prototypes can be found in Plato—the ascent and descent of human souls through the heavenly spheres in the *Timaeus*, the "flying chariot" of the *Phaedo*—and in Cicero's beautiful but propagandistic *Somnium Scipionis* (Dream of Scipio). The first real voyage-to-the Moon tales are to be found in the work of Lucian, his *True History* and *Icaromenippus*, which compared the lunar surface to "a shining island" (see Nicolson, 1948). During the Renaissance, thanks to translation, these and other stories about heavenly journeys gained a wide readership, and inspired contemporary versions. One of the most striking here, given its view of the lunar orb, is that presented in Ariosto's epic romance *Orlando Furioso* (1532), an enormously popular work that was translated into all the European languages. Its hero, Astolfo, ascends to heaven in a four-horse chariot, which bears its driver to a world ripe with dense familiarities:

> Swell'd like the Earth, and seem'd an Earth in size,
> Like this huge globe, whose wide extended space
> Vast oceans with circumfluent wave embrace . . .
> Far other lakes than ours this region yields,
> Far other rivers, and far other fields;
> Far other valleys, plains, and hills supplies,
> Where stately cities, towns, and castles rise.
> (quoted in Nicolson, 1948, p. 21)

Even before Harriot and Galileo, therefore, the image of the Moon as an Earth-like place was established as a literary commonplace and a popular belief; many people undoubtedly believed that this fiction was the truth. Seventeenth-century accounts of such voyages multiplied in the immediate wake of the astronomical discoveries of Galileo, Kepler, and others. Stories in this new tradition borrowed from three basic sources: Galileo's *Sidereus nuncius*; existing fantasy–utopian literature; and accounts of exploration in the New World, such as those of Hakluyt or Harriot's own *Briefe Report*. The first two influences have often been noted by scholars; the last, however, has not. Yet, as the passages quoted earlier from Galileo's work and from Kepler's *Dissertio cum nuncio siderio* aptly reveal, the colonial–geographic sensibility had even penetrated astronomy by this time. Given this truth, it could not but be taken up by the new genre of the cosmic voyage.

Indeed, one finds that the very first work to appear in this new genre was Kepler's own *Somnium*, which I have mentioned earlier. Named after Cicero's *Somnium Scipionis* (in which the heavens are laid out as a moral–theological tapestry, bidding final obedience to Rome), it was a fantasy work never finished, appearing four years after its author's death, in 1634. It was therefore something of a dream-work itself, one that only death could wake into public existence. As a form of literature, it suffers from this fitful gestation; it seems quaint, confused, not wholly successful, an attempt to use a fictional frame as an excuse for putting an astronomical eye, of musing intent, on the Moon. Kepler's narrator is transported upward by a daemon, who proceeds to speak, in astronomical and mathematical terms, about what a trained observer would see and experience once landed on the lunar surface. This daemon mentions hot and cold regions; the Earth and its cycle of days, months, and years; and the appearance of the Sun, the Zodiac, and the other stars. He states, at the outset, that "No inactive persons are accepted into our company . . . we choose only those who have spent their lives on horseback, or have shipped often to the Indies" (Lear, 1965, p. 103). This is because the lunar surface is a difficult, extreme environment, in earthly terms. The daemon remarks, "The chief alleviation of the heat [on the sunlit side] is the constant cloudiness and rains, which sometimes prevail throughout half the region or more" (p. 158). A footnote to this sentence notes that "Jose d' Acosta writes the same about the provinces of the New World."

Attached to the main text are an appendix, containing 34 axioms of lunar geography, and a Latin translation of (should we be surprised?) Plutarch's own *On the Face which Appears in the Orb of the Moon.* Kepler's "Geographical, or, if you prefer, Selenographic Appendix" is no doubt the most serious part of the work, the true essence of his "complete geography of the Moon," once promised to Galileo as evidence of Kepler's own sovereignty over the mapping of the lunar surface. Many of these axioms were based on his own telescopic observations; he perceives the "spots" (one said to resemble the "Austrian shield") as filled with varying amounts of water, such that "some are analogous to our swamps, some to our seas" (p. 171). In general, Kepler follows Galileo's lead in "earthifying" the Moon. But he takes one final step, in a direction Galileo would not have considered. In axioms 28 through 33, after offering what can only be called a remarkable geologic analysis of lunar processes ("Soil is eroded from the rocky subterranean ribs, valleys are washed out, so that mountains rise up, waters flow downward into the low areas marked by spots and there reach equilibrium . . . ," p. 173), Kepler's daemon takes up the perfect circularity of many lunar craters, along with the central uplifted area within them, as evidence of habitation: here, he says, "the descendants of Endymion [in Greek mythology, a beautiful youth loved by the Moon Goddess] make a practice of measuring out the spaces of their towns for the sake of guarding them . . ." (pp. 165–166). The daemon goes on: "Men do not make mountains and seas on the surface of the earth . . . but men do make on earth cities and fortresses, in which order and art are perceptible" (p. 174). The Moon, in short, has its parallel political geography too, a geography, in fact, that involves establishing and defending territory. Kepler even tries to explain what he surmises to be the precise method of construction of these fortresses, as if he were suggesting it as a possible model to his Hapsburg patrons. This may well have been a ruse; as an avowed Protestant, Kepler was constantly in need of ways to ingratiate himself to the Catholic emperor he served. Yet ruse or not, his "hypothesis" for the lunar craters embodied something critical to the sensibility of the time. The realities of geography in the early 17th century were realities of exploration, which included national defense. For Kepler, no less than for other astronomers, the Moon was something to be comprehended through the filter of "territory." The implications of Gilbert's map and of Galileo's images found a new specificity in the text of Kepler's fiction.

The *Somnium* was the first in the new genre of Moon voyages, but it was by no means the most successful or the most popular. Even before it had appeared in published form, another, more skilled and fanciful story had been written in England, though this would not itself appear until 1638, four years after Kepler's work. This book drew less on Cicero, and far less on technical astronomy, than it did on Harriot's *Briefe Report*, both in its brevity (a few dozen pages) and certain elements of its narrative. Francis Godwin's *The Man in the Moone: Or, A Discourse of a Voyage Thither by Domingo Gonsales, The Speedy Messenger* (1638), plainly reveals the three sources for the new genre—Galileo, utopian literature, and accounts of the New World—in its title. Like Kepler's, this was a posthumous work (Godwin had died five years earlier), probably written in the late 1620s or early 1630s. It no doubt had Francis Bacon's own *Sylva sylvarum* (1626) as an essential and immediate source, this being evident in the borrowed "flying chariot" with which Godwin has his Spanish "hero" drawn upward to the Moon by a flock of enthusiastic, migrating geese (see McColley, 1937). The book is a part-humorous, part-satirical account of an undeserving opportunist (tiny in stature, cowardly by nature) who, after being tossed from one life circumstance to another, across various and sundry parts of the globe, lands on the island of St. Helena and from there is transported to the lunar surface, where he meets with a variety of inhabitants and their rulers, who convert him to a better outlook, whence he returns to Earth, landing in China. The story ends with the narrator hoping he might one day return to Spain so that "by inriching my Country with the knowledge of hidden mysteries, I may once reape the glory of my fortunate misfortunes" (McColley, 1937, p. 48). With such hope for colonial fame, Godwin's Gonsales reports, as did Kepler's daemon, that the Moon has its seas, lands, islands, and so forth, and goes on to describe a catalog of items and experiences that act no less as a "map" of his discovery than did

Harriot's. In the preface, moreover, Godwin has him advise us that "in substance thou hast here a new discovery of a new *world*, which perchance may finde little better entertainment in thy opinion, than that of *Columbus* at first. . . . Yet [whose survey] of *America* [hath since been] betray'd unto knowledge soe much as hath since encreast into a vaste plantation. . . . That there should be *Antipodes* [people living on opposite sides of the Earth] was once thought as great a Paradox as now that the Moon should bee habitable. But the knowledge of this may seeme more properly reserv'd for this our discovering age: In which our *Galilaeusses*, can by advantage of their spectacles gaze the Sunne into spots, & descry mountains in the *Moon*" (p. 2).

Columbus and Galileo, therefore, are to be canonized on the same basic level of "discovery." Indeed, however fancifully, the Moon is here written as a territory to be actually visited, traded with, learned from, and possibly exploited, just like the "far side" of the Earth. Gonsales, like Harriot, describes such things as food, metals, precious stones, language, and ethnography. But he goes still further: he notes, too, that the Moon's inhabitants are able to discern at birth who among them will be wicked or imperfect, and that these undesirables are "vented" onto the Earth from "a certaine high hill in the North of *America*, whose people I can easily beleeve to be wholly descended of them, partly in regard of their colour" (p. 40). The inversions here are many, and not to be missed. In the mouth of England's enemy is placed the "discovery" that the natives of England's own newly claimed New World are, in fact, dangerous rejects who have been sent down to Earth. By way of the Moon, North America is proposed as a kind of colonial prison. Whatever the intended irony of this, the more literal implications, seen in the light of later historical reality, are both striking and disturbing—just as Harriot's sympathetic treatment of the Native Americans in his account must always be qualified by John White's picturing of them in Europeanized, that is, Renaissance poses, a fact that presages later attempts to Christianize and subdue them by other means.

Finally, Gonsales—actually Godwin here—performs another trick of reversal, this time without irony. He does this by using the new astronomical knowledge to turn one of the most ancient views of the Moon back upon the Earth:

Whereas the Earth according to her naturall motion (for that such a motion she hath, I am now constrained to joyne in opinion with *Copernicus*) turneth round upon her own Axe . . . I should at the first see in the middle of the body . . . a spot like unto a Peare that had a morsell bitten out upon the one side of him; after certaine [hours], I should see that spot slide away to the *East* side. This no doubt was the maine of *Affrike*. Then should I perceive a great shining brightnesse to occupy that roome, during the like time (which was undoubtedly none other then the great *Atlantick* Ocean). After that succeeded a spot almost of an Ovall form, even just such as we see *America* to have in our Mapps. Then another vast cleernesse representing the *West Ocean*; and lastly a medly of spots, like the Countries of the *East Indies*. (p. 22)

So the Earth, too, becomes an orb of "spots" and "brightnesses," the sign that the Moon itself has become a new vantage point for projected "seeing," a point from which one acquires, even in literal form, the colonial appetite for new territory, here delivered in the vision of Africa as a fruit with a large bite taken out of it. Kepler's daemon, too, had tried to describe the Earth's "spots," and had come up with interesting images: Africa as "a human head cut off at the shoulders," Europe as a "young girl with a long dress," South America as a "bell." But these were not given in the context of the Earth's own motion—the motion of the eye sailing across the terrestrial regions—and therefore lack the more suggestive dimensions of Godwin's hero, who spoke in the moment of discovery, actually perceiving them. It is the realm of projected experience that is missing from Kepler's work as a whole, a realm of literary power that Godwin evokes and that fills out, more fully, the mentioned reversal of geographically "seeing."

Godwin's work proved to be immensely popular in its day, both in England and on the Continent, and was continually reissued well into the 18th century in various translations (Bush [1945] mentions 25 editions in four languages by 1768). It was undoubtedly the inspiration for Cyrano de Bergerac's own *L'Autre Monde; ou, Les états et empires de la Lune* (1649), a much longer work in which French society, especially its monarchical, bureaucratic, and Catholic authorities, is satirized in some detail by being reembodied in various ingenious ways. Lastly, Godwin's book was also once included in a Hakluyt-type work, called *View of the English Acquisitions . . . in the East Indies* (assembled by publisher Nathaniel Crouch in 1686; see McColley, 1937).

The same year that *The Man in the Moone* appeared, another, more serious effort of popularization was published. This was John Wilkins's *The Discovery of a World in the Moone* (1638). Wilkins, who later became one of the founding members of the Royal Society and who launched an effort to construct a universal language for the purpose of "the new experimental philosophy" (see Chapter 2), is here more intent on spreading the Galilean–Copernican gospel. His treatment is earnest and pedagogic, yet it overlaps with Godwin's larger geography of lunar meaning. Not only does it promote the idea of land and sea on the Moon ("there are high mountaines, deepe vallies, and spacious plaines"), but it also claims that the Moon has "an atmosphaera, or an orbe of grosse vaporous aire," and points out that if the Moon has inhabitants, "as their world is our Moone, so our world is their Moone" (1973, pp. 212–213). Finally, Wilkins directly picks up on Kepler's own claim in *Dissertio cum nuncio siderio:*

> It is the opinion of Keplar, that as soon as the art of flying is found out, some of their nation will make one of the first colonies that shall transplant into that other world. I suppose his appropriating this preheminence to his own countrymen, may arise from an over-partial affection to them. But yet . . . whenever that art is invented, or any other, whereby a man may be conveyed some twenty miles high . . . then it is not altogether improbable that some other may be successful in this attempt. (p. 11)

No more clear statement of the realm of intentions behind creating a lunar geography were ever made. Unless, of course, it be that of Samuel Butler's elegant satire on the Royal Society of London, "Elephant in the Moon" (c. 1660s), which begins:

> A Learn'd Society of late,
> The glory of a foreign State,
> Agreed upon a Summer's Night,
> To search the Moon by her own Light;
> To take an Invent'ry of all
> Her Real Estate, and personall;
> And make an accurate Survey
> Of all her Lands, and how they lay,
> As true as that of Ireland, where
> The sly Surveyors stole a Shire;
> T'observe her Country, how 'twas planted;
> With what sh' abounded most, or wanted;
> And make the proper'st Observations,
> For settling of new Plantations . . .

Before the century was out, a host of new works on similar themes had appeared, most imitative of Kepler, Godwin, Wilkins, and de Bergerac, but some, like Butler's, satirical. A survey of this writing is offered by Nicolson (1948), who rightly points out how it helped define the possibility of the science fiction of Jules Verne, H. G. Wells, and others.

Wilkins and Godwin, as well as de Bergerac and Butler, all show that the Moon had undergone a revolution in public consciousness within a mere decade and a half after Galileo. The problem of longitude, meanwhile, was a problem of Earth geography, politically bound and nationally pursued, that depended entirely on the finding of a colonial-type landscape on the Moon. The longitude problem relied on exactly the type of consciousness that these literary works reveal—the consciousness of the Moon's own Earth-like character and therefore its capability of being claimed as an extension of existing powers on Earth. It is hardly a coincidence or an unrelated truth that astronomers who first mapped the lunar surface for determining longitude also believed it probable that the Moon was inhabited. Nor should it be thought insignificant that a man such as Wilkins, who would one day strive for a "universal character," would take up study of a geography that was now beginning, in the voice of the age, to cry out for a universal system of "places" and therefore names.

The Peiresc–Gassendi Project: Naturalism at Its Tragic Peak

The earliest of the new attempts to map the lunar surface was that undertaken by the team of Pierre Gassendi (1592–1655) and Nicolas-Claude Fabri de Peiresc (1580–1637). Gassendi was one of the most eminent French mathematicians and astronomers of the 17th century, an avid opponent of both Aristotle and Descartes, and a powerful influence on the thought and literature of the time. His lifelong friend, Peiresc, on the other hand, was more of an accomplished dilettante or *l'homme universel*, with avid interests in astronomy, archaeology, natural history, geography, and other areas. Peiresc had foreseen, at the same time as had Galileo, the possible importance of using celestial objects for estimating longitude: in January 1628, he and Gassendi made careful, painstaking observations of a lunar eclipse in Aix-en-Provence, in planned tandem with other astronomers in Paris. This permitted a precise determination of the difference in longitude between these two sites. According to Humbert (1931), this success encouraged Peiresc to try and organize a wide network of observers, in Rome, Tunisia, Egypt, Syria, and elsewhere (areas chosen so that weather would not be a problem), to conduct similar observations and thereby establish the relevant method and advance geography into the realm of geodesy. Somewhere within this project, too, was the hope of establishing a world reference base in France: Peiresc, long-term adviser to the local Aix-en-Provence government, preferred his city for this site (weather conditions were generally excellent there too). Political realities, however, might well have argued for Paris. In any case, the project could not proceed without a highly accurate lunar map, and the two men thus embarked on the attempt to construct one.

The project was underwritten to a high degree by the enthusiasm and funding of Peiresc himself. Gassendi, however, with the encouragement of Galileo (who even supplied telescopic lenses for the purpose), was eventually forced to take over the lead role, due to his friend's increasing illness and disability after 1634. Yet Peiresc, even in these years, had an insight: to surpass Galileo, it was now required that a trained artist be commissioned to produce images of the Moon. The very first of these, one Fredeau (noted in Gassendi's diary), was originally engaged first to do a portrait of a rare gazelle that Peirsec had shipped from Africa, destined as a gift to the powerful Cardinal Barberini in Rome. Being on site, Fredeau was asked to do a chalk drawing of the full Moon (July 10, 1634). Having drawn one distant object, he was thus asked to draw another. Even in this manner, the Moon could not escape the Earth.

Dissatisfied with the result, Peiresc then enlisted the services of a more renowned artist, Claude Sauve (or Salvat). On August 26 of the same year, a second image was produced, also unimpressive. It appears that the weather was somewhat foggy, but more, that Sauve, unaccustomed to looking through a telescope, was unable to draw what he saw very well. During the next several months weather remained a problem, but a number of images were produced. Then, in March 1636, a lunar eclipse offered a special opportunity to detail the advance of the shadow over various portions of the lunar surface. This time the results seem to have been better: the artist had essentially caught up to, and even surpassed, the level of recorded perception revealed in Galileo's own images.[15] However, no map resulted. Indeed, Sauve seems to have been let go, for when next we hear of the project a full year later, Peirsec has persuaded a still more famous artist, Claude Mellan—whom he calls "one of the great painters of the century and the most precise engraver in copper yet born"—to produce images of the Moon, whose worth would be "remembered for all time" (quoted in Humbert, 1931, p. 198). Due to weather once again, Mellan worked intermittently throughout the fall of 1636 before finishing enough drawings and paintings to yield three engraved images, two quarter phases and one full Moon phase. Gassendi must have been impressed, for he wrote to Galileo himself (who knew Mellan personally) in December, promising him copies of the final result.

One can well see, from Figure 8, the cause of his enthusiasm. Mellan's images are such a startling leap over the stylized, manneristic drawings of the 1620s and early 30s—even over Galileo's own pictures—that we are left almost stunned. Mellan's images, after all, are the first we have since Leonardo that were drawn by an artist (and engraver) of the first rank. More, the artist in this case had Galileo's work to serve as his beginning standard, Gassendi and Peirsec's years of observations to draw upon, and by this time, the whole toolbox of both Renaissance and baroque techniques to employ. One feels, looking at these images, Mellan had no agenda beyond sheer observation; they seem of a quality Van Eyck would have admired. Indeed, let us invoke him further, for at this point a new stage seems to be reached: Van Eyck shows us the Moon in the hands of "art"; Leonardo, art and science without separation; Galileo, likewise (though perhaps more science than art); but now, with Gassendi, Peiresc, and Mellan, we have art, as a separate domain of expertise, hired in the service of science. In exactly the same manner as geographers of the 17th century, astronomers were now realizing that the making of accurate maps demanded the combined efforts of different areas of knowledge and skill. To observe and to draw was no longer enough; one had to observe with learned precision and to draw with the highest level of craft. To be successful, lunar maps too had to be the result of trained ability. They had to be documents on which were inscribed the powers of high civilization.

In June 1637, Peiresc died, and with him much of the life went out of the project. Certainly the major funding disappeared, and with it, the ability to produce more than the three engravings already completed. Some years later, hoping that the work would not be wasted, Gassendi sent copies of two of these images to other astronomers, notably to Johannes Hevelius in April 1644, then in the midst of his own lunar mapping program, whose result would be the first true atlas of the Moon and would have an enormous effect on lunar astronomy. Indeed, news of this effort led still another interested party, Michael Florent Van Langren in Belgium, to rush his own map into print, thereby becoming the first to offer a complete nomenclature for the lunar surface. This activity reveals both the widely recognized value of such a map at the time and the consequent heat of competition to produce one. In such a climate, it was perhaps inevitable that, without any further work and promotion, the Gassendi–Peiresc–Mellan project would disappear into the shadows of "influence." One of the great failings of Mellan's images, after all, was also their very success, namely, their purely visual

FIGURE 8. Three lunar images produced and engraved by Claude Mellan during his work for Gassendi and Peiresc in 1636. Courtesy of Ewen Whitaker.

character, for his "maps" lacked any kind of text, especially names. As such, they qualified as raw material only, for the purpose of determining longitude meant that textuality (nomenclature) was now an absolute necessity. Without this dimension, they could be used as aids for advancing selenography, but not standards.

This does not mean, however, that a naming scheme was ignored. Gassendi's journal and correspondence contain the outline of such a scheme, never made official, but clearly viewed as an important part of the original project. Gassendi's designations are all in Latin and remain obedient to the concept of geographical "seas" and "land," though otherwise they have a unique character. Like Gilbert, he named a Northern Sea (*Boreum mare*), and like Harriot, he recognized a similarity of form (as then drawn on maps of Asia) between the Mare Crisium and *Caspia*; the Mare Humorum, which lies on the opposite side of the visible hemisphere and shares a similar shape, he called *Anticaspia* (after Ptolemaic convention). The Mare Vaporum he baptized *Hecates penetrale*, using the Greek name. On the list of titles, he included: *Vallis umbrosa, Rupes nivea, Amara Mons,* and *Lacuna,* thus calling directly upon terrestrial features (*Rupes* = "scarp"; *Mons* = "mountain" or "hill"). The combined area of the Mare Serenitatis, Mare Tranquilitatis, Mare Fecunditatis, and Mare Nectaris he lumped together as *Homuncio,* a region on the east side of Mellan's drawing that crudely resembles a human form slightly tilted backward (no doubt corresponding to the "Man in the Moone" so often mentioned in earlier writings, such as those of Harriot's friend Sir William Lower). The explosive crater Tycho, so prominently displayed on Mellan's image in the lower left center, Gassendi called *Umbilicus lunaris*, while the other such crater, Copernicus, located in the upper left, he titled *Carthusia*, either after the monastary or the area of its location, immediately north of Grenoble.

Gassendi's system of nomenclature, in its imaginative variability, was therefore intended to project onto the Moon a host of titles derived from classical myth, contemporary geography, and even human anatomy. His Moon was to be (as it largely is now) a Latinated surface, no less than the text of Virgil, the early 17th-century maps of North America, or the medical treatises of the time. All these were "geographic" too, concerned with locations, places, the contours of visibility. As it happens, in his correspondence with Lobkowitz, Gassendi states that he will happily leave to others a more detailed naming of the surface. The former wrote back to him to propose that "promontories, islands, and valleys" be given the names of contemporary *savants*: "All our friends will be there, yourself, Peiresc, Mersenne, and Naude" (quoted in Humbert,

1931, p. 200). To Gassendi's system, therefore, was to be added a memorial dimension, a canonizing not merely of "greats" but of a circle of friendships, whose geography of communal striving, achievement, disappointment, and ambition would thus find its final and eternal resting place in the rugged topography of another world.

History was to prove itself kind to this scheme. Where Van Langren would substitute his own memorial canon and Hevelius abandon such a scheme altogether, the final arbiter of lunar names, Giambattista Riccioli (1598–1671) would reinstate Lobkowitz's idea and, in 1651, give Gassendi (now in his declining years) and his departed friend Peiresc back to the orb with which so much of their lives had been entwined.

Van Langren: The First Textual Map and a Catholic Moon

The version of the lunar surface produced by Michael Van Langren (1600?–1675) has been the subject of several studies and is generally regarded as "the first true map of the Moon; that is, it depicted not only the surface shadings as seen at full Moon, but also a large number of topographical features (craters, peaks, mountain ranges) which only become visible at other phases" (Whitaker, 1989, p. 128; see also Prinz, 1903–1904; Bosmans, 1903, 1910; Wislicenus, 1901). In other words, Van Langren did not seek to produce an "observation," but a generalized plot of the total surface, unburdened by the demands of naturalism. His map is an abstraction, strictly cartographic (Figure 9). In the conventions of mapmaking of the time, Whitaker is therefore correct: Van Langren's is the first true textual Moon: surrounded, framed, supported, and invaded by writing, as were regions on terrestrial maps. Van Langren's textuality: his invocations of Theodorus, Cicero, Pliny, Seneca, and, yes, Plutarch; his long confessional narrative about his own struggles to conceive and finish the project; his title (*Lumina Austriaca Philippica*, "By the Light of Philip of Austria"); and his fulsome nomenclatural scheme, based largely on living political figures—all these aspects transform the Moon into a clear historical document. Van Langren's Moon, that is, is not the same Moon as that of Galileo or Gassendi and Mellan. It is a new Moon lately risen from its own moment of historical need.

Van Langren, about whom little is actually known, came from a family of royal cosmographers and mathematicians, and to these skills he added his own of engraving, engineering, and cartography. Intellectually, he was almost ideally suited to produce the first Earth-type map of the lunar surface. Personally, religiously, professionally, and politically, meanwhile, he was deeply committed to the fortunes of a Catholic Europe, led by Spain. His family had moved south out of what is now Holland, a center of long-term Protestant revolt, to the Spanish Netherlands, settling in Brussels. The period 1618–1648 saw the outbreak of the Thirty Years' War, the culmination of religious–dynastic conflicts generated during the late 16th century, and the period of final downfall for Spain as the dominant European power. Spain entered this war in 1621, on the death of Philip III and the ascent of his son Philip IV. By this time, Van Langren—possibly responding directly to Philip III's offer of monetary award, but more likely to the probability of royal favor and ensuing fame—was already at work trying to find a method for determining longitude at sea, and, like Peiresc, he focused his efforts on the Moon.

He presented this idea to Isabelle, Philip III's sister and princess of the Spanish Netherlands, as early as 1625. By this time, and during the years immediately following, he was at work on a number of other projects of immediate political and military utility: plans for a fortified port to be built at Mardyck; a map and similar plans for the harbor at Oostende; a map of a canal system linking the Rhine with the Meuse River in Spanish-controlled territory; other maps of Luxembourg, Mechlin, and the duchy of Brabant; early drawings for a three-shot cannon; and various astronomical and maritime tables (now lost, but which might have included lunar observations). Some of these projects were not published until the 1640s and 50s, yet it is likely that Van Langren at least mentioned them to the princess, for his more decided purpose was to take his ideas to the king himself. Impressed with these ideas and the evidence of his loyalty, Isabelle became a patron and agreed to write to the king (now her nephew, Philip IV) requesting an audience for the young cosmographer, who traveled to Spain in 1631, remaining there under royal auspices until 1634, when Isabelle died and when rumors of France's own imminent entry into the war began to circulate (this would mark the end of Spain's hopes for victory). Philip, too, appears to have been impressed with Van Langren, agreeing to finance his continued efforts and allowing him the new title of *Cosmographe du Roi* (royal cosmographer).

The death of Isabelle struck Van Langren a hard blow, and his work on a lunar map, which by this time seems to have included a large number of early drawings (perhaps as many as 30 or more), came to a halt. A friend and colleague, Erycius Puteanus, then became his immediate patron, encouraging him in various projects during the 1640s. Sometime between 1643 and 1645, it seems, Van Langren learned of other efforts to map and name the Moon, and this news prompted him to quickly assemble and publish his own definitive version, thereby laying claim to a long-held and now reawakened ambition, which amounted to both a promise and a means of securing royal favor. By February 1645, Van Langren had an early manuscript version of his map, complete with nomenclature, and applied to the king's Privy Council for permission to publish it. In his request, he notes that he has "adorned [the Moon] with the names of eminent persons, as His Majesty has wished." He then makes the following request: "As the supplicant fears that some other person might well change the mentioned denominations and by such means alter and place in confusion the observations here achieved, and moreover prevent the supplicant from attaining his rightful compensation for such work, he hereby humbly requests it please your Majesty to expressly order his subjects to change nothing on said figure . . . that he issue a decree regarding this measure and privilege to be sent with all possible speed to a certain individual, in order to prevent him from advancing further with his own design, to the disadvantage of His Majesty" (quoted in Bosmans, 1903, p. 113).

It is not known exactly who this "certain individual" was, but undoubtedly Hevelius (see below) is an excellent candidate, for his lunar mapping project was well under way by this time and had become common knowledge among European astronomers. It seems, in fact, that Van Langren learned of these efforts at a relatively late date, no doubt due to reasons of war and possibly to his concentration on nonastronomical subjects during the late 1630s and early 1640s. In any case, one finds him now calling on the king to secure, by royal fiat, a new realm of scientific discourse. This, we recall, was not a new thing: Galileo had named the four "Medici stars" in the hopes of securing the patronage of that powerful family, and had even insisted that God himself had guided the choice.[16] Van Langren, meanwhile, received the king's permission and the general decree was issued (though whether it was sent to the intended "individual," a Protestant, is not known).

Between February and the end of May, Van Langren completed at least four maps of the lunar surface, two of which are identical (see Wislicenus, 1901; Prinz, 1903–1904; Bosmans, 1903, 1910). These are of varying quality and detail, the most carefully finalized being the one shown in Figure 9, carrying as many as 322 names (it is this image of which two copies survive). The other maps have fewer names and lack the specific date, author signature, and long inscription at the bottom.

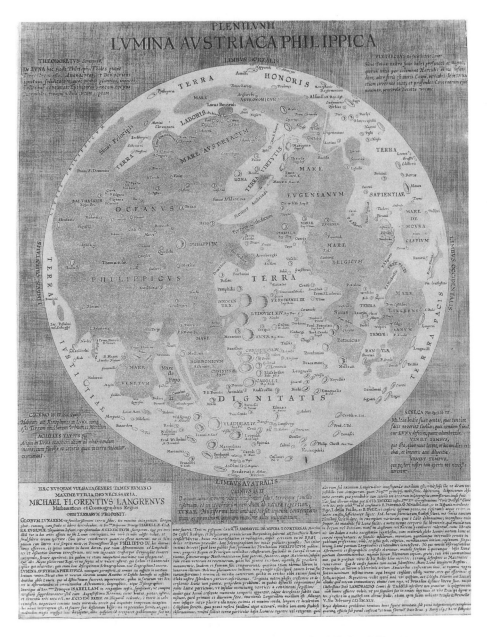

FIGURE 9. Map of the Moon published in 1645 by Michael Van Langren, showing nomenclature and lengthy inscription. Courtesy of Ewen Whitaker.

Moreover, the nomenclature tends to shift a bit between images. It seems clear that Van Langren was in a hurry and was trying out different approaches to creating a suitable map, but also that he was consciously and cleverly directing his eye to suit different audiences and needs. The version he sent to the Privy Council, for example, he knew had to make an immediate and powerful effect: this is the only one of the maps to be illuminated. According to Bosmans (1903), the colors are beautiful and highly artistic, with the lunar disc a pale yet vivid yellow, the surrounding page tinted in blue, the "seas" a deep aquamarine, and the craters a violet-brown with traces of gilding. This is the simplest of the maps: its features are smallest in number, their outline more generalized, and the craters show the least relief of any of the images. Such simplicity was obviously calculated to allow for the best effects of color and to avoid overwhelming his unscientific viewers with detail. Another map (now located in Strassburg) is very similar to the final, published version, but lacks several detailed features, names, and inscriptions, and thus seems a sort of "draft." The version of Figure 9 most fully obeys the general conventions for a published map of the time. In the mid-17th century, it was common to publish maps rich in textual decor, with quotations in each of the four major corners; titles above and below; descriptions of various voyages and discoveries placed within the map to cover the blank space of uncharted areas; and small biographies of a mapmaker's purpose, abilities, and experience, engraved in Latin in some lower portion of the image. Van Langren obeyed most of these conventions in his final map, which he presumably intended for a scholarly audience.

The commonness of these features, however, cannot diminish the unique and very remarkable qualities of this document. The nomenclature itself (which I will discuss in a moment) is one reason for this; another, however, no less striking, is the content of the long inscription given at the bottom, a writing Van Langren rendered as part confession, part personal history, part paranoia, and part propaganda, an honest and undisguised portrayal of its author's intentions and weaknesses. Indeed, it is very much worth quoting at some length, for it provides one of the best examples in any age to show how the personal ambitions and failings of a particular scientist become the immediate material for a new language (in this case of names):[17]

> *These things, remaining until now unpublished despite their great utility for mankind, despite even their necessity, Michael-Florent van Langren, mathematician and cosmographer to the king, offers to the entire world.*

The lunar globe is at once the most obvious of the stars and the least known. Its geographic description was undertaken by me with great care and effort for her most Serene Highness, Princess of Belgium, Isabelle-Claire-Eugenie, Infante of Spain, and I have finished the task. The great love of this princess for the sciences urged her to command my assistance for her observations of the moon; still more, she visited me in order that we might contemplate together the secrets of that star. She understood their great importance. Charging me therefore with letters in her own hand, she sent me to Spain, there to abide by the all-powerful king Philippe IV, to whom I was to present my observations, to publish them in his name, and to therefore furnish a reliable astronomical method for determining longitudes and distances of terrestrial locations and, through these, to correct enormous geographical errors. Marine navigation might also make use of them.

In all this, the great king took a keen interest. He had me often called to his side, in order to assist him in observing the sky and moon with the aid of the telescope. He further allowed me to give this selenographic description or lunar geography the title *By the Light of Philip of Austria* and to make it known to the world under the auspices of his name. He also approved setting down the names of famous persons for the luminous and brilliant mountains and islands of the lunar globe. These will serve to distinguish [such features], and it will be possible henceforth to make use of them in observations and corrections of an astronomical, geographical, and hydrographical nature. In his letter of response to her most Serene Highness, Princess Isabelle, [the king] ordered that the required expenses be paid. Yet, upon my return from Spain, in 1634, this august heroine, whose goodness, justice, piety, and clemency were known to all the world, was returned to the earth en route to the skies, to there witness these great marvels at close hand.

It pains me to recall what misfortune of time my work, already in progress, then suffered interruption and, deprived of its support, vacillation. It began little by little to be disclosed to others, to such a point where the danger arose of seeing another make off with it and publish it under his name. Finally, the excellent don Emmanuel de Moura y Cortereal . . . governor of the Belgian provinces and Burgundy, etc. intimate adviser on finance, was informed as to the uniqueness and utility of the work. He foresaw the eternal glory which would be showered upon the king and gave permission for the publication of this selenography . . .

Encouraged by this decree (that it might be for the benefit of the general public!), we have begun with the publication of this Philippian Full Moon, adorned with proper names. These are of kings and princes (who reign at present in Europe, as the patrons, promoters, and supporters of the mathematical sciences), more still of ancients and moderns, eminent in this field, who have acquired commendation and glory by the excellent monuments of their genius. We will also publish a book in their honor. To our profound regret we have been unable up till now (but hope to correct this soon) to learn the names and achievements of those foreigners of excellence in the sciences, in order to inscribe them equally on our brilliant globe.

We possess already thirty images of the waxing and waning phases of the moon. These appear incessantly. We show distinctly all the details of the lunar surface, such as the islands and summits of the mountains. Removed very often from the continents, they appear instantly during sunrise on the moon and disappear suddenly during sunset. They provide the means which are best, and of an almost daily employ, for the determining of longitudes. . . . Such are among the things that, by order of that most Serene Princess, we demonstrated, in 1631, to scholars very knowledgeable, illustrious, and renowned in this science: E. Puteanus and G. Wendelin; then, in Spain, by order of the king, J. della Faille and B. Petit . . .

To avoid, however, the confusion that would result in astronomical and geographic observation from changes that someone might effect on these denominations of the moon, we have distributed a large number of copies of this engraving, without expense and very widely, to all those whom we venerate as being today the benefactors, defenders and patrons of these studies. We therefore dedicate this image of the moon most humbly to the kings, the princes, and the noble amateurs of these sciences. We request of them to approve our nomenclature of names, to change nothing, and to agree in the greater part of what we offer them here . . .

5th of the Ides of February, 1645, Bruxelles

Prohibited by decree of the king: any change in the names of this figure, under pain of indignation; any counterfeit copies of said image, under pain of confiscation and three florins fine.

Given this day, 3 March 1645

There are many things one might say about this writing, which, as it happened, turned out to be the epitaph to all its author's ambitious plans for his work on the Moon. The following year, 1646, saw the death of Puteanus, and Van Langren, deprived of his second patron, abandoned the lunar surface for more mundane and profitable subjects, especially of a military sort (fortification plans and the like), seeking to follow royal favor during the last, particularly intense years of war.

In this writing, which he chose to leave the world as a legacy to his vision for the Moon, Van Langren shows himself, as in all his activities and writings, exceedingly dependent—financially, but even more, emotionally—on the favor of his patrons. Such favor, it appears, was his lifeblood. And this lifeblood spilled itself across the surface of his map. Indeed, Van Langren's Moon was exactly that: a personal orb of gratitude, faith, and obsequiousness. Could it be any surprise, given the sentiments expressed in his text, that his nomenclature would be, above all, oriented toward preserving forever the remains of a Catholic Europe even then about to disappear? Again, the title itself, claiming Austria as a terrain of Philip, is a direct expression of the Spanish Hapsburg's own move to join forces with the Austrian branch of the family to fight the Protestants in Germany. As for the "seas," Van Langren gave them titles such as *Oceanus Philippicus, Mare Austriacum, Mare Borbonicum, Mare Popo* (i.e., Pope), *Mare Eugenianum* (for the princess), *Mare Belgicum, Fretum Catholicum,* and so on. The intervening continents, or terrae, he denotes with names such as *Honoris, Sapientiae, Dignitatis, Pacis, Virtutis,* and (again) *Philippicus,* thus implying the plenary virtues that exist between and among the referents just mentioned. The major craters, meanwhile, are all named after Catholic kings and princes, both living and dead: Philip IV, Ludwig XIV, Innocent X, Alfons XX, Carl I, Ferdinand III, Christian IV, Ladislaus IV, and so forth. Each of these is also followed by a specific royal title, for example, Carl I, *Reg. Britt.* (Charles I, king of England). A few other large craters, of somewhat smaller size, are named after queens and princesses, along with their titles. Saints' names—Augustine, Bede, Dominic, Francis, Ignatius, and so on—are reserved for headlands and capes. Finally, the minor craters are usually given the names of famous scholars, mainly mathematicians and astronomers, both ancient and contemporary—Hipparchus, Aristarchus, Kepler, Copernicus, Galileo—but a few are reserved for Van Langren's personal friends: Puteanus, Wendelinus, Lafalli, and himself, Langrenus.

The hierarchy of this scheme is obvious and was not lost on astronomers of the day. The effort to claim the Moon as Catholic territory, to make Pythagoras and Euclid the semantic subjects of Philip and Isabelle, is only too clear. The scheme far too overtly placed "science" at the very bottom of coined importance, and fawningly bowed to the imperial ambitions of a few select nations. Indeed, the king of Spain himself appears no fewer than four times on Van Langren's Moon. The lunar surface, according to this scheme, did not belong to astronomy: astronomy belonged to royal power. Thus, Van Langren's map also inscribed a certain truth, namely, the relative fragility of contemporary science in real-world terms, its vulnerability to systems of patronage, its status as a sometime protectorate of individual rulers.

In the months just before his map was published, Van Langren spoke of his choice of names with Puteanus. These letters, in more than one instance, come eerily close to Gassendi's own memorial impulse toward friends and colleagues; it is as if the scent of death were in the air. Puteanus, for example, now less than a year away from his own demise, expresses his gratitude that he might see himself "written in your moon and destined to live there always with you, as we have been together here below, on our mortal earth." He advised Van Langren, "Do not use many ancient names. . . . Use those of our contemporaries, in order that they gain such renown that they will then work to fulfill their reputation" (quoted in Bosmans, 1903, pp. 123–124). Elsewhere, he agrees with the use of princes, kings, and royal families, and he also recommends expanding the pantheon of scholars, even to include such non-Catholics as Huygens. "It must be done," he says, "in a manner such that our enemies should have no pretext for producing a new map of the moon, in their own image" (p. 130).

A sense of mortality pervades these words, a mortality that includes the project itself. The war in Europe, which lasted through much of the lives of these men, was not simply fought on the Moon they inscribed, it was lost there as well.

Hevelius: A Moon of "Higher" Origins

It seems both ironic and significant that Van Langren mentioned, in his inscription, a desire to include "foreigners of excellence in the sciences," yet admits that he does not know their names. His own scheme, of course, left out a fair number of his living (non-Catholic) contemporaries, whose names he probably did know, or could easily have learned. One of these was Johannes Hevelius (1611–1689), Protestant merchant, astronomer, and city official of Danzig (Gdansk), who became Van Langren's main rival with regard to mapping and naming the Moon, and against whom, in the last years of his life, the Belgian bitterly and ironically complained for having ignored his own work. This, however, was characteristic of Hevelius, who often placed himself above all other living astronomers, saying even of Galileo that he "lacked a sufficiently good telescope, or he could not be sufficiently attentive to those observations of his, or, most likely, he was ignorant of the art of picturing and drawing" (quoted in Winkler and Van Helden, 1992, p. 195).

Hevelius, in truth, was favored more by circumstance, skill, and worldliness than genius and originality. He was the eldest son of a wealthy brewer of German ancestry, who, like other middle-class families with recently acquired means and high ambitions for their children, provided an excellent education for Johannes before involving him, at an early age, in the family business. The financial success of the brewery, whose ownership eventually passed entirely to him, freed Hevelius from the burdens of patronage, which so many scientists of his day had to obey and endure. It also permitted him the means to construct the largest and most sophisticated observatory in all of Europe, one to which scholars, diplomats, and even royalty all made pilgrimage (see Beziat, 1875; North, 1972). Beyond this, Hevelius's education and interests gave him considerable skill in several crucial areas: engraving, optics, instrument making, lens grinding, and observing. Less a scholar than Van Langren, he was a far more able craftsman. His observatory, built in the early 1640s, contained not only a wide range of astronomical instruments of his own design and manufacture, but an art studio, all the equipment required for high-quality engraving, and possibly a printing press as well. Indeed, Hevelius's success depended upon his ability to largely control, with a minimum of intermediaries, the means of production of his own science—from raw observation to published result. In some ways, it was this liberty from most ordinary social restraints on the making of science that is most impressive about Hevelius.

Most biographies relate that the young Johannes grew deeply attached to his first tutor and later friend, Peter Kruger, a well-known mathematician and astronomer. Kruger gave private lessons to Hevelius and encouraged him to take up drawing, engraving, and instrument making (MacPike, 1937). After traveling to Leyden, London, Paris, and Avignon, during which he met many of the century's most famous astronomers, Johannes was called home in 1634 to study law and to thereafter take his place in his father's business and in the administration of his native city, both duties incumbent upon a prominent merchant's son. The story then goes that, having neglected astronomy for several years (to his own private chagrin and longing), he turned to it once again and with final resolve in 1639, when called to the bedside of his old mentor, who, now fatally ill, expressed in his dying words the wish that his favorite pupil not abandon the subject that had bound them so closely in life (Westphal, 1820; Beziat, 1875). Kruger's death worked a conversion upon Hevelius: he decided to devote his life to a higher purpose. During the next few years, he gave all his free time and considerable expense to building astronomical instruments; to observing the planets, the Sun, and especially the Moon; and to starting up a prodigious correspondence with other astronomers, scholars, diplomats, and officials through whom he was able to learn in detail of other efforts like his own, and of others' discoveries.

Through these contacts he created a kind of network to transmit throughout Europe the news of his own work and successes. This network, when seen in light of his total astronomical operation, suggests that Hevelius applied to his science many of the lessons he learned from his commercial activities. Shrewd in producing his own science, he seems to have been equally aware of a need to enter the "invisible college" of his scientific contemporaries, to trade and bargain for information if need be, and to change course when advisable. Beziat tells us, for example, that upon learning of Gassendi's project for a lunar map, "Hevelius wanted to renounce his own work; and it was only the pleading of Gassendi, and the assurance that he had abandoned his intention, that [Hevelius] reconvened his own study of the Moon" (1875, p. 504).

Gassendi, in fact, as Beziat reveals, actually opposed publication of Mellan's images in order to send them to Danzig in 1644, thereby "leaving to Hevelius the honor of the first lunar map" (p. 504). The French astronomer, in other words, traded away to another the culmination of a project his friend and patron (Peiresc) had pursued for more than a decade. The reasons he did this were no doubt connected to his desire to see the project actually brought to fruition, but there is no denying that Hevelius had positioned himself well for the reception of these images at this early date (less than 4 years into his astronomical work). Indeed, seeing two of Mellan's engravings helped convince Hevelius to broaden his own work considerably, from a single map of the full Moon to a complete atlas of all the lunar phases. On this he appears to have labored night and day for a year or more thereafter, observing after sunset and drawing in daylight. With the publication of Van Langren's map in May 1645, he expanded the project yet further to include more than 40 detailed drawings, with several of the full Moon showing the limits of libration, and others mapping the waxing and waning phases (to which he gave individual names), along with detailed engravings showing the construction of his instruments, as well as many optical sketches, and nearly 500 pages of explanatory text, including a glossary of some 275 named features of the lunar surface.

This work, one of the grandest publications in all of 17th-century European astronomy, was entitled the *Selenographia sive lunae descriptio* (Selenography; or, A Description of the Moon). Appearing in 1647, its title was apt: it attempted nothing less than the most complete description in text and image of the lunar face, as well as nearly all other areas of telescopic observation of the day. It did not venture deeply into theoretical, mathematical, or computistic territory; indeed, the first seven chapters largely repeat what were by then well-established arguments disproving the idea of the Moon as mirror. They also offer a treatise on lenses and lens making, the construction of telescopes, observations on the planets, the moons of Jupiter, and sunspots. As such, they provide a kind of guide to "the practice of observation" itself (Winkler and Van Helden, 1993, p. 106).

But the true core of the work is its discussion of the Moon, and in particular its images. Though much inferior in naturalism to those of Mellan, these nonetheless present the Moon in so many phases as to offer the reader (who is as much a watcher) a kind of time-lapse succession of the full lunar cycle (one can, in fact, flip the pages to gain this very effect). Those drawings showing equipment used in the observatory, meanwhile, present their machines and instruments not as they would appear to a visitor but instead in completely disassembled form, with pieces lying about so that the eye is given access to everything, to every last detail. Letters placed on each element are used in the text to refer to assembly and operation, but the reference is still visual. As in Galileo's *Sidereus nuncius*, the narrative is itself focused on delineating things, mostly observations. But here, the relationship of effect is reversed: the text is unequal to the images themselves, which seem to strive for their own power of explanation, effected through a merger between naturalism, cartography, and baroque decoration (see Figure 10). The title page to *Selenographia*, meanwhile, unlike that of Galileo's *Sidereus*, is replete with terms such as *accurata, delineatio, observationes, figuris accuratisime aeri incisis,* and *experiendi.* The claim is for a superior kind of eye and vision. It is therefore less for an "unfolding of great and very wonderful sights" before the reader, or for making him into a virtual witness, than for overwhelming him with the evidence of Hevelius's *own* witness, the testimony of his eye. Visually, his images lock one into what amounts to an individual world of seeing and picturing. As his title page says: "questions of astronomical, optical, and physical importance are both propounded and resolved" (*astronomicae, opticae, physicaque quastiones proponuntur atque resolvuntur*), with emphasis, one suspects, on the final claim. In contrast to Van Langren, Hevelius did not draw the Moon as an overtly political idea; he did not wish to celebrate a royal patron but instead to make visible, for other astronomers, his own royal powers of observation.

Hevelius had a tendency to belittle his contemporaries, because his true ambition was to enter the ranks of the greatest naked eye celestial seers of the past: Hipparchus, Ptolemy, Al-Battani, Ulugh Beg, Tycho

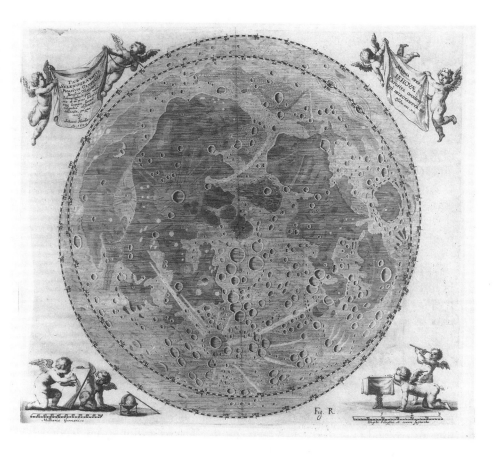

FIGURE 10. Image of the full Moon engraved by Johannes Hevelius and published in his *Selenographia* (1647). Courtesy of Ewen Whitaker.

Brahe. These luminaries, for example, he had gathered together and engraved as a council of sages on the frontispiece of his final work, *Prodromus astronomiae* (Astronomical catalogue, published posthumously), where they surround Urania, muse of astronomy, to whom Hevelius himself is shown presenting his work for judgment. The image is truly a minor masterpiece of baroque engraving, with its myriad details of clouds, cherubs, cartouches, animals, and unfolding scrolls—all there to be seen and lingered over, a kind of "vision" or *somnium* whose center is Hevelius's own appeal to join the pantheon of the canonical eyewitnesses of the skies.

The frontispiece to the *Selenographia*, on the other hand, is even more rich in meaningful iconography (Figure 11). This, one should perhaps note, was not drawn and engraved by the author himself (in the lowermost left and right corners we see this was done by Adolf Boy and S. Falk, respectively), but was probably designed by him. It shows Contemplation, telescope in hand, parting the clouds of ignorance, through which also shine the images of Hevelius's Moon on the left and his Sun (complete with sunspots) on the right. Immediately below, two putti (cherubs) hold up a scroll on which is written: "Lift up your eyes to the heights and see who has created these things" (*Attollire in sublime oculos vestros, et videre qui creaverit ista*). The words are from Isaiah 40:26 (Winkler and Van Helden, 1993). Together with the image, they seem to say that the true astronomer is he who does not merely observe, but who contemplates, reflects, and finally, worships. Only in this way can he become (as presumably Hevelius has) the messenger of divine truth.

Below the two putti, meanwhile, stand two figures on pedestals, holding a banner with the title of the work. One of these figures, on the left, is the famous 11th-century Islamic astronomer, mathematician, and expert on optics, Al-Hasen, author of a detailed commentary on Ptolemy (critical of the latter's ideas on the Moon). On the right is none other than Galileo, a telescope in one hand, the other grasping the title banner with care, fingers pointing subtly but noticeably downward toward Hevelius's name. Beneath Al-Hasen, in a framed niche, is the pate of a head (meant to indicate the higher portion of the brain) under which is written the word *ratione* (reason); Galileo stands above a niche inscribed with a human eye and the word *sensu* (perception). The overall message is therefore clear and merges neatly with that of the figures above. The powers of Al-Hasen and Galileo, of Arabic and European astronomy, though great in achievement, are nonetheless limited in separation and must be joined in any true selenography worthy of the name. Again, mind and eye must be joined, and the promise is that Hevelius has been the one to do this.[18]

It is a cliché of astronomical history that Tycho Brahe was the final significant observer of the heavens to rely solely on the naked eye. This, however, is untrue. In a field renowned during the 17th century for so many "firsts," Hevelius fits this "last" more than anyone. Most of his career was devoted not the the Moon but to making precise positional measurements of the stars and constellations, comets, and planetary motions. For this the telescope was simply inadequate, technically speaking, until the late 1660s when the use of calipers or cross hairs became widespread. Earlier on, the telescope was a tool of *discovery*, not of measurement. With regard to the Moon, Hevelius participated in this

FIGURE 11.   Frontispiece to *Selenographia* by Hevelius. Reprinted by permission of Harcourt Brace & Co.

contemporary exploration. But for the rest, he employed only the naked eye as his instrument. "Like Tycho Brahe and his predecessors," Hevelius worked "by measuring the positions of stars through slits adjustable by screws" (Lambrecht, 1967, p. vi). He appears to have been blessed with excellent vision; the equipment in his observatory represented sophisticated versions of what were often ancient apparatuses, or else implements that were rapidly becoming outmoded everywhere else in Europe by use of the telescope. In this sense, Hevelius was the last of his kind, and (as the frontispiece to his *Prodromus astronomiae* strongly suggests) he knew it.[19] His desire was to become equal to the greatest of the ancients, to carry on, in modern guise, the world of classical astronomy. Even when the telescope had advanced to become a measuring tool, Hevelius did not switch over. He had become too invested in the past.

As with other aspects of his science, he did not want to employ intermediaries here either. To a degree, Hevelius lived in an astronomical world much of his own making: he was, as I have implied, a kind of singular monarch who reigned expertly over every territory of his realm, in complete control of the borders and products of each. This analogy gains a literal dimension, moreover, when one considers the fact that several kings, princes, and princesses visited him at his observatory; that his expenses in building this intellectual fortress were "deemed worthy to excite the emulation of rulers" throughout Europe (Lalande, quoted in Beziat, 1875, p. 661); that King Louis XIV of France, on the advice of Colbert, included Hevelius among the 60 non-French scholars to whom he granted a lifelong pension; and that, following a tragic fire in 1679 that destroyed the observatory and many of Hevelius's manuscripts, the kings of England, France, and Poland all contributed funds for its immediate rebuilding (Westphal, 1820). Royalty of another type was also drawn to Hevelius's majesty over the heavens. No less a figure than Pope Innocent X, upon being shown a copy of the *Selenographia*, is said to have remarked that "such a book would have no equal were it not written by a heretic" (Hevelius was a Protestant).

One returns to Hevelius's view of himself, which, in addition to the frontispiece drawings mentioned above, also gained expression in the opening literary portion of the *Selenographia*. The work begins not with a normal prose address to the reader, expressing humility or hope regarding the efforts to be found within. It opens instead with a series of lavishly (again the adjective "baroque" comes to mind) honorary poems about the author himself, written in elaborate Latin by various friends who were scholars, teachers, officials, and ministers. The tone is distinct in its courtliness: Hevelius is the muse of the Moon; his virtues surpass those of Columbus, Daedalus, and the Greek heroes; he has performed a total eclipse of Galileo; he is the new glory of his city; his name shall live forever, beyond time, and so on. It seems a dedication, in short, more appropriate to a patron prince, yet it was one commissioned and published by the author himself.

Royalty thus recognized in Hevelius something admirable, something of power and reputation akin to themselves. He, too, after all, had gained control over a sought-after territory and had presented this to the world. More, he had done this by means of the eye most of all, providing detailed surveys of a domain he claimed to explore, map, and name better than anyone before him. Like any monarch whose feet were planted too deeply in the soil of the past, his reign was not to last; history itself would defeat him. But for the moment, he was king of his world, which sought to include all the heavens. As the pope's words suggest, his achievement undoubtedly had a political dimension within the context of the religious wars of the time—less overt, certainly, than Van Langren's polemical Moon, but still constituting a "claim" in which those opposed to Hapsburg Spain and Austria could take some degree of pride.

🖅🖅

Finally, one comes to Hevelius's naming scheme for the lunar surface. This, too, reveals a very special loyalty to the past. It was a scheme founded on an analogy, geographic in nature, but carried to a literal level. Simply put, Hevelius perceived in the Moon's surface, turned 90 degrees sideways (counterclockwise), an immediate resemblance to the Mediterranean region—in brief, the classical world as known to the Greeks and the Romans. His nomenclatural image is therefore completely, strikingly, an Earth-type map, showing the lunar surface entirely conquered by the conventions of a classicist terrestrial cartography (see Figure 12). One finds on the Hevelian Moon such features as the Mediterranean, Adriatica, Hyperboreum, Black, and Caspian Seas, all in their relative positions; such islands as Majorca, Malta, Sardinia, Sicily, Crete, and Cyprus (and many others); the Italian and Peloponnesian peninsulas; Africa, Libya, Arabia, Armenia; the Alps and the Atlas, Tarus and Caucusus Mountains; and any number of more local features named for cities such as Byzantium, Herculaneum, Athens, for classical gods and heroes (e.g., Letoa, Neptune, Hercules, Cadmus), or for specific mountains and valleys on Earth (e.g., Sinai, Olympus, Ida, Hajalon).

This by no means exhausts the full range of Hevelius's scheme, but it gives some idea of its orientation. A full glossary of the names he used,

FIGURE 12. Hevelius's nomenclatural map of the Moon, showing his interpretive scheme for observed features. Courtesy of Ewen Whitaker.

including some 275 titles, claims to offer the titles of the Moon's chief "seas, bays, islands, continents, peninsulas, capes, lakes, swamps, rivers, plains, mountains, and valleys" (p. 228). Each of these, moreover, is listed in alphabetical order and described in terms of its reference location on Earth, along with other names that have traditionally been attributed to it. The attempt to discover a terrestrial geography in the lunar surface here reaches an end point, with every type of earthbound geomorphic phenomenon identified and analogically named. Indeed, one *does* find the Nile drawn in, with three stream-like lines ("inaccurately" debouching into the Bay of Sirte). The craters, Hevelius says, are actually valleys and are therefore not round as they seem, but only appear to be so due to their great distance, which obscures their irregularity (see Delambre, 1969). The map is shaded and hatched according to cartographic conventions of the time, with watery areas darkly outlined and a great number of mountains and mountain ranges drawn in the form of small rocky protrusions, casting shadows to the lower right (sunrise). Scholars have often expressed dismay or confusion as to why Hevelius did this (see, e.g., Strobell and Masursky, 1990), or why, for instance, he drew the bright rays of Tycho and Kepler/Copernicus as long narrow lines of mountains. Any such confusions vanish, however, when one understands the degree to which he quite literally "saw" the Moon as another Earth. As a detailed example, he drew Sicily (Kepler on today's maps) as a large volcanic island, and named its center Mt. Aetna, after the most well-known and active volcano in Europe: the manner in which this is drawn approximates the general morphology of a strombolian-type volcano, with radiating ridges resulting from lava flows. In his text, furthermore, Hevelius speaks of another feature he calls *Mons Porphyrites* (today, the formation Aristarchus) as being "without doubt" a volcano in "the midst of continual eruption," similar in its red color to "those known to us as Aetna, Heckla, Vesuvias, etc." (p. 353; see also Beziat, pp. 511–512).

The relevant map here is therefore not merely a nomenclatural image, as is often suggested. It is the result of an analogy taken into the realm of actual perception.[20] It is, in fact, the third in a series of pictures labeled P, R, and Q, with the first of these (P) showing the surface as an ornamental design of flat spots and bright areas, with the second (R) adding topography to the craters and more complex shading, thereby coming closer to a naturalistic view. The map of image Q (see Figure 12), then, is related to the other two as an interpretation to "data": it is Hevelius's "theory" of the Moon's reality.[21] Regarding whether or not the Moon had inhabitants, Hevelius believed that it did. But his rationale seems completely ironic: "Just because we do not perceive any beings there, it does not follow there are none. Would a man raised in a forest, in the midst of birds and quadrapeds, be able to form an idea of the ocean, and the animals which live there?" (quoted in Beziat, 1875, p. 512). Yet it is Hevelius himself, isolated in his magnificent self-sufficiency, who gave the Moon necessary life—for how could a world made literally in the image of a select portion of the Earth, a portion "highest" in civilization, complete not merely with seas, plains, and mountains, but with rivers, swamps, and volcanoes too, *not* be inhabited? Hevelius is using text and image in a combined manner that asks us to merge reading and seeing into one experience. We should consider, that is, what kind of cognitive act is actually involved in "recognizing" (and inscribing) a feature such as the Nile River on the lunar surface. Had Godwin a copy of Hevelius's map, his hero Gonsales might well have explored the upper reaches of this lunar stream, perhaps even to its headwaters (the Mountains of the Earth?), centuries before its correlative in terrestrial Africa. The satires of Godwin and de Bergerac simply postulated, as a literary device, such possibilities; Hevelius, on the other hand, transformed them into a phenomenology.

In the chapter devoted to his naming scheme, the author says he turned away from using the names of astronomers or other persons "to avoid human vanity and petty jealousies." An often-quoted statement, this seems unconvincing. Whether it involved a direct criticism of Van Langren's nomenclature is not known. Certainly, it was a turning away from tradition within science itself.[22] It would seem clear, however, that Hevelius believed his scheme would win out, and that, instead of seeking royal decrees to such effect, he felt a need to find some neutral, harmonious ground that might avoid the signs of conflict then raging throughout Europe, and through the scientific community as well.

Hevelius sought a "higher" vision for the Moon, a second charting of the classical world that might serve the unifying purpose of monumentalizing the (presumed) origins of Western scholarship. The Moon would be a surface on which every scientist, of no matter what nation or religion, could perceive his own origin. The domain Hevelius himself had built, had ruled, and through which he had gained fame, was one deeply linked to intellectual traditions looking back to this origin, and it was this world, in its "original" form, that he sought to resettle on the Moon. Thus, the effect was problematic: where Van Langren placed "science," in the form of its heroes, at the bottom of his nomenclatural hierarchy, Hevelius eliminated it altogether. For him, the Moon was a possession of European classicism but not of science, which was forbidden a single specific claim. In trying, perhaps, to create a vision for unity, Hevelius made invisible the very community he was trying to reach. Even more, his scheme enthroned the classical world at a time when the ancients themselves were being increasingly rejected as models for understanding the physical universe. The entire 17th-century movement away from classicism in the sciences, beginning with the challenges to Aristotle, are completely denied by the Hevelian nomenclature. The moderns, in short, who by then *were* science, are nowhere to be found. The fact that Hevelius did not include his own name anywhere on the lunar surface becomes a sign of the inevitable failure of his scheme. Indeed, if Van

Langren had aimed at the preservation, for all time, of a Catholic Europe even then in the throes of defeat, Hevelius, with his adherence to ancient ways of seeing, sought to embody an ancient seat of origins, yet one whose means and thought were now, finally, being superseded by those of the moderns. An eternal honorarium to Greece and Rome was not well suited to an age eager for new confidences of its own.

Riccioli: The Moon as a Conflictual Community

For nearly a century after it was published, the *Selenographia* remained the standard reference on the Moon in northern Europe. Hevelius's maps and his nomenclature held sway, while those of Van Langren, and of others who came after, were ignored or forgotten. The reasons for this have a great deal to do with the politics of the time: continued religious and national conflict (including the decline of Spain and Italy as major centers of intellectual culture), the ill-fated choice of names Van Langren had made, and Hevelius' own overarching fame, his great network of contacts, and the scale and general accuracy of his work. Yet his Moon, too, did not survive the 18th century. The state of religion and religious war helped ensure that even as early as the 1650s and 60s, it would fall from the sky in much of southern Europe, being replaced with a new version produced by Giovanni Battista Riccioli (1598–1671).

Riccioli was a Jesuit priest and professor at Bologna. A flamboyant intellectual, he taught not only literature and theology, but philosophy, mathematics, and eventually astronomy as well. His was a restless intellect in search of a center, a nexus on which he could focus his ambitions. During his own lifetime and for two centuries thereafter, he was renowned for his ability to digest incredible amounts of reading material and to move among disciplines with an ease denied most of his contemporaries. His talents allowed (or drove) him to publish a work of even greater scope and magnitude than Hevelius's, indeed a work that sought nothing less than a comprehensive discussion of all the heavens and whose ambition is adequately revealed by its title, *Almagestum novum*.[23]

Riccioli sought to become the Ptolemy of Catholic astronomy. His work is a monumental attempt to "save" the Christian version of the cosmos against Copernicus, Kepler, and Galileo. And yet, despite this, it is a work that ends up in admiring support of these same "enemies" of the old order. The *Almagestum* provides an encyclopedic history of earlier work and thought on astronomy, especially on the Moon, in literature, philosophy, and science, from classical antiquity to the present, and for this reason it served as a standard historical reference well into the 19th century (see, e.g., Bailly, 1779). It tried to synthesize, to reorder, and to give critical meaning to the entire corpus of Greek, Roman, and European writing dealing with astronomy, including writing that fell outside the realm of "science." The book represents the culling of thousands of texts in a half-dozen languages. Even if one takes note of all its failures, it remains a work of spectacular energy and ability. Yet among its many efforts, the only one whose result is still recognized in the 20th century is Riccioli's lunar nomenclature, which in fact composes the Moon as we know it today. The reasons for this, too, have everything to do with the type of community he placed on the lunar orb.

Riccioli's work appeared in 1651. From the moment it appeared, it was seen as the culmination of attempts at lunar mapmaking that followed in the wake of Van Langren (Whitaker, 1989).[24] As noted by Kopal and Carder (1974), most of these maps show the direct influence of Hevelius (his images P and R). As maps, however, these attempts lacked names, and though they show no obvious regression in their aesthetic form, as do the drawings made immediately after Galileo, they are nonetheless manneristic stylizations based on copies of existing work. Their general degree of accuracy varies little; their decorativeness is much more in line with Hevelius than with the naturalism of Mellan or Galileo. This is true for Riccioli too, but here the debt to Hevelius is openly admitted. Moreover, regarding his nomenclatural image, Riccioli clearly sacrificed artistry and design to the necessity of letting the names themselves stand forth (see Figure 13). The librational limits are entirely copied from Hevelius's images, as are a fair number of the features (compare Figures 10 and 13). The only real advance, again, other than a slightly higher degree of accuracy, is Riccioli's nomenclature.

This scheme avoided Van Langren's Catholicism, yet followed his and Gassendi's lead by repopulating the Moon with the names of famous astronomers. For this reason, in part, it soon attracted attention and approval within the scientific community; it was favored by Cassini, Hooke, Huygens, Mercator, and Wren (who constructed the first lunar globe), among many others. A century later, one finds the great French astronomer Lalande full of praise for Hevelius's observations but dismissive of his naming scheme, which he called "rather bizarre" (Beziat, 1875, p. 506). By this time, the Riccioli nomenclature had become nearly standard, and was finally made so by the pathbreaking lunar atlas of Johann Schröter in 1791, which thereafter set the pattern for lunar maps

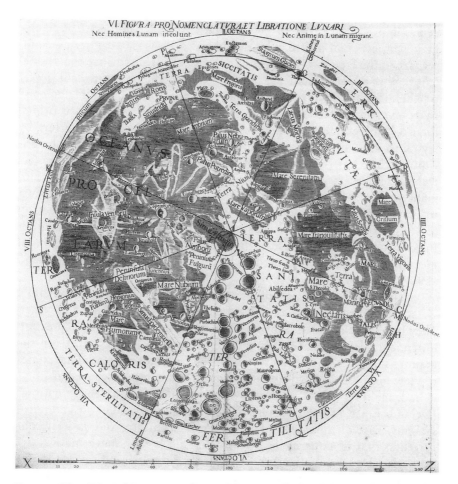

FIGURE 13. Riccioli's nomenclatural image of the Moon, showing the naming scheme that remains in use today. Courtesy of Ewen Whitaker.

down to the photographic era (Batson, Whitaker, and Wilhelms, 1990; Strobell and Masursky, 1990).

To understand Riccioli's scheme of names—the scheme that eclipsed both Van Langren and Hevelius, and even the conflict between them— one needs to understand something of the *Almagestum novum* and its context. Its many difficulties, contradictions, and ironies are a direct result of the Counter-Reformation, in which the Jesuits played no small role. It is evident from the title, as I have noted, that the author conceived himself as a type of neo-Ptolemy. Indeed, another of Riccioli's major projects (unfinished at his death, however) was to compose a counterpart to the latter's *Cosmographia*, "a single great treatise that would embrace all the geographical knowledge of his time" (Campedelli, 1975, p. 411). And if Ptolemy had originally sought to "save the phenomena" by re-explaining the classical cosmos on the basis of ecliptical orbits, then Riccioli was again attempting a modern version of his effort by trying to uphold Aristotle and the literal word of the Scriptures by critiquing the Copernicans and offering his own substitute theory. As Delambre describes it, "The Copernicans complained that the theologians, completely ignorant in mathematics, had hurled decrees against their system without reason. The desire to respond to them urged Riccioli to study astronomy; thus it was not strictly a love of truth, but rather the desire to plead the cause of the theologians and to defend the literal meaning of Scripture, that rendered our author an adversary of Galileo" (1969, vol. 2, p. 275).

The problem for Riccioli was that as an able mathematician and a gifted amateur scientist, as a devout Catholic and a Jesuit pledged to obey his Church, and as a lover of the truth, he soon found his position untenable. He attacks Galileo and Kepler, and Copernicus too, but the manner in which he does so seems, in the end, half-hearted. One comes to view him, "as an advocate who, unable to abandon a cause he knows is bad, makes every effort to show it up to defeat and condemnation" (Delambre, 1969, p. 275). In his discussions of the various cosmological systems, Riccioli finds none so elegant, so accurate, and so well conceived as that of Copernicus, yet he calls this "only a mathematical hypothesis." He tries to divorce it from probable truth, and then proposes a scheme he knows is more complex and less plausible, in which Jupiter, Saturn, and the Sun revolve around the Earth; Mercury, Venus and Mars about the Sun; and the Moon with no motion at all. As Delambre notes, "Doubtless he imagined this hypothesis [which was closer to that of Tycho Brahe] in order to persuade the reader that he believed in the immobility of the Earth; yet, despite such efforts, one sees that without his cowl he would have been a Copernican" (p. 279).

All this is wonderfully revealed in different form by the frontispiece to the *Almagestum novum* (Figure 14). In its overall structure, the image is clearly modeled after Hevelius's engraving for his *Selenographia* (see Figure 10). It lacks, however, the latter's elegance, balance, and restraint, being much more crowded, overdrawn, and visually confusing. Indeed, Riccioli's image, like his book, is absolutely crammed with all manner of mythological, biblical, scientific, classical,

FIGURE 14. Frontispiece to Giambatistta Riccioli's *Almagestum novum* (1651). See text for discussion. Courtesy of Owen Gingerich.

and seemingly eccentric (personal) allusions. In its upper portion, for example, it is jammed with all the planets, the sun, and a comet, each carried by a separate putti, the whole arranged on either side of a radiant divine hand whose three outstretched fingers are labeled "number," "measure," and "weight" (*numerus, mensura, pondus*). The face of the sun is drawn to resemble the common representation in ancient Rome, while the Moon is given its modern, "realistic," cratered surface. Below stands Urania on the right, holding a balance on which are weighed against each other the systems of Copernicus and Riccioli, with the latter the more weighty of the two. "It will not be tipped for all time," says the muse of astronomy (*Non Inclinabitur in saeculum saeculi*), enigmatically. On the ground lie Ptolemy and his system, discarded; from Ptolemy's lips emerge the words "I am raised up while I am corrected" (*erigor dum corrigor*). Copernicus's system is the lighter, yet it is also the higher, closer both to Riccioli and to God's hand, whose fingers point downward toward it. It is nearer the light of heaven, and it rests immediately above Ptolemy himself, who seems to be looking upward at it, suggesting that it was this system, not Riccioli's, which has corrected and replaced the ancients.

Opposite Urania, meanwhile, is the figure (one assumes) of Riccioli himself, barely clothed, like a classical statue, in a loin wrap. Urania's form is covered with stars, and these are matched by a host of eyes that appear all over Riccioli's naked body. He is gazing and pointing upward; from his mouth unfurl the words "I will look upon your heavens, the work of your hands" (*Videbo caelos tuos, opera digitor tuor*). It is with his entire body and soul, therefore, both ever-watchful, that the author seems to be dedicating himself to the work of observing divine creation. And yet, it is likely that few scholars of the time would have looked upon this image without calling to mind several phrases from Virgil's *Aeneid*, where the subject is Fame (*Fama*): "A dreadful monster and vast, with so many feathers on her body as there are watchful eyes beneath . . . By night she flies between heaven and earth . . . often just as tenacious of falsehood and wrong as a messenger of truth (Book IV, lines 180–189; *horrendum monstrum ingens, quot plumae sunt cui corpore, tot vigiles oculi subter . . . Nocte volat medio caeli terraeque . . . tam tenax ficti pravique, quam nuntia veri.*)

Riccioli, in the very content of his science, personifies the confusions and contradictions often suffered by the scholarly world during the Counter-Reformation. By training and employment, as a professor of theology, literature, and philosophy, he was bound to Rome and to Roman Catholic traditions of teaching and learning. As a self-taught scientist of no small ability, however, he was propelled into a domain where such traditions were in the process of being questioned and overturned. The Inquisition had forced Galileo to recant, and Riccioli seems to have agreed with this decision, for he refused to read his works. Yet his own skillful experiments proved and even refined Galileo's law of falling bodies and led to a profound advance in using the pendulum to measure time (as Galileo had tried and failed to do). If the Tuscan had even been brought near the flame, Riccioli would undoubtedly have felt the heat himself.

The *Almagestum novum*, meanwhile, is a magnificently complex but narratively unstable work. And this, too, is exactly due to its author's intellectual and spiritual librations, not only between science and theology, but between literature and astrology, classical myth and biblical stories, history and biography. It is a work of many voices, often at odds with one another. Much of it is a compendium of existing knowledge, established methods, and historical observations, a narrative in the encyclopedic tradition. But it has many other elements as well. In its early pages, one finds all of the following: a straightforward historical discussion of astronomers and commentary on their works; the story of Prometheus; (who he says was abandoned to the Caucus in order to discern the laws of heaven); the "true" tale of a dead saint whose blood liquifies on the date of the autumn equinox; an elegant method for calculating the Earth's curvature; and a discussion of the Inquisition's lack of an official decree regarding the Earth's immobility ("however, as Catholics, we are obligated by prudence and obedience . . . to teach nothing which is absolutely contrary"; see Delambre, 1969, p. 277).

Riccioli wrote the work while serving as prefect of instruction at the University of Bologna, and he was granted leave from teaching in order to complete it. The *Almagestum* was thus known to the authorities and was considered by them as a highly significant effort to help stem the tide of rising heretical views on the heavens. Riccioli himself seems to have produced it in a fever pitch of work spanning but 3 years. And while it was used thereafter quite widely, it was also used variably, on the one hand by Jesuit professors as far away as Scandinavia intent on disproving Copernicus, and on the other hand by scholars who had long accepted the heliocentric view. Indeed, its greatest employment was as a standard reference text for every survey of the physical and mathematical sciences written during the 18th and 19th centuries (see, e.g., Bailly, 1779; Libes, 1810). Yet authors in this category rarely fail to censure the *Almagestum novum* for its rejections of the "new astronomy."[25] The long-term reception of this work included praise for its erudition, qualified appreciation for its science, but condemnation for its basic philosophy, as well as disinterest in its use of literature and myth. Thus, in the response it engendered, as in well as the contents it offered, Riccioli's book represents a complex cultural document, torn by differing allegiances.

⬚⬚

Riccioli seems to have had at his disposal the lunar images of both Van Langren and Hevelius. He states that Van Langren had sent him a number of his unpublished drawings, and that, after comparing them with those of Hevelius, he decided that while Van Langren had been first with his published map, Hevelius had been the superior in every other way, in terms of his overall conception, in drawing skill, and in the production of a great work such as the *Selenographia* (Delambre, 1969, p. 281). The problem with Hevelius, however, was his nomenclature. Riccioli did not speak of Van Langren's naming scheme, which no doubt pleased him more, as a prayer to a Catholic Europe, but which he rejected just the same. Discussing it probably would have cast an ill omen over his own, far more secular scheme, at least in the eyes of the Jesuits. Riccioli's nomenclature, as we shall see in a moment, was loyal to many masters, and thus historically very shrewd. Hevelius, meanwhile, was unacceptable for a different type of zealotry. His scheme showed he had gone too far in treating the Moon as another Earth (Delambre, 1969, p. 282). His Moon was too small: not only was it confined to a cramped region beyond which the known world had been greatly expanded, but this region was itself denied its own modern history, "frozen" in its pre-Christian state. Hevelius had sought, in other words, to make the Moon the property, the reflecting mirror, of the ancients only. Such could not be accepted in a period when the moderns—such as Hevelius himself—had gained so much ground.

The images of the Moon in Riccioli's book were not drawn by him but by one of his most talented students, Francesco Maria Grimaldi (1618–1663), who was a collaborator on much of the *Almagestum novum* and who constructed the maps by comparing his own observations against many earlier versions, refining and updating all markings over a period of years to a higher standard of accuracy. Scholars also credit Grimaldi (well known for his researches on light) with having been the one to propose using the names of famous scientists and philosophers for lunar craters (e.g., Eastwood, 1972). In any case, the engravings he produced (six in all), though less ornate than those of Hevelius, were more exact and useful, showing much less of a tendency to smooth out features and to add nonexistent craters. Less aesthetic, the images are more scientific; they indicate that even at this early date, art and science had begun to discover conflict, or at the least, competition.

The images appeared on two separate pages, one with a full-moon drawing, lacking nomenclature and surrounded by four smaller images of various phases. The other drawing, complete with nomenclature, is shown in Figure 13. The first page of engravings is notable for making direct acknowledgment in its title to the work of Van Langren, Hevelius, Divini, Sersale, and other moon artists of the 1640s, all of whose efforts are said to be "partly adapted, partly corrected." A half-moon image, on the other hand, is attributed directly to Fontana.[26] This type of citation, with its direct admittance of debt to others, was a quiet yet extraordinary addition. It had never appeared before, on any Moon image. The need to seem original—however impossible, in the wake of Galileo, Mellan, and their imitators—was no doubt important to those seeking some degree of fame from the new discoveries and growing public interest in the Moon (printers also wanted their books to sell well, and the semblance of originality was important to them too). Any glance through the many images of the day, say, those published between 1630 and 1660 (see, e.g., Kopal and Carder, 1974), would reveal at once how many were simply stolen or slightly modified from the work of others (especially Galileo and Hevelius).

The issue of attribution is not a minor one. Riccioli and Grimaldi provided in their art what science would soon come to adopt in text: a reference list, a calling on a community of like investigators. In a certain sense, this holds a key to the Riccioli nomenclature, too, which has come down as the nominative language of lunar science.

This naming scheme consisted of two basic parts, one devoted to major land and water forms, the other to the craters. The seas, continents, lakes, peninsulas, and so forth were all named after the effects and influences that had been attributed to the Moon, both astrologically and in more general cultural terms, down through the millennia, by various peoples and eras, mainly in the West. The seas in particular (as we know them today) show this directly in their invocation of fecundity (*Mare fecunditatis*), serenity (*Mare Serenitatis*), madness (*Mare crisium*), storminess (*Oceanus Procellarum*), mist and vapors (*Mare Humorum, Mare Vaporum*), rain (*Mare Imbrium*), cold (*Mare Frigoris*), and so on. Continents were also named (these are no longer identified), usually in opposition to their watery relatives: heat (*Terra caloris*), sterility (*Terra sterilitatis*), healthfulness (*Terra sanitatis*), liveliness (*Terra vitae*), cheerfulness (*Terra vigoris*), and so on.

It is in his naming of the craters, however, that Riccioli was most inventive. These, he decided, would embody the progress of scientific thought about the Moon. Riccioli conceived a hierarchy that would distribute famous astronomers and philosophers vertically on the Moon's surface according to era, philosophy, and collegiality. Those of ancient Greek ancestry appeared toward the top, in octants I, II, and III (see Figure 13); those of Rome came just below, toward the center, in octants IV, V, and IV (upper parts). These were followed downward by medieval European and Arabic thinkers in the top lower half of the image. These names, it should be noted, are not limited to astronomers per se but also include the authors and translators of important pedagogic works that proved crucial to medieval learning, such as Martianus Capella, Manilius, Bede, Alcuin, Gerber, Alfraganus, and Dante (of them, only the last was later removed). Below all these "ancients," in the bottom portion of octants V, VI, and VII, and in the whole of VIII, the names of the moderns were placed, the most contemporary at the very outer edge. Included here were a great number of Riccioli's colleagues and—most importantly —a large number of his heliocentric adversaries. In a stroke of melded spite and wit, Riccioli cast the leaders of heliocentrism far out into the Sea of Storms (*Oceanus Procellarum*): Copernicus was kept closest in, within sight of land at least; Kepler, however, was banished to a forbidding and windswept volcanic island (*Insula ventoru*), while Galileo was exiled still further toward the far margin, as distant from the ancients as possible. All three, that is, were left in a state of floating remove for having unjustly tried to make the Earth mobile. And for readers who knew something of the lives, works, and personalities of this holy trinity of the new astronomy, Riccioli's choices had yet deeper (and more satirical) meaning (Copernicus had been a local canon, remaining closest to the "true Church"; Kepler, meanwhile, was a known Protestant; and Galileo rarely acknowledged a debt to anyone). There is subtlety here, however: Kepler's debt to Tycho Brahe's observations is manifest in his crater being positioned along one of the large rays extending toward the middle left from Tycho (just beyond the *Peninsula Deliriorum,* one might note). In addition, all the other heliocentrists are grouped nearby, an exiled community. Hevelius, for example, was granted a sizeable crater on land, just beyond Galileo, in *Terra caloris.*

On the other hand, Van Langren was given a similar position, but on the opposite side of the Moon, in the Sea of Fertility, surrounded by his "regional contemporaries Vendelinus, Petavius, Snelius, and Furnerius" (Whitaker, 1978b, p. 381). This was actually a crater that Van Langren had named after himself, on his own map. Riccioli was thus, at one and the same time, citing this earlier attribution and framing it with a new, appreciative significance. Van Langren was granted the "fertility" of his vision as well his "place" within a community. Hevelius, meanwhile, found himself in a heated terrain, no doubt for incurring the wrath of the pontiff himself.

Clustered around the Sea of Nectar, meanwhile, one finds a number of holy saints (St. Catherine, St. Theophilus, St. Cyril, etc.), all with some connection to astronomy, whose stories, sometimes of a mystical nature, appear in Riccioli's text of the *Almagestum*. (These names remain on lunar maps of the present—though, interestingly, stripped of the "St." designation.) At the same time, lest this be mistaken for pure religious provincialism, there are a still greater number of Arabic names to be found just west of the Sea of Nectar (Alzophi, Albategni, Arzachei, Alphonsus, Abunezra, and others), in the Land of Healthfulness (*Terra sanitatis*). Once again, that is, Riccioli proves himself worthy, indeed exceptional, in his ecumenical citation of the past. Such inclusion of Arab astronomers is extremely significant from a cultural standpoint, since, within European universities, the names and works of these authors had gradually declined in mention and influence from the 14th century onward. The importance of Arab astronomy to Europe was (and is), however, beyond all reckoning: as early as the 7th and 8th centuries, the greater part of Greek science had been translated into Arabic and thereafter nativized and greatly augmented by the technical advances of Islam. When it was eventually translated into Latin during the 11th and 12th centuries, this augmented science became that of Europe (the Arabs did far more than merely "preserve" or "transfer" Greek science, as is so often and incorrectly said). Eventually, the ascent of Aristotle to the pinnacle of the intellectual canon, coupled with the military threat posed to Christendom by the Ottoman Empire and, somewhat later, the religious fervor of the Reformation and the Counter-Reformation, conspired to delete the names and works of Arabic thinkers from the common stock of technical literature (see Southern, 1962).

In a profound reversal of this long historical trend—whose effort, in effect, was to erase a portion of history—Riccioli put the Arabs back into the European sky, indeed into the very center of European astronomy, as inscribed on the lunar face. He placed them not just anywhere, that is, but shoulder to shoulder with some of the most hallowed names in classical science: Ptolemy, Hipparchus, and Agrippa. One cannot help but wonder at Riccioli's audacity, his symbolic boldness—again so reflective of his conflicting loyalties—in his "overthrowing" of Scholasticism by granting Albategni a larger, more central, and exceedingly prominent crater than that given to Aristotle himself, who, it turns out, floats largely alone in the chill waves of the *Mare frigoris*!

Finally, at the absolute outer west edge of octant VIII, immediately below Hevelius, Riccioli baptized two craters with his own and Grimaldi's names. His own feature, being the smaller is also in a more marginal position than that of his friend and collaborator.[27] Indeed, it is the last and final feature to be seen along the entire western edge of the lunar face. In all likelihood, this was a bow toward required humility; after all, the *Almagestum*, as I have noted, was hardly humble in its scope or in the referential claim of *its* title. Yet, however rhetorical, this modesty has gained over the centuries a degree of added symbolism. As the most "Western" representative, Riccioli was also the one to put the "East" back into the Moon.

⑤⑤

On the whole, then, Riccioli's nomenclature calls upon a far wider spectrum of astronomical culture than those of Van Langren and Hevelius combined. The names on his map are of persons with an enormous variety of connections to astronomy—scientific, astrological, literary, biblical, mythical, and philosophical—yet that also leave "science" preeminent. The Riccioli and Grimaldi map contains, in effect, a vast curriculum of voices that largely defined the Moon of Western culture—including its Eastern ingredients—down through the millennia. The title of the map ("Moon with Nomenclature and Libration") carries beneath it the sentence: *Nec Homines Lunam incolunt, Nec Animae in Lunam migrant* (Neither do humans inhabit the Moon, nor do spirits migrate there)— Riccioli believed that Holy Scripture did not provide support for the idea of lunar inhabitants, whether of flesh or of spirit. Divested of literal life, the lunar surface could be peopled with the textual–historical community of which Riccioli viewed himself a part, a college of the ages that his work sought to gather, teach, convert, and animate.

This encompassing quality should not be thought complete, of course. Despite his fastidiousness, Riccioli left out a great range of thinkers of non-Western origin whose names he could well have known. One in particular, Ulugh Beg, builder of a huge observatory in Samarkand during the 15th century and compiler of a highly valuable star catalogue, had been brought to light and popularized by Hevelius (who included Beg among his own sages of astronomy). No doubt the writings of Matteo Ricci, the brilliant Jesuit scholar who spent so many years in China, and whom Riccioli had read, held the names of famous Chinese astronomers; other works must have existed on Persian and Indian thinkers. The Riccioli and Grimaldi nomenclature, one realizes, is no less Eurocentric than that of Van Langren or Hevelius. Even the names of its Islamic luminaries, after all, were Latinized in spelling and confined to those authors who had proven useful to European scholars. In this, as in other things, the Jesuit professors revealed themselves men of their time.

But neither is it quite the case that their nomenclature succeeded because it directly flattered the belief that Europe was the center of everything. Yes, it greatly complimented the self-importance and ambitions for fame of the very community that had already gained possession over the Moon in an epistemological sense. But not this group alone. Its flattery would have to be seen in broader terms. The seas, for example—today, the most prominent and well-known features on the lunar surface—call directly upon common beliefs, many of them truly ancient in origin, existing in many parts of Asia and Africa as well as in Europe, which relate the Moon to such things as agriculture and the harvest; to fertility, birth, and growth; to the vicissitudes of weather; to the crises of human psychic life; and to the traditions of magic contained within astrology. All these things cross many boundaries of class, education, and sectarian faith, as well as culture and time; no doubt they seemed wholly embracing. Moreover, Riccioli's vertical hierarchy of personal names cast a certain degree of irony over the promise of possible immortality inherent in any such canon: for the lucky few who might one day join this list, after all, they could expect nothing more than to appear at the bottom, or else near the outer limits of visibility. Moreover, the several layers of meaning involved in placing such controversial figures as Copernicus and Galileo in a region of "storms," or, for that matter, Aristotle in a frigid sea (*Mare Frigoris*), can hardly have escaped the notice or appreciation of Riccioli's contemporaries. In this sense, the map reads (and was no doubt read) as a kind of code, a historical document merging private and public sensibilities, full of biases and, at the same time, ripe with loyalties to the Moon as an object of Western cultural imagining. If a form of flattery, therefore, inscribing a "textual community," this map also had the ability to include a much greater membership, well into the future, than simply astronomers alone.

In short, the Riccioli–Grimaldi nomenclature recommended itself on a number of levels. Textually, it suited the age perfectly, by both incorporating and exceeding the politics of science at the time. Though it included partisan elements, it also appealed to a "higher" and more harmonious realm. With its poetic, historical, and scientific qualities, it could be said to have nearly personified the scholarly gentleman of the late

17th and 18th centuries. Its political content, though no less reflective of the period than that of its competitors, neither denied nor centralized the conflicts of the Counter-Reformation, but included them in a manner subtle enough to allow "astronomy" and "scholarship" to show through more brightly. Though the "moderns" are given a low standing on this map, spatially speaking, one should not overlook the fact that there are so many more of them than the ancients—indeed, more than the ancients, the medievals, and the saints combined. Despite Riccioli's conservatism, in other words, his map ends up documenting the perception, dear to Galileo, Bacon, and the whole trend of 17th-century science, that more of worth had been done in the last 100 years than during the whole of the preceding millennia. In contrast to Hevelius, the Riccioli nomenclature incorporates the classical world, but reveals it as limited, fallible, and superseded. The Moon here—our Moon today—finally becomes an embodiment of its own history in the West.

Conclusion

The eventual success of the Riccioli–Grimaldi nomenclature meant that science would inherit a mid-17th-century lunar discourse. This was finally made doctrine in 1935, when the International Astronomical Union released its report standardizing the names of nearly 600 lunar formations (Blagg and Muller, 1935), remaining true, with several exceptions (Dante being one of them), to Riccioli's scheme. Further updated and greatly expanded during the 1960s (see Whitaker et al., 1963), the official nomenclature continued to build on this scheme but narrowed it to the names of deceased scientists only. All nonscientists were forbidden any further place in the lunar canon. At the same time, the use of Latin was strictly adhered to, with many new terms being added to designate different types of observed features (e.g., *tholas,* for hill, *rupes,* for scarp, etc.). These were also applied to the other planets and their moons as well, effectively casting the whole of the solar system eventually into a frame of Latinated discourse. Since the 1980s, a great many non-Western names—of artists, writers, musicians, gods and goddesses, towns and cities (ancient and modern)—have been added to the pantheon of astronomical titles. Not, however, for the Moon.

And yet, despite the attempts of modern astronomy to restrain this surface nomenclaturally, Riccioli's scheme preserves a kind of magnificent topography of association. Though it has been amended and expanded, its names reveal the myriad voices and the complex range of historical beliefs, superstitions, theories, and ambitions that went into the making of the Moon as a cultural object of the imagination. Such associations remain alive today, just beneath the armored skin of contemporary jargon; they are indelible to Western astronomy. The Moon may be "stuck" in the 17th century, but its reflection of the shadows and brilliance of that age reveal it, too, as a kind of linguistic peak, bringing into direct view a rugged landscape of history, language, and culture that the speech of astronomy today would urge us to ignore, forget, or disavow.

The oldest earthly geography of the skies remains a historical document of the age that "discovered" it. One can view this fact in a number of different ways. In addition to what I have just said, one could also point to the colonial history of the lunar surface. One could say that after the Moon was rendered visible as an object by Van Eyck and Leonardo, Gilbert was free to impose upon it an exploratory ambition in the most primitive sense, claiming it as a nameable, possessable territory. Harriot, meanwhile, rendered it part of traditional astronomical representation, while Galileo gave it, finally, a pictorial rhetoric that transformed it into a true geography of Earth-type origin. All three of these were essential to yielding a modern Western concept of "planet"—the wandering star that came to be fixed by being mapped and named into scientific study. In the detailed mapping schemes that followed, one finds the Earth expanded in new ways. Van Langren and Hevelius show two very different approaches to the creation of place, reflective of different, mature nation-state sensibilities. Van Langren, like the Spanish conquistadors in the New World, impelled to sanctify the imperial court-based systems of worldly and spiritual patronage that gave them their own power, named the Moon mainly after kings, queens, and saints. Hevelius, on the other hand, followed an approach more in line with Protestant efforts of exploration and settlement, in which features of a new environment were often named after places left behind in the homeland. The first reflects a condition of empire, seeking centripetal aggrandizement; the second, a hope for new provincialisms, centrifugal in nature. The semiotics of loyalty are very different in each case: to overt exploitation on the one hand, and to memorialization on the other.

Riccioli, by contrast, made of the Moon a territory neither imperial nor memorial but something like a combination of these: pedagogical. If Van Langren represents the conquering impulse, and Hevelius the discovery of a new homeland (in the image of the old), then Riccioli arrives to people the new territory, to give it a chronicle, a textual space, an ideal history.

The Moon of today represents a combination of "observation" and "colonization." United in the form of the map, these two aspects of 17th-century European experience were just as much a part of science as they were of politics. The world-encircling voyages of Magellan and Drake, the pressing technical problems of navigation, the innumerable tales and objects of curiosity brought back from distant places, and the rivalries between newly consolidating nation-states that made of the New World a domain where Old World power relations were mapped and imposed— all of these things helped create "geography" on the lunar surface as they did on Earth. The alien qualities of the Moon, at once expressed by the swirl of distant ocean and land, could also be tamed and overcome by this same presumption, whose disposition was pursued into detailed perceptions of earthly forms, of mountains, islands, and valleys, which therefore made of the lunar surface a true "landscape" of familiar (even classical) aspect. One sees that the voiceless, atextual naturalism of Galileo and Mellan was not preferred: only mapping the Moon, through the passion of names, could "annihilate the space between near and far" (Livingstone, 1992, p. 98). Yet naturalism, too, had won the day. For just as colonial explorers had documented things and places the ancients has never dreamed of, and had openly cast doubt on the primacy of classical wisdom, so exactly did the new lunar astronomy record a territory in the same mold. "It is a New World," Amerigo Vespucci is reported to have said, "because the ancients had no knowledge of it" (Glacken, 1967, p. 362). As of 1610, such a statement had equal relevance to the Earth and the Moon.

The map, the image, and the name were the means by which the Old World of Europe could be selectively re-created in the new. What the historian J. H. Elliot has said of America and the Indies applies equally well, or even better, to the lunar surface:

> The reverence of late medieval Europeans for their Christian and classical traditions had salutary consequences for their approach to the New World, in that it enabled them to set it into some kind of perspective in relation to themselves. . . . But against these possible advantages must be set certain obvious disadvantages, which in some ways made the task of assimilation appreciably harder. . . . Christendom's own sense of self-dissatisfaction found expression in the longing for a return to a better state of things. The return might be to the lost Christian paradise, or to the Golden Age of the ancients, or to some elusive combination of both these imagined worlds. (1970, pp. 24–25)

Van Langren inscribed on his Moon the yearning for a lost empire, Hevelius for a Golden Age. But the Moon, as an entity in space and in time, demanded far more than this. It required integration into Europe's evolving mental image both of the natural world and of the historical process. It had to be a new and unknown geography with a recognizable, heroic past, a writing of poetry and progress.

The map implied travel, and the vessel by which science journeyed to the Moon, and took control over it, was the telescope. It was with this instrument, and its complement, the microscope, that the representational imagination gained a central place in science during the early 17th century. These instruments, as extensions of the eye, demanded a reporting of "sight." Yet they did not merely strengthen human perception, they reconstituted and transformed it as well. They did this not only by broadening the eye's horizons, by creating a host of new visibilities whereby the world and the heavens could be re-formed into a universe with much greater detail and much greater expanse; they also did it, with regard to the Moon especially, by making ancient concepts visible or apocryphal. The idea of the Moon as another Earth, complete with mountains, seas, and islands, was a vision that had to be grasped, literalized as perception, and entered into the finality of scientific imagery. More than this, it had to become itself a means by which an evolving present could claim superiority: Galileo's little book *Sidereus nuncius* was a formidable act of conquest in its own right, for it proved that a living generation of thinkers could achieve unprecedented advances in knowledge and thus enter the geography of fame previously reserved for the ancients. It was now evident, that is, that neither Greece nor Rome, nor the Church Fathers, had known as much about the physical universe as did contemporary thinkers. As the world and the heavens were drawn closer by means of documented perceptions, the ancients receded. This is revealed, only too clearly, by an attempt such as Riccioli's, to compile a new *Almagestum*, surpassing Ptolemy, and to textualize the Moon with an astronomical community that culminated in a wealth of "moderns."

The nomenclatural systems of Van Langren, Hevelius, and Riccioli were all published in a very brief period of only six years between 1645 and 1651, during which the entire balance of power in Europe shifted away from Spain and Catholic southern Europe to Protestant northern Europe. It is ironic, if revealing, that these three authors, a Catholic, a Protestant, and a Jesuit, demonstrate a very similar kind of transition. From Van Langren to Hevelius, and finally to Riccioli, "science" in the form of canonized names advances from background to foreground, while

religious figures retreat to a position below even that of "Saracen" thinkers. Indeed, Van Langren's map did not merely slight the moderns by
making them so secondary; in its abject obedience to royal privilege and
its troubled personal confessions, it provided a declaration of political
dependence for science generally, and thus of fragility. This was entirely
accurate: astronomy in the early 17th century was not yet a professional
discipline of its own; it lacked institutional support and was a pursuit of
individuals deeply indebted to their sponsors. The unfortunate fate of
the Peiresc–Gassendi project and of Van Langren's larger ambitions, and,
conversely, the unique success of Hevelius, all testify to this. A major
exception was the Society of Jesus in Italy (see Heilbron, 1989); their
missionary campaign to win back territory, spatial and intellectual, for
the glory of Rome impelled the Jesuits to support the teaching and study
of scientific subjects, and to grant sabbaticals to those such as Riccioli
who promised to produce works that would serve as weapons for the
cause. The success of Riccioli, vis-à-vis the Moon, therefore has both
ironic and straightforward significance: ostensibly interested in defeating the heliocentric "moderns," his work was nonetheless supported by
a crucial forerunner of modern institutional support.

Van Langren's scheme too obviously stained the Moon with the
existing world of politics and patronage. Its own favoritism, which be-
littled science in more than one way, could not hope to succeed. Hevelius,
on the other hand, made of the Moon a billboard for an outmoded classicism, really his own eccentric vision, yielding another geography vacant of science. It was Riccioli who finally took fuller possession of the
lunar surface for astronomy. Yet this "science" was neither a unified nor
a wholly stable thing, for it brought together ancient, medieval, heliocentric, mythological, and religious representatives. Riccioli, like Van
Langren and Hevelius, included allegiance to the patron who gave him
support; his "treatment" of those such as Copernicus, Kepler, and others makes this evident. Yet he did not shrink from something far broader
and more appealing to his time. Writing the Moon as a text of its own
ecumenical past was a great act, and we are fortunate to have it preserved today, for it reveals directly that any "science," be it physics or
astronomy, is a discourse of revealed history as much as of theory, fact,
or hypothesis.

Linguistically, the Moon is a complex, geographic document of
European culture. Though studied and investigated today by people
throughout the world, it will always speak of Europe first, of the politics of European science in the mid-17th century. This, however, should
be cause not for dismay or simple denouncement but instead for histori-

cal perception. The Moon of today is the result of a troubled adventure of the eye and hand and mind that went far beyond the limits of the terrestrial realm, only to discover them once again, in rarefied form.

Notes

1. This is likely to be the case for Oriental art, too, at least in China and Japan, where the Moon was also portrayed either as a luminous disc without features, a disc inscribed with an animal (commonly a rabbit), or as a crescent.

2. Leonardo drew a "Man in the Moon" image in one case, with the dark areas depicting a jack-o'-lantern type-face (done sometime c. 1505–1508); but on another page he produced a more detailed rendering in chalk of the western half, with the lunar maria plainly visible (drawn a bit later, c. 1513–1514). A text that accompanies the latter is as follows: "Others say that the surface of the Moon is smooth and polished and that, like a mirror, it reflects in itself the image of our Earth. This view is also false, inasmuch as since the land is uneven [*sic*] when the Moon is in the east it would reflect different spots from those it would show when it is above us or in the west; [but] the spots on the Moon . . . never vary in the course of its motion" (Reaves and Pedretti, 1987, p. 57). Notwithstanding that these comments largely repeat what had been said in antiquity, Leonardo shows himself a true observer of the lunar surface, if not the very first to draw it.

3. This is sunlight reflected from the Earth, to the Moon, and back to the Earth again, giving the night hemisphere of the Moon a faint luminosity. Earthshine had been depicted on Moon images from at least the 12th century onward, but always in the form of a face, dimmer and less distinct then the crescent. An example appears in the west rose window, Sainte-Chapelle, in Paris (14th century). I am indebted to Ewen Whitaker for pointing this and other examples out to me.

4. While there is no single definition of "map," we can see that Gilbert's image is more abstract than either Leonardo or Van Eyck's drawings, more of an idealized projection than a naturalistic portrayal. In addition, its major features are all named. Together, these two factors argue strongly for the "map" designation.

5. Ptolemy's text, compiled in A.D. 160, had been brought to Italy from Byzantium in 1406, translated into Latin, and first published in the late 1470s with maps of the various portions of the world known at the time. A total of no less than 8,000 placenames were included on these early maps, all in Latin.

6. The "chorographicall" of Camden's title refers to the discipline of chorography, a legitimate branch of geography at this time (derived from Strabo's works), which involved the textual description of places and their

delineation on a map or chart. Camden's efforts, meanwhile, were very close to those of Hakluyt and the movement to strengthen and vitalize England. Sketching out the traveler's experience of surrounding lands, in detailed "scientific" fashion, he was essentially arguing that their differences be subsumed under a single, higher authority, one of "improved civilization," capable of uniting them in power and purpose. When James I claimed the throne 17 years later in 1603, uniting Scotland and England, Camden's wish had come to pass: the new king was crowned ruler of "Great Britain."

   7. These drawings deserve a central place in any discussion of the history of relationships between art and science. They are magnificently detailed, and show a definite command of the techniques of Renaissance chiaroscuro, foreshortening, and so on (see Hulton and Quinn, 1964). To say, for example, as Edgerton recently has, that "Britain in 1609 [the year Harriot did his first drawing of the Moon] was still medieval as far as the visual arts were concerned" (p. 237) is simply not true, and worse still, misleading. Theodor De Bry's engraved versions of White's drawings, published in 1590 along with Harriot's *Briefe Report* (as the first part of De Bry's own collection *America*), depict numerous Indian men and women in the distinctive bodily sway of Renaissance (actually reclaimed Hellenistic) poses, for example, those of Michelangelo's *David* or Botticelli's *Graces*. This has been pointed out before, perhaps most eloquently by Howard Mumford Jones (1964) and J. H. Elliot, the latter of whom states that "readers [of] De Bry's famous engravings . . . could be forgiven for assuming that the forests of America were peopled by heroic nudes, whose perfectly proportioned bodies made them first cousins of the ancient Greeks and Romans" (1970, p. 23).
   Studies like that of Edgerton (1991), Winkler and Van Helden (1992), or Kemp (1990), despite their excellence in other respects, nonetheless give privilege to an old prejudice which emphasizes the presumed naturalism and overall superiority of Italian art vis-à-vis northern European art. It is a simple thing to show how mistaken this is; indeed, I have done so above with regard to the Moon (one of Edgerton's main subjects, as it happens) and Van Eyck. It would in fact be just as simple to show the pronounced *unnaturalism* of many natural phenomena—rocks, hills, rivers, mountains and so on—in Italian art, from Giotto through Leonardo. For all the hyped abilities of these artists to portray things realistically, especially following the invention of linear perspective, it is utterly clear to any geologist, for example, that these features of the natural world are props, the visual equivalent of constructed stage sets. The oft-quoted line by Vasari, regarding Giotto's capacity to "imitate and reproduce nature," and his bucolic portrait of the young genius, learning to sketch sheep from "nature" by scratching out portraits in stone, seems still to bear its burden of overstated heroicism and propaganda 450 years later. Art history has inherited Vasari's emphasis on humanistic realism. Yet Vasari's "nature," is not our "nature"; he is speaking particularly of design and form (especially human form), not of "observation" per se.

Moreover, as has been noted before, linear perspective does not reproduce the world as an observer sees it, but distorts a scene in favor of geometric projection. The human eye is stereoscopic and does not act as a point source for vision (contemporary cognitive theories of human sight, such as those of James Gibson, are highly complex and involve a range of shifting perspectives). The so-called distance-point method of northern art is much better suited to the portrayal of realistic landscape. In any case, it should be recognized that one is dealing with two very different, if sometimes overlapping, approaches to representation, and that each, as a collection of solutions to various related problems, bears its own heightened capabilities. If "realism" or "naturalism" is the issue, it needs to be partitioned and more precisely examined. Though not my purpose here, the bias in favor of Italy, already dealt a death blow more than 10 years ago by Alpers's excellent study (1983), should be put to rest in any future writing on the subject.

8. There has been some discussion regarding whether Harriot actually had received a copy of Galileo's work or had merely heard about it, possibly via Kepler's *Dissertatio cum nuncio sidereo* (A conversation with the messenger from the stars), published in May 1610. John North, however, indicates that at least one other English scientist (Sir Christopher Heydon) had procured and read a copy of Galileo's book by early July, and assumes that Harriot had the book in his possession by about the same time (1974, pp. 136–137).

9. A number of these pages are reproduced in Shirley (1978, pp. 291–300).

10. The effects of this on planetary astronomy, down to the present day, should not be ignored. Literalizing the Moon as another Earth has had its long-term influence in the realm of technical conception. Features observed on other planets, that is, have been thought of by contemporary astronomers in just those cognitive terms set out for the Earth, whether directly or by predisposition: seas created by "floods" or "great outpourings"; mountains by "convulsive" action; "valleys" by erosive work; "plains" by long periods of inactivity. Moreover, when, in the last 50 years, it became apparent that these surface realities were relatively unchanging on the lunar surface, that they had been there for eons without modification, it could only be concluded—again in earthly terms (due to the demands of difference)—that the Moon was a "dead world," whose "evolution" had already reached its end. The literalization of analogy in the 17th century between Moon and Earth meant the gain, in the end, of a set scheme for describing and even explaining other worlds by modern astronomical science.

11. The lunar surface does, however, strongly resemble in general form certain volcanic provinces in desert areas (Craters of the Moon National Park, in south-central Idaho, for example, is an aptly—if a bit too literally—named example). However, it is unlikely that Galileo ever visited such places. At the same time, as we will see below, another early observer of the Moon—Hevelius—noted and inscribed on one of his images a similarity

between the "explosive" area of Copernicus-Kepler and the Mediterranean volcano Mt. Aetna. Galileo, however, who was probably familiar with the latter, and could no doubt see this feature on the Moon, left it out altogether. Indeed, the one main comparison he makes to an actual terrestrial site involves using the large, circular valley (actually a basin) of Bohemia as a model for the exaggerated crater (Albategnius) previously mentioned. No doubt Galileo had seen this portrayed on maps, for it is not visible to the eye. This is one more argument, then, that his perception of the lunar surface was in part guided by terrestrial cartographic forms.

12. Note, for example, this statement by Alpers: "In spite of the Renaissance revolution in painting, northern mapmakers and artists persisted in conceiving of a picture as a surface on which to set forth or inscribe the world rather than as a stage for significant human actions" (1983, p. 137).

13. The larger role of images in Galileo's book is a rich topic for study. In addition to his engravings of the Moon, he included mathematical (geometric) diagrams, mainly optical in nature; sketches of the fixed stars in several constellations, with no artistic overlay of pictures, only the stars themselves depicted in symbolic six-point fashion; and as many as 60 versions of Jupiter and its moons, arranged horizontally, inserted directly into the text, with Jupiter itself as a small open circle and each satellite as an asterisk. No attempt is made to render these images the least bit naturalistic. Discovery, with regard to their content, involved the "mere" existence and position of new entities, not their actual visual appearance. Showing their presence, symbolically, was sufficient to document that an observation had taken place—though in the case of the Jovian moons, Galileo obviously felt that a great number of examples, by the weight of sheer excess, was needed for persuasion as well. In terms of effort, then, the Moon drew a vastly greater amount of representational energy: it is clearly the visual center of the book. Here, that is, Galileo recognized that to be diagrammatic was not enough: to demonstrate witness, he had to offer a picture "stained with the real" (a phrase I borrow from Alpers, 1983).

14. During much of the 16th and early 17th centuries, it was thought that changes in the angle (declination) of the compass needle might show systematic patterns over the Earth's surface, therefore providing a means for estimating east-west position. This proved not to be the case, however, since the Earth's magnetic field does not vary systematically enough and also changes over time. With regard to time as a means of determining longitude, this was surmised on the basis of the following: as the Earth rotates 360 degrees in 24 hours, this translates into 0.25 degrees of longitude every minute. Therefore, if two observers at different locations on the Earth could somehow perceive the same phenomenon coming into view at the same instant, and note down the exact time they did so, then the difference in these logged times could be converted into distance.

15. This should make us all the more impressed with Galileo, who

seems to have studied the Moon and produced his drawings of it in less than 3 weeks, between November 30 and December 18, 1609. See the "Introduction" by Van Helden to his recent translation of *Sidereus nuncius* (Chicago: University of Chicago Press, 1989), especially pp. 4–9.

16. Indeed, with regard to naming the heavens generally, there is probably no more eloquent expression of the mixed aspirations for fame and flattery in the here and now, and immortality later on, than in the brief introductory address Galileo aims at "Most Serene Cosimo II de'Medici." The author begins by stating how all ages have tried "to preserve from oblivion and ruin" the names of "excellent men" by carving statues, building pyramids, erecting cities, writing great works. Then, in a sudden reversal, he writes:

> But why do I mention these things as though human ingenuity, content with these [earthly] realms, has not dared to proceed beyond them? Indeed, looking far ahead, and knowing full well that all human monuments perish in the end . . . human ingenuity contrived more incorruptible symbols against which voracious time and envious old age can lay no claim. And thus, moving to the heavens, it assigned to the familiar and eternal orbs of the most brilliant stars the names of those who . . . were judged worthy to enjoy with the stars an eternal life. . . . This especially noble and admirable invention of human sagacity, however, has been out of use for many generations. . . . But now, Most Serene Prince, we are able to augur truer and more felicitous things for Your Highness, for scarcely have the immortal graces of your soul begun to shine forth on earth than bright stars offer themselves in the heavens which, like tongues, will speak of and celebrate your most excellent virtues for all time. Behold, therefore, four stars reserved for your illustrious name, and not of the common sort and multitude of the less notable fixed stars, but of the illustrious order of wandering stars, which, indeed, make their journeys and orbits with a marvelous speed around the star of Jupiter, the most noble of them all, with mutually different motions, like children of the same family, while meanwhile all together, in mutual harmony, complete their great revolutions every twelve years about the center of the world, that is, about the Sun itself. Indeed, it appears that the Maker of the Stars himself, by clear arguments, admonished me to call these new planets by the illustrious name of Your Highness before all others. For as these stars, like the offspring worthy of Jupiter, never depart from his side except for the smallest distance, so who does not know the clemency, the gentleness of spirit, the agreeableness of manners, the splendor of the royal blood, the majesty in actions, and the breadth of authority and rule over others, all of which qualities find a domicile and exaltation for themselves in Your Highness? (1989, pp. 29–31)

I have quoted this at length to help show that through the haze of required rhetorical tribute, Galileo is making a cunning plea that would have Cosimo II's acceptance of this gift of immortality linked to a recognition of the Copernican system. The skill with which this is done is striking, and involves the clever use of metaphor at several levels. Nor should we ignore, on the other hand, a heavy quality of irony hovering in the last part, an irony Galileo was prone to indulge in some of his other writings as well (see Chapter 2). Jupiter, that is, was hardly known to Renaissance readers as a figure of such uniform grandness and majesty. On the contrary, one of the most widely read and often-quoted works of Latin literature—a work that no educated reader of the 16th and 17th century would have gone without was Ovid's *Metamorphoses*, in which the god was often portrayed in a manner exactly opposite to Galileo's own catalogue of virtues, as capricious, unforgiving, lustful, and self-centered. Such were the qualities of power gone bad. Was Galileo therefore offering a veiled warning to his would-be patron? Probably not. It is more likely that he was commenting, in elliptic fashion, upon the very traditions of discourse that forced him to adopt such a fawning tone and language. But the point is clear: to name these new "stars" as he did was a calculated move with political, economic, and professional intent. That they are named today the "Galilean satellites" is perhaps the final, and just, irony of this episode.

17. What follows is the major part of Van Langren's inscription, taken from a French translation of the original Latin (Bosmans, 1910, pp. 251–254). The first paragraph corresponds to the title lines shown on the left side of the inscription (see Figure 9).

18. Both figures, meanwhile, look upward toward Contemplatio and the noted scroll. The reader, finally, is being advised to cast his eyes within these pages, and there see who has re-created for ordinary sight the *rerum Caelestium* itself.

19. More than a few scientists at the time were skeptical of Hevelius's abilities as an observer. The Royal Society, for example, sent the young Edmund Halley to Danzig in order to check the accuracy of Hevelius's naked eye measurements. This was done by order of Robert Hooke, then secretary to the Society, whose own frequent and productive use of both telescope and microscope had rendered him highly suspicious of the power of plain sight. It seems likely, too, that Hooke regarded Hevelius as representing a type of vestigial science, a former stage whose time had passed. Halley reported that position measurements made by him with a telescope and by Hevelius with his instruments agreed to a very high degree of precision. This seems to have allayed Hooke's distrust for the time being. Yet, as the telescope was itself the standard for this comparison, Hevelius's success only strengthened the general idea that the limits of the unaided eye had been reached, and that to advance in any profound sense, astronomy would have to bring the heavens closer in one form or another—either through a strengthening of human vision or a theoretical system that could explain, in mathematical terms, what could be seen. The year of Hevelius's death was the same in which Newton's *Principia* appeared: a coincidence that both frames and interprets the significance of each vis-à-vis the century as a whole.

20. This can be shown, too, interestingly enough, where this naming scheme exhausts itself, toward the top of the map. Here, that is, Hevelius could conceive of no other major names than those beginning with "Hyperborea," the ancient Greek denomination for anything of the "extreme north," beyond the known limits of civilized exploration. On Hevelius's map, one finds seas, bays, mountain ranges, lakes, and other features, as well as the entire "continent" (*regio*) within the topmost libration limits, all named with this designation. Claiming full exploration of the lunar surface, Hevelius nonetheless ends up reiterating a provincialism of the classical world.

21. Most works on 17th century astronomy have given this map (Q) only a glancing mention, preferring instead to discuss at length the more naturalistic drawing (R) which can be directly compared with the maps of Van Langren, Mellan, Fontana, and Galileo, and can be judged in terms of the spatial accuracy of its features (mostly excellent). One sees that the naturalism of this image is vastly inferior to Mellan's, or for that matter Galileo's. It has a decorative feel, suffering from too many effects of design (many of its features are smoothed out, made symmetrical; craters are added for visual balance). One therefore has less of "observation" in its pure state, than in Mellan's images, and more of an attempt to both instruct and please the eye at the same time. Surrounded by *putti* (winged cherubs) in each corner performing some task of measurement or announcement, the map is an aesthetic object first, a scientific result second. It seems to me Hevelius's work, his description of the Moon, culminates not here but in his true cartographic image, with its interpretive vision.

22. During the whole of the 17th and most of the 18th centuries, it was an expected part of scientific literature to discuss the thinking and thinkers of the past, often in little potted histories that serve as textual portraits, and that placed an author's own efforts and claims for originality in connection with canonical luminaries of the past. This, as discussed above, was exactly what Hevelius himself did, pictorially, in his *Prodromus astronomiae*.

23. The word *Almagest*—"the greatest work"—was the title of Ptolemy's synthesizing work on classical astronomy, originally called *Syntaxis mathematica* (it gained the later title via its translation into Arabic in the 8th century and then into Latin 400 years later). Riccioli was thus claiming, in effect, that his work might stand as a grand synthesis for as much as a thousand years. Ironically, much of the Almagest was by this time accepted as disproven by Copernicus, Kepler, and Galileo, and would soon be dealt its final blow by Newton. Like Hevelius, then, Riccioli wanted a place among the ancients.

24. With the exception of Van Langren himself, these attempts were all made by Germans and Italians, for example, by Hevelius and Anton Schyrlaeus (a Bohemian friar and optician), or those such as Eustachio Divini (a Roman optician), and Gerolamo Sirsalis (a Jesuit priest).

25. The most sympathetic dismissal I could find was offered by A. Libes, who said of Riccioli: "One owes him some recognition . . . finally, for having worked constantly in a useful manner, if not for science, at least for those who cultivate it" (1810, p. 94).

26. This image is reproduced as Figure 8.13 (p. 136) in Whitaker (1989).

27. In contrast to this, Riccioli's list of the constellations includes none of the traditional Greek or Latin names. Instead, it gives only those of the so-called "Christianized heavens," recently coined or assembled by Julius Schiller in his lavishly produced celestial atlas, *Caelum stellatum christianum* ("Stars of the Christian heavens," 1627). This work, which must be counted among the most elaborately artistic and technically accurate depictions of the century, replaces the signs of the Zodiac and most of the circumpolar constellations with the names of various saints (a comparative list can be found in Delambre, 1969, pp. 298–99). The fact that Riccioli did not impose this type of nomenclature on the Moon, as had Van Langren, speaks to a wider ambition. His training in, love of, and fastidious knowledge about literature and philosophy, as well as the Bible and science, encouraged him to see the lunar surface as a far larger canvas, culturally speaking, than the heavens.

References

Alpers, S. (1983). *The Art of Describing.* Chicago: University of Chicago Press.

Ariew, R. (1984). Galileo's Lunar Observations in the Context of Medieval Lunar Theory. *Studies in the History and Philosophy of Science, 15,* 213–216.

As Sufi. (1954). *Suwaru'l-Kawakib* (Uranometrica). Decan, India: Oriental Publications Bureau.

Bailly, J. S. (1779). *Histoire de l'Astronomie Moderne.* vol. 2. Paris: Chez DeBure.

Batson, R. M., Whitaker, E. A., and D. E. Wilhelms. (1990). History of Planetary Cartography. In R. Greeley and R. Batson (Eds.), *Planetary Mapping* (pp. 12–59). Cambridge: Cambridge University Press.

Beziat, L. C. (1875). La vie et les travaux de Jean Hevelius. *Bullettino di bibliografia e di storia delle scienze matematiche e fisiche, 8,* 497–558, 589–669.

Blagg, M. and K. Müller. (1935). *Named Lunar Formations.* London: Percy Lund & Humphries.

Bloom, T. (1978). Borrowed Perceptions: Harriot's Maps of the Moon. *Journal for the History of Astronomy, 9,* 117–122.

Borrell, A. P. (1967). Historia de la cartografia lunar. *Urania, 266,* 1–69.

Bosmans, H. (1903). La carte lunaire de Van Langren conservée aux archives générales du royaume, à Bruxelles. *Revue des questions scientifiques, 3d ser., 4,* 108–139.

Bosmans, H. (1910). La carte lunaire de Van Langren conservée a l'Université de Leyde. *Revue des questions scientifiques, 3d ser., 17,* 248–264.

Brown, H. (1973). Nicolas Claude Fabri de Peiresc. *Dictionary of Scientific Biography,* vol. 8 (pp. 488–492). New York: Scribner's.

Bush, D. (1945). *English Literature in the Earlier Seventeenth Century, 1600–1660.* New York: Oxford University Press.

Campedelli, L. (1975). Giambattista Riccioli. *Dictionary of Scientific Biography,* vol. 8 (pp. 411–412). New York: Scribner's.

Cherniss, H. and W. C. Helmbold (1957). *Plutarch's Moralia.* Loeb Classical Library, vol. 12. Cambridge, MA: Harvard University Press.

Cyrano de Bergerac (1978). *L'Autre Monde; ou Les États et Empires de la lune.* Paris: Éditions Sociales.

Delambre, J. B. (1969). *Histoire de l'Astronomie Moderne.* vol. 2. New York: Johnson Reprint Corp. (Facsimile of original work published in 1821.)

Dhanens, E. (1980). *Hubert and Jan van Eyck.* New York: Fine Arts Press.

Drake, S. (1970). *Galileo Studies.* Ann Arbor: University of Michigan Press.

Drake, S. (1972). Galileo Galilei. *Dictionary of Scientific Biography,* vol. 5 (pp. 237–249). New York: Scribner's.

Drake, S. (1976). Galileo's First Telescopic Observations. *Journal for the History of Astronomy, 7,* 153–168.

Eastwood, B. S. (1972). Francesco Maria Grimaldi. *Dictionary of Scientific Biography,* vol. 5 (pp. 542–545). New York: Scribner's.

Edgerton, S. Y. (1991). *The Heritage of Giotto's Geometry: Art and Science on the Eve of the Scientific Revolution.* Ithaca, NY: Cornell University Press.

Elliot, J. H. (1970). *The Old World in the New.* Cambridge: Cambridge University Press.

Galileo. (1989). *Sidereus nuncius.* Transl. by A. Van Helden. Chicago: University of Chicago Press.

Gilbert, W. (1958). *On the Magnet.* Transl. by S. P. Thompson. New York: Basic Books.

Glacken, C. J. (1967). *Traces on the Rhodian Shore.* Berkeley and Los Angeles: University of California Press.

Goetz, W. (1937). Die Entwicklung des Wirklichkeitssinnes vom 12. zum 14. Jahrhundert. *Archiv für Kulturgeschichte, 27*, 33–73.

Harbison, C. (1989). *Jan Van Eyck: The Play of Realism*. London: Reaktion.

Harriot, T. (1986). A Briefe and True Report of the New Found Land of Virginia. In A. L. Rowse (Ed.), *Richard Hakluyt, Voyages to the Virginia Colonies* (pp. 107–136). London: Century.

Hartner, W. (1975). Terrestrial Interpretation of Lunar Spots. In M. Righini-Bonelli and W. Shea, *Reason, Experiment, and Mysticism*. (pp. 25–38). New York: Science History Publications.

Heilbron, J. (1989). Science in the Church. *Science in Context, 3*, 9–28.

Hevelius, J. (1967). *Selenographia sive Lunae Descriptio*. New York: Johnson Reprint Co.

Hevelius, J. (1968). *The Star Atlas*. Tashkent, USSR: "Fan" Press/Academy of Sciences of the Uzbek SSR (in Russian and English)

Hulton, P., and D. B. Quinn. (1964). *The American Drawings of John White, 1577–1590*. 2 vols. Chapel Hill: University of North Carolina Press.

Humbert, P. (1931). La première carte de la lune. *Revue des questions scientifiques, 20*, 193–204.

Jones, H. M. (1964). *O Strange New World*. New York: Macmillan.

Kelly, S. (1965). *The "De mundo" of William Gilbert*. Amsterdam: Menno Hertzberger.

Kelly, S. (1972). William Gilbert. *Dictionary of Scientific Biography*, vol. 5 (pp. 396–401). New York: Scribner's.

Kemp, M. (1990). *The Science of Art*. New Haven: Yale University Press.

Koestler, A. (1959). *The Sleepwalkers*. New York: Macmillan.

Kopal, Z., and R. W. Carder. (1974). *Mapping of the Moon: Past and Present*. Dordrecht: D. Reidel.

Lambrecht, H. (1967). Preface. In J. Hevelius, *Selenographia sive Lunae Descriptio*. New York: Johnson Reprint Co.

Lear, J. (1965). *Kepler's Dream*. Berkeley: University of California Press.

Libes, A. (1810). *Histoire philosophique des progrès de la physique*. vol. 2. Paris: Immerzeel.

Livingstone, D. N. (1992). *The Geographical Tradition*. Oxford: Basil Blackwell.

Lohne, J. (1973). Thomas Harriot. *Dictionary of Scientific Biography*, vol. 6 (pp. 124–129). New York: Scribner's.

MacPike, E. F. (1937). *Hevelius, Flamsteed and Halley*. London: Taylor and Francis.

McColley, G. (1936). The Seventeenth Century Doctrine of a Plurality of Worlds. *Annals of Science, 1*, 385–430.

McColley, G. (Ed.). (1937). *The Man in the Moone* and *Nuncius Inanimatus* by Francis Godwin (1638). *Smith College Studies in Modern Languages* 19(1).

Montgomery, S. L. (1994). The First Naturalistic Drawings of the Moon: Jan Van Eyck and the Art of Observation. *Journal for the History of Astronomy*, xxv, 317–320.

Nicolson, M. (1948). *Voyages to the Moon*. New York: Macmillan.

North, J. (1972). Johannes Hevelius. *Dictionary of Scientific Biography*, vol. 6 (pp. 360–364). New York: Scribner's.

North, J. (1974). Thomas Harriot and the first telescopic observations of sunspots. In J. W. Shirley (Ed.), *Thomas Harriot: Renaissance Scientist* (pp. 129–165). Oxford: Clarendon Press.

Panofsky, E. (1953). *Early Netherlandish Painting*. 2 vols. Cambridge, MA: Harvard University Press.

Parry, J. H. (1981). *The Age of Reconnaissance: Discovery, Exploration and Settlement, 1450–1650*. Berkeley and Los Angeles: University of California Press.

Pepper, J. V. (1974). Harriot's Earlier Work on Mathematical Navigation: Theory and Practice. In J. Shirley. (Ed.), *Thomas Harriot: Renaissance Scientist* (pp. 54–90). Oxford: Clarendon Press.

Porter, R., and M. Teich. (Eds.). (1992). *The Scientific Revolution in National Context*. Cambridge: Cambridge Univesity Press.

Prinz, W. (1903–1904). L'original de la première carte lunaire de Van Langren. *Ciel et Terre*, 24, 99–105, 149–155.

Reaves, G., and C. Pedretti. (1987). Leonardo da Vinci's Drawings of the Surface Features of the Moon. *Journal for the History of Astronomy*, 18, 55–58.

Revel, J. (1991). Knowledge of the Territory. *Science in Context*, 4, 133–161.

Rochot, B. (1972). Pierre Gassendi. *Dictionary of Scientific Biography*, vol. 5 (pp. 284–290). New York: Scribner's.

Rosen, E. (1950). The Title of Galileo's *Sidereus Nuncius*. *Isis*, 41, 287–289.

Rosen, E. (1974). Harriot's Science: the Intellectual Background. In J. Shirley (Ed.), *Thomas Harriot: Renaissance Scientist* (pp. 1–15). Oxford: Clarendon Press.

Rowse, A. L. (Ed.). (1986). Introduction. In Richard Hakluyt, *Voyages to the Virginia Colonies*. London: Century.

Shirley, J. W. (1978). Thomas Harriot's Lunar Observations. *Studia Copernicana (Science and History: Studies in Honor of Edward Rosen)*, 16, 283–308.

Shirley, J. W. (1983). *Thomas Harriot: A Biography*. Oxford: Clarendon Press.

Smet, A. (1973). Michael Florent van Langren. *Dictionary of Scientific Biography*, vol. 8 (pp. 25–26). New York: Scribner's.

Southern, R. W. (1962). *Western Views of Islam in the Midde Ages*. New York: Vintage.

Stevens, H. (1900). *Thomas Harriot, the Mathematician, the Philosopher and the Scholar*. London: Privately printed.

Stock, B. (1983). *The Implications of Literacy*. Princeton: Princeton University Press.

Strobell, M. E. and H. Masursky. (1990). Planetary Nomenclature. In R. Greeley and R. Batson (Eds.) *Planetary Mapping* (pp. 96–140). Cambridge: Cambridge University Press.

Van de Vyver, O. (1971a). Original Sources of some Early Lunar Maps. *Journal for the History of Astronomy*, 2, 86–97.

Van de Vyver, O. (1971b). Lunar maps of the XVIIth century. *Vatican Observatory Publications*, 1, 71–114.

Van Helden, A. (1989a). Introduction. In Galileo, *Sidereus nuncius*. Transl. by A. Van Helden. Chicago: University of Chicago Press.

Van Helden, A. (1989b). Galileo, telescopic astronomy, and the Copernican system. In R. Taton and C. Wilson (Eds.), *Planetary Astronomy from the Renaissance to the Rise of Astrophysics, Part A: Tycho Brahe to Newton* (pp. 81–105). Cambridge: Cambridge University Press.

Von Euw, A. (1993). Die kunstlerische Gestaltung der astronomischen und komputistischen Handschriften des Westens. In P. Butzer and D. Lohrmann (Eds.), *Science in Western and Eastern Civilization in Carolingian Times* (pp. 251–288). Basel: Birkhaüser.

Westfall, R. S. (1985). Scientific Patronage: Galileo and the Telescope. *Isis*, 76, 11–31.

Westphal, J. H. (1820). *Leben, Studien und Schriften des Astronomen Johann Hevelius*. Koenigsberg: Hallervord.

Whitaker, E. A. (1978a). Galileo's Lunar Observations and the Dating of the Composition of *Sidereus nuncius*. *Journal for the History of Astronomy*, 9, 155–169.

Whitaker, E. A. (1978). Why is the Brightest Lunar Crater Named Aristarchus? *Sky and Telescope*, November, 380–382.

Whitaker, E. A. (1989). Selenography in the Seventeenth Century. In R. Taton and C. Wilson (Eds.), *Planetary Astronomy from the Renaissance to the Rise of Astrophysics, Part A: Tycho Brahe to Newton* (pp. 119–143). Cambridge: Cambridge University Press.

White, L., Jr. (1947). Natural Science and Naturalistic Art in the Middle Ages. *American Historical Review*, 62(3), 421–435.

Wilkins, J. (1973). *The Discovery of a World in the Moone*. Delmar, NY: Scholars' Facsimiles and Reprints.

Winkler, M. G. and A. Van Helden. (1992). Representing the Heavens: Galileo and Visual Astronomy. *Isis, 83,* 195–217.

Winkler, M. G. and A. Van Helden (1993). Johannes Hevelius and The Visual Language of Astronomy. In J. V. Field and F. A. L. James (Eds.), *Renaissance and Revolution: Humanists, Scholars, Craftsmen and Natural Philosophers in Early Modern Europe*. Cambridge: Cambridge University Press.

Wislicenus, W. (1901). Ueber die Mondkarten des Langrenus. *Bibliotheca Mathematica, 3d ser., 2,* 384–391.

Younes, T. (1991). Closing Remarks. In D. L. Hawksworth (Ed.), *Improving the Stability of Names: Needs and Options* (pp. 337–338). Taunus, Germany: Koeltz Scientific Books.

5

Science by Other Means

Japanese Science and the Politics of Translation

Anywhere one cares to tread in the landscape of contemporary knowledge, one's steps are guided, albeit invisibly, by the realities of translation. The worlds of discourse associated with modern scholarship may seem, at first, magnificent in their monolingual accessibility. They appear, even naturally, our own. Yet a moment's thought is enough to reveal that they would not be possible, as they are, unless their pathways had been often paved with stones brought in from elsewhere, quarried by foreign hands, and then replanted in ground not native to their bedrock.

One knows this to be true, of course, with regard to literature, philosophy, the humanities in general. The very idea of Western culture, especially what was once known as "high culture," is fundamentally based on the sharing and appropriation of materials across borders, oceans, and centuries. The metabolic function of "culture," by which such materials are absorbed, digested, and eventually assimilated to the national sensibility, could not happen without the constant transfer of words from one language to another.

If we tend to forget this truth, it is perhaps because such forgetting is the very aim and highest achievement of translation. But while we only "forget" the cultural significance of translation in the case of poetry or philosophy, we do much worse with regard to the natural sciences. Here, forgetfulness has tended to be joined to disinterest. An enormous library of writings can be found on the subject of literary translation. Opinions and observations reach down the millennia, from Roman times to the present. Nearly every great author in the humanistic canon, from Cicero to Augustine, from Goethe to Derrida, has written something on the topic, in some form. This is not to include, moreover, the constant attention of literary critics, historians, anthropologists, linguists, and, of course, translators themselves. Today, there is even a new academic discipline entitled "translation studies" (see, e.g., Bassnett-McGuire, 1991), focusing on the theory and practice of this ancient craft—surely the second oldest profession in the realm of authorship.

But where, amid all this writing, does the topic of technical translation arise? So far as I can tell, after years of reading in the field, it doesn't—at least not in any significant or consistent way. There are barely a handful of studies that deal with technical translation at all, and when they do it is almost always in tones of boredom or dismissal. A recent study, for example, handles the topic like this:

> In moving factual information about from tongue to tongue, the notion of word-for-word fidelity reigns, and its ideal practice should attain the perfection of the high quality reproduction of photocopy machines. Such . . . is proper for information transfer . . . but obviously not for art. At the same time, consider how unjust it would be, how disastrous, were an expressive fidelity to be applied to ordinary information transfer. (Barnestone, 1993, p. 33)

The notion that modern science—no less than literature, and perhaps even more—owes an incalculable debt to translation does not attract much interest. Science is science, the idea goes, whether it takes place in French, in German, or in English. It is a specialized discourse, formal in its structure and vocabulary, and the process involved in transferring it from one language to another must be far less demanding, less interpretive, certainly less "expressive"—and even less real (as true translation)—than the rendering of "art," whether this be novels or poems.

Anyone with a deep curiosity about the history, methods, or theory of technical translation—not to say, its larger effects—is left to a small number of excellent but unfortunately often overlooked studies in the history of medieval science (see, e.g., Lindberg, 1978; Hourani, 1972; O'Leary, 1949; d'Alverny, 1982). Otherwise, he or she is left largely to a vacant field. What should be a rich province of scholarship, stretching across all periods, remains a confined space, trod by few. Despite the obvious role technical translation has played in the building of the modern world, in the spread of science and technology between various nations, and determining how they have been nativized, its significance has been hugely ignored. How can this be when the powers of "science" in the last two centuries have been commensurate with those of translation? How can this disinterest in technical translation continue despite long-standing debates about the nature of scientific language itself, debates that began in the days of Francis Bacon? Surprisingly, prior to the 20th century, one looks in vain even for the very idea of "scientific translation." This oversight is all the more curious and ironic considering the many efforts made over the last two centuries to define a "science of translation."

In effect, the role of linguistic transfer in the creation of modern science, and therefore the influence of various nations on other nations, is for all practical purposes completely virgin territory. Diplomats, politicians, and scholars speak often, and with no small concern, of "technology transfer." The movement of knowledge from "first world" to "third world" cultures, or, via trade and espionage, between "first world" competitors such as the United States and Japan, has attracted great attention and aroused much anxiety. In the postwar era, it has played an ever increasing role in government circles. Yet how central to any of this discussion has been the issue of translation? In 1989, the U.S. Congress, inspired by fears of rising trade imbalances, took a reportedly "bold step" forward by allocating a total of $1 million for translation of Japanese scientific information into English (this to be distributed over several years). In an era when federal grants for research and development have broken the $100 billion mark, such a gesture barely qualifies as symbolic.

I would like to offer a brief anecdote to reveal how this prejudice has manifested itself in scholarly circles, even among scientists. The 1950s, one might recall, were a heyday of early excitement over the possibility of machine translation, the replacement of human (and therefore fallible) translators, whether largely or altogether, with programmed computers. Under the influence of new mathematical theories of communication pioneered by Claude Shannon and Bernard Mandelbrot, the ancient dream of discovering the "natural laws" of human language, laws believed to be "as stable as those governing the motions of heavenly bodies" (Oettinger, 1959), seemed finally to be within reach. The fundamental problems were felt to be understood; only the technology remained to be developed. Those who pursued this hope were by no means simpleminded utopians. Nor were they mere technicians. They included linguists, theoretical mathematicians, physicists, and literary scholars, most of whom were multilingual and had done translations themselves. Within this group, there was ready recognition of the difficulties involved in any system of automatic translation, especially where sophisticated, expressive writing was concerned. Literature, it was felt, or any sort of written discourse rich in idiomatic expression, would never in a thousand years yield itself to something so crude as "lexical transfer," the word-for-word (brick-by-brick) delivery of a text in one language into a text in another by means of a dictionary-based program. The results of such an effort would produce either gibberish or comedy (or both); "literature" would be destroyed.

Not so "science," however. When it came to this area of language, a different opinion dominated. The questions were asked: Could technical material really be rendered in this way, one word at a time, by the crudest process imaginable for any translation? Would the result be at all intelligible, legitimate, useful? The experts all seemed to agree: "experimental results suggest that for scientific texts the answer is yes." Moreover, "a monolingual reader, expert in the subject matter of the text being translated, should find it possible in most instances to extract the essential content of the original . . . often *more accurately than a bilingual layman*" (Oettinger, 1959, pp. 257–258, italics added).

Buried in these statements were a number of significant assumptions. One was the belief that scientific discourse comprised something enormously more simple than literature, something universal in its properties. To speak it in any single language automatically gave one more access to it internationally than knowledge of other human languages. Within the realm of science, it was assumed, the physicist or the geologist is, by definition, more global an individual than the multilingual person. This drew on another basic idea, namely that scientific speech constitutes a form of discourse that exceeds all others in terms of its fixed rules, its grammatical and syntactic purity, its reduction of expression to information. "In this uniformity," the linguist Leonard Bloomfield had written a few decades earlier, "the differences between languages (as English, French, German), far-reaching and deep-seated as they are, constitute merely a part of the communicative dross. We say that scientific discourse is translatable, and mean by this that . . . the difference between languages . . . *has no scientific effect*" (1939, p. 4, italics added).

Such belief in the supreme internationalism of scientific language prevailed in Bloomfield's time and remains a commonplace today. Indeed, it defines one of the most ubiquitous beliefs about science both within and outside the scientific community. It is an invested political belief (handed down from the Enlightenment) that bears within it an ideal model of boundless sharing, cooperation, and world peace. It is the dream of recovering, finally, the speech that united humankind before Babel. Such a view has been important to the 20th century: during a period when older value systems utterly collapsed, and when war, fostered by national rivalry, led to suffering and dehumanization on a monstrous new scale, the idea of "science" as an embodiment of universal progress, human welfare, rationality, and internationalism became, especially after World War I, enormously appealing.

The philosophical therapy of this faith, however, did not prevent it from downplaying or ignoring certain realities of science and its language(s). Many of these have been taken to task since the late 1960s, when science might be said to have lost its halo. But the issue of translation has yet to be tackled. Most scholars have skipped over this terrain, vast as it is, leaving the belief intact that French science is linguistically the same as German science, and both the same as English or Russian or Japanese science. Indeed, scholarship as a whole has regularly contributed to this belief by always speaking of Science, the grand and universal entity, the work of only one or two writers of a single nationality. As commonly employed, there is only one Science, with one universal tongue, spoken by its loyal members throughout and beyond the diversity of their individual cultures, as the grand intellectual Internationale.

I

Such ideas, clearly, deserve some scrutiny. How true can they be? How far can one accept on a literal, everyday basis, Einstein's idea of the "common language of science"? No one can doubt that national differences exist in many areas of scientific practice. Obviously French science and English science, for example, are different with regard to the detailed public and private realities of research, the specific institutional structures or other public theaters which they inhabit, the manner in which they accept and integrate ideas from abroad. All these realities, and many others arising from day-to-day behavior and the context of labor (intellectual or otherwise), bear the immediate imprint of the culture in which they occur. They, too, are science and have their own epistemological effects.

Many examples might be found to document national influence upon the content of scientific understanding. The reasons why certain fields have evolved rapidly and thrived in some countries but not in others have sometimes been the subject of historical study. It is well known, for instance, that early centralization of French science in the form of L'Académie des Sciences led to a strong (if narrow) consensus regarding the basic principles of experimental philosophy and the primacy of the Cartesian mechanical view, which held that no motion could occur without the transfer of some substance ("No action at a distance" was a common slogan). This acted as a filter for outside ideas, with the result that Newtonian physics, with its emphasis on an unexplainable force (gravity), was viewed as a return to occultism and was resisted by French scientific circles for more than 50 years after its introduction.[1] More generally, one could easily point to how certain disciplines have been emphasized over others in particular countries, due to national political priorities (e.g., areas of physics linked to military uses during the cold war, or fields of biomedicine relevant to research on cancer and AIDS).

But how do these realities, these scientific traditions, relate back to Bloomfield's point about language? Can it be said here, too, that science as written text undergoes an important transformation across national boundaries? To what degree might there be inevitable differences, given that language is perhaps the most localized of universal human tools? To begin an answer is, at this point, to call for evidence. Let me briefly offer two excerpts for comparison, one in French, the other in English, both written by prominent contemporary researchers in the area of African geology. These selections are taken from technical articles published within the last decade and concern the Early Paleozoic period (roughly 450–600 million years ago) in the formation of the African continent:

L'histoire de l'Afrique au Paleozoique se déroule toute entière dans le cadre du super-continent de Gondwana. Celui ci, on le sait, s'était constitue comme tel aux environs de 600 Ma, au cours de l'évenement thermotectonique panafricain. Il ne se disloquera qu'au cours due Mesozoique, 450 Ma plus tard. On doit, par conséquent, penser l'histoire de l'Afrique, durant cette periode, comme déterminée à la fois pars [quelques] facteurs principaux, [qui inclues]: l'héritage panafricain, la destruction des chaines formées, leur péneplanation, puis le "souvenir" de ces chaines. (Fabre, 1988, pp. 2–3)

(The history of Africa in the Paleozoic unfolds entirely within the frame of the supercontinent of Gondwana. This, one knows, was formed somewhere in the neighborhood of 600 million years before the present, during the pan-African thermotectonic event. It broke apart only during the Mesozoic, 450 million years later. One should, as a result, conceive the history of Africa during this period as determined simultaneously by [several] principal factors, [including]: the legacy of the pan-African event, the destruction of the mountain chains that had been formed, their peneplanation, and the preserved "memory" of these chains.)

A review of the Palaeozoic stratigraphical record of western Gondwana reveals that for the purposes of rigorous reconstruction the information from many localities is sparse, and some of the relevant interpretations appear to be inconsistent or contradictory. Palaeozoic sedimentary rocks of western Gondwana are preserved along the western margin and in the interior of South America, and on the broad northern and narrow southern margins of Africa. In describing these deposits we refer to basins principally as the geographical location of the record, without implying a particular Palaeozoic tectonic framework. (Van Houten and Hargraves, 1987, p. 345)

These two passages, while obviously not representative of all geologic discourse in French and English, do reveal certain differences in the types of expression ordinarily allowed or even expected (in my translation of the first excerpt, I have tried to remain as faithful as possible to the original, without scientizing any of the structure or wording). To the attentive eye—certainly to the eyes of a contemporary geologist—there are, in fact, striking contrasts in style, tone, and even cognitive content between the two texts. Compared with the English passage, the French has a decided "literary" quality. The variation in sentence length, revealing a sensitivity to drama and effect, the type of word usage, the conscious turn to metaphor (even if placed in quotations)—all are elements that give the French a touch of "writerliness" almost entirely lacking from the more functional, monotonic English. To those versed in scientific prose, the French excerpt has a certain archaic, 19th-century feel to it. One notes, for example, that the word *histoire* is used by itself—not, for example, as "*histoire géotectonique.*" This not only postpones any dominance of jargon; it also permits a double reading of the word, which encompasses both "history" in the broad sense and "story" in the narrative sense, as something put together, even fictionalized, and then recounted by a particular speaker interested in both persuading and entertaining a group of listeners. This sense of putting together "Africa," as a creation of one type of history, has its own richness of relevant meaning. It has deeper, philosophic connections to the modern geologic view of Africa as a unique entity with a unique "fit" in the larger assembly of continents, known as Gondwanaland, that prevailed through Paleozoic time. "Histoire" implies a community of readers interested in the ways in which this entity is created by contemporary narrative and its exploits. Geologic history and the data that builds it together become the field of tale-telling interpretation. Nothing like this can be gleaned from the utilitarian "reconstruction" of the English passage, with its engineering (i.e., intrascientific) connotations.

There are, therefore, differences in both sense and sensibility between these two passages. My own research in the area of geology tells me that such expressive differences carry pretty much across the board in French scientific discourse. Even more definitive examples, in fact, can usually be found in technical abstracts, where the writing is intended to be as dense and functional as possible and where the ideal of a one-to-one correspondence between languages is a clear goal. Here, too, that is, such differences as those pointed out above are equally apparent.[2]

These remarks, brief and introductory though they are, suggest that one might pursue such a line of inquiry in more detail. A full study that attempted to say something about the "character" of science in a particular language, and how this differed from science in other languages, would constitute a unique type of effort, for this topic has been ignored by scholars interested in technical communication. Such study could do worse than to look too at how variations in scientific expression can not be divested from the larger literary traditions in various countries and the ways in which these traditions have been shaped and hardened by such things as education, social class, political power, the development of publishing, national self-image, the place of authorship, and so forth.

The history of scientific discourse in any particular language is also a history of continued coinage, the creation of vocabularies. Each discipline must therefore have its own "histoire" and each term within it too. Single terms in fact are always the result of conscious choice, and such choice must always, by definition, bear the marks of larger influences, characteristic of their time and of the men and women who made them. A nomenclature is built from thousands of such choices; it leaks history at every pore. Tracing the progress of such a vocabulary, even at a simple level, would be a revealing task. Yet the nationalizing of science via language is only one area for inquiry. Another, no less intriguing, concerns how different languages have influenced each other—how, for example, scientific discourse in English or German or French has affected the development of technical speech in non-European, even non-Western cultures. It is now estimated that between 60 and 70% of all technical literature in the world is first published in English. English has become, more than any other language, the international *lingua scientia*. Yet this is largely a recent (mainly postwar) phenomenon. Prior to the 1930s,

German was the major language of science. Much science was also published in French and Spanish, and even in Dutch and Portuguese. Certain nations have experienced a historical succession of linguistic influences with regard to the evolution of scientific terminology. One other aspect to "nationalization," therefore, would be the individual ways in which different countries have metabolized these influences, as expressed in their discourse of science today. Here, one feels, is a new area, where the very meaning of "translation" takes on a different, possibly Frankensteinian character.

II

Without doubt, one of the best languages in which to pursue such an inquiry is Japanese, and this for a number of reasons. Modern scientific discourse is as advanced in this country as anywhere in the world, yet it was born relatively late and grew up under strong foreign influences. Until the late 19th century, Japanese science combined ideas and terms from both traditional Chinese and Western natural philosophy, in a shifting proportion that moved away from the former and toward the latter. The tale of this historical change in loyalty is written in the nomenclature of certain fields today, which therefore contain memorials to their own evolution. The most rapid period of scientific modernization, which began in the late 1860s, took place as part of a deliberate governmental plan, whose basic idea was to model different fields after their most successful counterparts in the West. During this period, Holland, which had served Japan as the major conduit for Western science for more than two centuries, was abandoned for countries such as Germany, England, France, and the United States. Yet the process of adoption and adaptation was neither smooth nor fully under the government's control. Certain disciplines, such as chemistry, developed conflicting attachments to more than one country and therefore manifested a split in terms of the languages they followed (German and English in this case). The change in such attachments over time were shaped by changing historical and political realities, especially during the present century.

The linguistic effects of all this require a larger context, that of the Japanese language itself. More than Western languages, Japanese reveals its history on its skin. It is composed of three different writing systems: two syllabaries (called *katakana* and *hiragana*) and one ideogram system (*kanji*). The latter was imported from China sometime in the fifth or sixth century, when contact between the two countries encouraged the Japanese nobility's conversion to Buddhism and an interest in Chinese philosophy. Prior to this time, Japan appears to have had a purely oral culture. This culture, with its traditions of myth and poetry, was therefore made literate—"translated" to itself—through an alien system of symbols, developed for another language. Chinese writing could not, however, be used without some changes, and these took place in two basic ways that remain visible today. The first was to use Chinese characters semantically, purely for their meaning, and to give them Japanese sounds (called *kun* reading); the second was to do the very opposite, to employ the characters phonetically; that is, to use Chinese sounds for writing Japanese words, with no regard to the original meaning of the symbols themselves. The sound given each character was an approximation in Japanese of the original Chinese sound (this was first called *manyo-gana*, after a famous collection of poems, *Manyo-shu* (The gathering of ten thousand leaves), compiled around A.D. 759; later, however, it came to be known more simply as the *on* [literally, "sound"] reading of a character). Both these forms involve the use of Chinese writing to record Japanese speech. "Record," however, does not convey the depth of adaptation involved. This, rather, is suggested by the common practice among linguists and scholars when they speak of *kun* reading during the early and medieval periods as "translation readings" (see, e.g., Habein, 1984, p. 22).

Alongside these forms of nativization, Chinese itself became—and remained for more than a millennium—the language of the literary and scholarly intelligensia, of government edicts, chronicles, and other institutional areas of writing. This long-term presence of Chinese proved of great importance to the further development of written Japanese. Priests, especially, when deciphering religious works out loud, commonly slipped from Chinese into Japanese readings. Over time they developed a system of diacritical marks that they placed alongside the *kanji* to aid in this translation process. Such marks often took the form of simplified characters or parts of characters, and came to be known as *kana* (literally, "temporary name" or "interim name"). By the 11th century, such marks had developed into their own system for writing Japanese sounds, eventually called *katakana* (*kata* meaning, originally, "imperfect" or "half-formed").

The other syllabary, *hiragana*, was developed at about the same time, also out of a direct connection to Chinese, but in an entirely different way. It evolved not through reading but through writing, as a kind of aesthetic end point to the art of calligraphy practiced among the nobility, who regularly wrote poems in Chinese. During the Heian period (9th–12th centuries), when such writing reached its height, the tracing of certain characters, often used for their *on* (phonetic) readings, became highly stylized and was finally reduced to a few flowing, cursive strokes. These were called *onnamoji* ("woman's writing"), since women were not allowed to study Chinese (though many did so) and thus were forced to write in Japanese, through these aesthetically simplified *kanji*.[3] By the end of the Heian epoch, *hiragana* (literally, "smooth" *kana*), woman's writing, had become the syllabary used in combination with Chinese characters to form the main portion of the Japanese writing system that has continued down to the present. *Katakana*, meanwhile, was assigned a variety of special uses, including phonetic spelling of Japanese words for emphasis (similar to Western italics), and more importantly, the phonetic rendering of foreign words, especially, beginning in the 16th century, those derived from Western languages.

Katakana was a creation of the ascetic world of the Buddhist priesthood. In actual form, it is simple, angular, unadorned (see below). *Hiragana*, by contrast, is sensual and curvilinear, reflecting its origin in the rarefied cosmos of the Heian aristocracy, where luxury and decorativeness defined an everyday sensibility. It was derived not as a reading aid, for those who sought to absorb and teach "wisdom," but as a type of writing for those required to produce "art," in the form of sophisticated, stylized expression. Modern Japanese writing, therefore, in combining these two scripts with *kanji*, offers in its very graphic qualities a visual representation of historical origins—indeed, of history itself, inasmuch as this history combines wholesale importation of another writing system with several styles of nativization.

Despite the development of the *kana* systems, however, Chinese remained the official language of scholarship and government for centuries thereafter. Most new words imported from abroad, whether from China or the West, were written phonetically in characters (including such terms as *pan*, for "bread"). Others, such as *tabaco*, were pronounced in imitation of the Western term but written semantically in Chinese (with the *kanji* for "smoke" and "grass"). Moreover, orthography re-

あ い う え お ね   Hiragana

フ カ エ オ ス ニ   Katakana

mained entirely faithful to the Chinese standard, starting on the right side of a page and occurring in vertical lines. These and other uses of Chinese were given great stimulus in the early Edo period (17th century) with the establishment of the Tokugawa shogunate and the influence of neo-Confucianism (the Chu Hsi School), which was sanctioned as state doctrine. Chinese was taught to the samurai class, who comprised the top of the social hierarchy, as a foundation of their schooling. It was also promoted by the advent of printing, by private education (through which the lesser classes sought to compete with the samurai), by public lectures, and by the generally increased status of scholars as leaders of official and sometimes popular sentiment.

This is not to say that Chinese was entirely dominant throughout Japanese society. It was, instead, the formal language of learning and social privilege. It formed the roof to a vernacular revolution of sorts that began from the 18th century onward and proceeded in several directions. One group of scholars and writers, for example, the so-called *kokugaku-sha* or "intellectual nationalists," rejected all foreign influence and tried to reinstate ancient styles of writing used in the so-called Japanese "classics" of the early medieval period. Among the rising merchant class, meanwhile, there developed a number of popular types of writing that employed Japanese, even colloquial Japanese, in developing such genres as satire, farce, folktales, children's books, pornography, and *feuilleton* fiction. By the end of the Edo period in the middle 19th century, the native orthography had split into such a diversity of styles, many associated with specific literary uses, that literacy became difficult to adequately define and reading Japanese could often be confusing: single characters and simple character combinations, for example, had acquired so many different readings (phonetic and semantic both) that they often could not be accurately deciphered without some sort of reading aid.[4]

During the whole of the Edo period, Japan had strongly restricted relations with other countries, especially those of the West (the standard view that the country closed off nearly all such contact is significantly exaggerated). The Dutch, however, were permitted to carry on limited trade with Japan, and through this intercourse introduced books and ideas from Europe into the country. A new language of scholarship—Dutch—sprang up and became important (mainly in the sciences) by around 1800. Following Perry's visit in 1853 and the restoration of the imperial government in 1867, the country as a whole turned more fully toward the West for models of modernization. This turn, in combination with the forces of vernacularization mentioned above, spelled the downfall of Chinese. From the late 19th century onward the use of Japanese began to penetrate all levels and kinds of scholarly writing, though often in the wake of Western languages (Dutch early on, German and English somewhat later). The influence of Western languages convinced many Japanese thinkers of the need for standardizing their language and for uniting its written and spoken forms. A number of proposals were put forward. Some advocated using Western writing as a model, by which was meant either employing only *katakana* (the "Japanese alphabet") or else doing away with all native forms and adopting the Roman alphabet directly. A more measured, and eventually successful response, put forth by the famed leader of the "Meiji Enlightenment," Fukuzawa Yukichi, was to limit the number of *kanji* for general, public use and to standardize all readings according to a scheme that would incorporate both *on* and *kun* versions, along with any others that approached a similar level of wide usage. This reform, however, was long in coming, and was not achieved until after World War II, when the Ministry of Education put into effect a series of sweeping changes that have acted (not without difficulty) to create the semblance of a modern, relatively stabilized and uniform written language.

At the turn of the century, however, written Japanese had become an amalgam of all the influences just mentioned. Japanese writers often used Western-style punctuation, including commas, periods, question and exclamation marks, and, depending on their intended audience, could choose among several common orthographies: vertical, from right to left (traditional Chinese-style); horizontal, from left to right (Western style); or horizontal, right to left (mixed style). Since World War II, Japan has been heavily influenced by the United States, and it has become more and more common to write horizontally and to put Western loan words in the Roman alphabet, especially if these words are borrowed from English. This is particularly true in the sciences, where it is now a regular practice to insert English nomenclature in the published articles of such fields as astronomy, physics, computer science, and biotechnology. But in other areas, such as literature, older forms persist: novels are still published in vertical form (right to left), as are most newspapers and magazines, books on culture and the arts, and so forth.

Taken as a whole, contemporary written Japanese represents a field where overlap occurs between a host of graphic–historical deposits. This is true, moreover, no matter what the subject material, and no matter what the level of writing.[5] Indeed, to say that the visual realities of this language, as written today, wear their history on their sleeve is to state only half the truth, for they also reveal much about the actual processes behind this history, as I have said. Graphically, the different types of writing—*kanji*, the two *kanas*, the Roman alphabet—might be said to compete with each other. Despite all its recent reforms, Japanese remains a battleground between ancients and moderns, a space where different eras, places, philosophies, and origins have all left their historical trace.

III

What has just been said is perhaps more true of scientific discourse in Japanese than other forms of expression. To understand this, one must also understand something else fundamental to this discourse.

Scientific Japanese differs from technical language in Western countries in one very significant way. Because it is written in ideograms known to almost any college-educated person, nonscientists can readily decipher the basic meaning of many of the most complex scientific terms. They can, in a sense, "read" scientific jargon far more easily than their counterparts in the West (who are no longer schooled in Greek and Latin). For example, the terms "pyroclastic" and "protoxylem," no doubt obscure to the average literature major in the United States, are written in Japanese with the ordinary characters for "broken by fire" (*kasai*) and "original living wood part" (*genseikibu*)—in this case, *kanji* that any 7th grader knows. Such intelligibility is not maintained for all technical terms, however. A great deal of contemporary scientific jargon is written in *katakana* and represents direct phonetic importation from Western languages, especially English. However, in certain fields where the use of *kanji* tends to dominate (e.g., botany, mathematics, geology, and astronomy), general public access to technical language is extraordinarily high by the standards of Occidental linguistic reality.

What, then, of style? How "literary" might scientific discourse be in Japanese? It is my strong impression, as a regular translator of technical articles, that despite profound linguistic differences in grammatical structure, Japanese scientific writing is far closer to English in its basic intended aesthetic qualities than to other Western languages, such as German, French, or Spanish. This seems true in light of such things as general tone, length of sentences, use of first-person pronouns, manner of citation, organization, and, most of all, employment of jargon as the syntactic center of expression. Japanese scientific writers appear to follow the English example in keeping things as direct and unadorned as possible. The following excerpt provides an example:

> Beppu Bay lies on the eastern margin of the Beppu-Shimagara gra-
> ben(1) (Matsumoto, 1979) and Hohi volcanic zone ([written in En-
> glish]; Kamata, 1989b), which developed as a result of volcanic ac-
> tivity and graben(2) formation in Pliocene-present time (approxi-
> mately 0–5 Ma [million years before present]). Within the Hohi vol-
> canic zone, andesitic volcanism was associated with normal faulting
> and caldera subsidence involving 2–3 km vertical subsidence of pre-
> Tertiary basement rocks. (Kamata, 1993, p. 39)

Such writing has all the straightforward dullness and efficiency one expects from contemporary geological discourse in English. It is characterized by an almost pure functionalism. At the same time, it would be
difficult to prove such stylistic influence, at least in any direct way. We
should perhaps beware of making everything in Japanese science seem
the result of foreign "influence."

But let us dwell a moment on some other features of this passage,
and note how it bears out what I said earlier. One sees, for instance, that
the term "Hohi volcanic zone" appears in English amid the flow of *kanji*
and *hiragana*. All numbers are also given in their Western (Arabic) symbols. The term "graben," on the other hand, is written in two ways within
the same sentence: the instance labeled (1) appears in *kanji* and represents the adopted Chinese term (absorbed during the late 19th century);
the second instance (2) uses *katakana*, thus indicating a distinctly Western origin (in this case German "*graben*," meaning "ditch" or "trench"
originally, but in geologic parlance indicating a fault-bounded, rectan-
gular or rhomboidal area of sinking). This kind of double writing is not
done for effect. Instead, *Shimagara-chiko* (*chiko* = *graben*) is part of a
proper name, coined either late in the last century or early in this one.
Juxtaposing it against "gu-rah-ben" in *katakana*, which is contemporary usage in geo-discourse, is to place different eras or tendencies of
nomenclatural coinage—different cultural loyalties—immediately against
one another. An earlier era when Chinese models sometimes still dominated is placed side by side with the postwar tendency to phoneticize
terms in English. On the other hand, "Pliocene" and "Tertiary," both
proper names of Western origin, are rendered in *kanji*, with the characters, respectively, for "bright new era" and "third period." The first of
these translations represents something of a nativized elaboration (the
Greek root means, simply, "more new"); it has a dynastic ring to it. The
second, meanwhile, is a direct translation from the English. Thus, again,
two very different sensibilities placed side by side; two different realms
of choice revealing of historical influence.

Such complexity can be found everywhere hovering at the surface of scientific writing in Japanese. No doubt it is not perceived as such by its users, Japanese scientists (else, one wonders, how would any "science" get done?). As insiders, they have been trained to ignore or (my earlier term) "forget" its reality, just as Western scientists have (though their discourse lacks the graphic dimensions of its Japanese equivalent). It comes immediately into focus, however, when one turns a lens in this direction and makes this discourse the subject of inquiry.

Another interesting tendency found in contemporary Japanese journals is to publish articles in Japanese with an abstract only in English. Far more striking, however, is the labeling of illustrations, tables, charts, or graphs entirely in English as well (particularly true in geology). The effect, for the English reader, is both remarkable and useful. Floating in a sea of otherwise incomprehensible signs, there are here maps and diagrams that are entirely legible to the Western eye, islands of valuable, summarizing information. Why this use of English? Since early in this century, Japanese scientists have been acutely aware of their linguistic isolation and have striven to internationalize the products of their labor. This isolation, however, also allows them an ingenuity. A demand to caption illustrations in English effectively requires Japanese scientists to learn English. It also recognizes (inadvertently?) that scientists everywhere often "read" the latest research in exactly this way, by skimming the visual information offered, even before perusing an abstract. (Indeed, the deeper internationalism to scientific discourse as a whole may lie not merely in mathematics or in certain modes of writing but in styles of drawing and practices of reading.)

In the face of all this, one perhaps feels an urge to declare the equal complexity of technical speech in European languages. Haven't terms there too derived from Greek, Latin, Arabic, English, French, German, Russian? The 17th, 18th, 19th, 20th centuries? And so forth? Certainly they have, and this truth has many implications with regard to the content of "science" and the ability to read its historical evolution as inscribed in such origins. In the case of Japanese, however, the emphasis seems different (I do not say "exotic"), due to the ideographic nature of the language. To understand the unique Japanese case, you need to imagine a different kind of written English, one in which terms from classical Greek or Arabic would be written in the Greek or Arabic alphabets, and other terms dating back to English's origins as Anglo-Saxon would be written using original Anglo-Saxon spellings and some of its special symbols. Moreover, you would have to imagine the simultaneous use of medieval and modern forms of certain words, each with their own graphic identities. In short, all derivations would have to be kept *visible*.

If this sounds exaggerated, we need only turn to the names of the elements for a definitive example. Indeed, as a linguistic phenomenon, these offer a riveting instance of how diverse the mixture of Chinese, Japanese, and Western influences is in Japanese technical language. I quote here from a recent study along these lines by Sugawara Kunika and Itakura Kiyonobu (1989, 1990a), who begin their discussion with a coy note: "The names of the elements in Japan," they write, "are rather complex in their construction." They then go on as follows:

> In addition to those elements such as hydrogen (*suiso*), oxygen (*sanso*), and carbon (*tanso*) that form their names by attaching the character *so* [literally, simple, principal], there are others [based on Chinese example] that use only a single *kanji*, for example, gold (*kin*), silver (*gin*), copper (*doh*), and iron (*tetsu*). Thinking this of interest, we might also recall those names such as mercury (*suigin*, from the characters for "water" and "silver"), zinc (*aen*, literally, "next to lead"), and platinum (*hakkin*, "white gold"), which are formed from two characters on the basis of their literal or analogical meanings. Then there are those names written solely in *katakana*, such as aluminum, sodium (written: *natrium*), tungsten, etc. Finally, there are also names whose original language was Latin, German, and English, or that mingle these together. All of this constitutes, moreover, a noted contrast to Chinese, in which the names of the elements today are all expressed with a single character. In Japan, the christening of this nomenclature has been based on a variety of circumstances, and in its contemporary mixture . . . can be said to surface something of the process by which chemistry in Japan was formed. (1990a, p. 193)

In short, the Periodic Table in Japanese represents a zone of linguistic and cultural collision. Moreover, the above passage speaks only of the final result, as it exists today. During the 19th century, however, when these names were first entering the language, they were anything but fixed and agreed upon. Many of them, along with such basic terms as "atom," "molecule," "reaction," were proposed and argued over in various other forms. As historians of science have pointed out, even the word *chemistry* itself had two variants until the last years of the century. One of these, *seimigaku*, written phonetically in *kanji*, had been adopted earlier from Dutch translations of French chemical works; the word *seimi*, in fact, was meant to simulate the French word *chemie*, which Dutch writers sometimes used (the more common Dutch equivalent is *Scheikunde*, the first three letters being pronounced similarly to the French "ch"). The other term, *kagaku*, had been more lately imported from

Chinese. Thus, both words were essentially foreign. The choice between them was a choice between traditions: the more "modern" of the two, involved using a Dutch-derived word to designate a Westernized science; the other, the more traditional called upon still older intellectual loyalties to China. The conflict therefore drew upon attitudes dipped in complexity. During the 1880s, the decision between these two terms became a focus of acrimonious debate before finally being ended by fiat during a meeting of the Tokyo Chemical Society—whose very name, one might note, was at stake (Hirota [1988] provides a detailed discussion of this episode).

At any point in its evolution, the standardized nomenclature of a scientific field is like the surface of a smooth sea. It stretches out, nearly flat, calm, easily crossed by those who know their way. Yet it covers a vast submarine landscape of looming forms, of mountains, canyons, plateaus, and formations of enormous scale, sometimes rising up to just below the surface. Sailing over this ocean, a few meters in any direction, one crosses great vertical distances of time and influence. To anchor and explore the depths below means to enter a realm normally thought to be removed from "science."

IV

The terrain of modern Japanese science has been created almost entirely since the later decades of the 18th century. This was a period that followed the lifting of a strict ban on Western books, a policy which had been in effect since 1630, when contact with the Western nations was severely restricted by the early shoguns, due to perceptions (not entirely unfounded) of political and cultural threat from European colonialism and missionaryism. "Most favored nation" status was granted to Holland, which was perceived to be interested only in economic matters and was allowed to maintain a trading station on the island of Deshima (literally, "Departure Island"), in Nagasaki harbor. Any and all books unloaded there had to be inspected by official censors before being allowed to touch mainland soil. Most Western works considered for admittance "were concerned with geography or nautical subjects. They had already been translated into Chinese, and any mention of the Christian religion in them had been expunged. By the 1760s, after the relaxation of the censorship ban, a number of Dutch books on medicine and various almanacs were being imported too, and among the items shipped to Japan were included many scientific instruments (especially telescopes), which had the effect of quickening an interest in Western learning (MacLean, 1974).

At the time, this was calculated to appeal to the tradition of natural philosophy in Japan, which was concentrated in two areas: botany (actually pharmacology, linked to herbal medicine) and astronomy (for the purposes of calendar making). Both areas were deeply joined to the Confucian social order put in place by the shogunate, which imposed a rigid class structure whose intellectual professions included physicians, Confucian scholars, and priests. All of these were hereditary callings and all were shaped by the great resurgence of neo-Confucian study and scholarship that the government supported. In a sense, this resurgence made Chinese scholarship a required fundament to any branch of intellectual endeavor. The Chinese classics were regarded as the source of wisdom and a path to virtue. Pharmacology and medicine in general drew on detailed study of these works: the plants and animals described within them, their uses, and the theory and illustration of the human body, were all discussed in terms of a moral order related to universal balance and harmony. Similarly, astronomy depended on literary study of chronology related to the seasons, constellations, planets, and so forth, as expressed in these ancient texts, such elements expressing the cyclical moral order as well. New reference works were also produced that catalogued these phenomena, that commented upon their occurrence and meaning within the classics, and that, somewhat later on, began to develop "theories" or "hypotheses," usually of a metaphysical nature, based on points of debated interpretation. All of this seems to have been generally known to the Dutch traders, who tailored their imports of "science" accordingly.

The Chinese literary tradition, however, though preeminent, did not enjoy a complete monopoly in intellectual life. Even by the late 17th century, the Japanese classics were also being invoked as important sources for intellectual example and knowledge (e.g., native plants, descriptions of astronomical phenomena in poetry, etc.). Japanese dissatisfaction with the inaccuracies of the Chinese calendar led to reform efforts and to a greater readiness to examine Western astronomical methods and ideas. With the lifting of the censorship ban in 1720 by Shogun Yoshimune (who himself seems to have been interested in the techniques of Western calendar making), Japan began slowly to turn away from China as the major seat of knowledge and toward Europe. Still, the Chinese classics continued to remain influential up through the early 19th century; they were regarded as a reservoir for higher forms of truth and the firm basis for higher levels of education. For the sciences too, they provided a metaphysical, vitalist framework or filter through which Occidental ideas and language were often absorbed, in somewhat re-constituted form. Science in Japan thus had literary origins, and emerged from a series of canonical works whose understanding was restricted to an elite group of scholars.

The vast majority of European books that came into Japan before about 1770 were Chinese translations or commentaries in Chinese, usu-ally written by Jesuit missionaries living in China. The sensitivity of the Japanese censors to Christian religious reference meant, in effect, that those books most likely to pass inspection were of a scientific nature. These were examined by a "college of interpreters" in Nagasaki who had been trained in the Dutch language at government command. Such training was often of a very rudimentary nature at best, but helped pro-vide the basis for a new realm of scholarship "Dutch Studies," *rangaku* (see, for example, Keene, 1968), that proved extremely significant later in the 18th century. Earlier on, however, after passing the censors, books were then shipped to the government library in Edo (modern Tokyo), where they were stored and kept out of general circulation. According to MacLean (1974), some Western books might also have entered Japan via the private libraries of Dutch officials on Deshima and thereafter found their way into private hands. In any case, what Nakayama (1969) says of astronomy was true for other fields of Western science as well during this early period, namely, that they had little success in penetrat-ing the existing neo-Confucian system of learning and inquiry. And yet, before long, it was this system itself that helped established the pro-found worth of technical knowledge from Europe, to such a degree that *rangaku* studies expanded tremendously (by the early 19th century) and nearly all interest in the West became concentrated in this area. Thus, during this critical period when the country was experimenting with a new opennness, "the knowledge that the Japanese managed to glean from and about the West was almost entirely confined to the natural sciences" (Blacker, 1969, p. 14).

In Tokugawa Japan, neo-Confucianism meant the reformist or Chu Hsi School, whose founder had lived and written in late 12th century China, during the Sung dynasty. Within this philosophy, as adopted by the Japanese, there existed two fundamental terms for helping explain the nature of the universe and the place of human beings within it. One of these terms was *ri* (*li* in Chinese), denoting ultimate ontological and moral order. According to Chu Hsi, perception of *ri*, either within or behind the material world, brought one in touch with the governing principle of the cosmos and thus led one to higher wisdom and ethical conduct. Posed alongside this, meanwhile, was *ki* (*ch'i*), whose description by Chu Hsi is at best vague, having to do with physical appearances of *ri*, but whose definition was ardently taken up by 17th- and 18th-century Japanese scholars interested in the natural world (Saigusa, 1962). One of the first of these was Hayashi Razan (1583–1657), perhaps the most influential of the new neo-Confucian scholars. For Razan, *ri* signified spirit, the fundamental essence; *ki*, on the other hand, was equivalent to the actual movements and workings of the material world. *Ki* was the materialization of *ri*, brought it into the realm of everyday life and action (Saigusa, 1962, p. 53).

Razan's scheme was modified and reconstituted by a number of thinkers over the next century and a half, yet, dualistic view remained intact. More conservative scholars, holding to a literal reading of Chu Hsi, tended to privilege the *ri* side of the equation, with its embodiment of virtue. Others, however, focused more on the methods by which the perception of *ri* could be accomplished, and this is where *ki* tended to play a role. Chu Hsi had created a number of aphorisms regarding this kind of inquiry, among the most important of which were: *kakubutsu kyuu-ri* (investigate things and exhaust, or penetrate, the *ri*), and *sokubutsu kyuu-ri* (probe into things and exhaust, or penetrate, the *ri*). In each case, the things (*butsu*) to be investigated or probed or penetrated were associated with *ki*, the actual physical world. The goal was spiritual; the means, potentially at least, material. All of which, as noted by Craig (1965, p. 139), constituted a "this-worldly mysticism with rationalistic implications."

Such implications came to the fore in the era when Western science began to be imported after 1720. Ideas of *ki* began to be associated with the physical workings and actual phenomenology of nature, and a shift in emphasis toward this side of the system took place. At this stage, that is, a purely abstract philosophy of Chinese origin, based on moral spiritualism, was increasingly rejected. "Those who investigate natural things," wrote Miura Baien (1723–1789), "must not, like all prior followers of Confucius, remain dogmatic in their adherence to the words of the sages. For mastering truth, nature itself is the better teacher" (quoted in Saigusa, 1962, p. 55). Baien represented the new type of *rangaku* scholar who emerged in the late 18th century, deeply schooled in the Chinese classics yet intrigued by, and open to the influence of, the newer knowledge from the West. He made a careful study of books on European science and came to view them as offering pragmatic "tools" for pursuing deeper questions. At the same time, he denounced or turned away from the speculative ruminations of more orthodox neo-Confucians, whom he saw as backward and closed (see Nakayama, 1964). Baien was thus one of the key "moderns" to argue against the "ancients." He followed up his thinking with actual work: he seems to have been the first to build and employ such instruments as a celestial globe, a telescope, and a microscope—all devices for taking the eye and mind closer to nature, further into its true workings. In his scheme of thought, *ri* was tantamount to Plato's *noumena* or Kant's *a priori*—the perfect or ideal form, what "should be." *Ki* signified the reality of what existed, of how things occurred. To investigate it meant to study a range of other componential realities: *sei* (quality or character), *ryo* (mass), and *shitsu* (substance) among them.

Baien thus represents a crucial turning point in the development of modern Japanese science. In a sense, he stands close to its origin. His framing of concepts such as those just mentioned helped provide Japan with the very first vocabulary relating to concepts of physical process and its analysis. Such basic terms as "matter" (*busshitsu*), "mass" (*shitsuryo*), and "body" (as used in physics, *buttai*) go directly back to him. They are terms that had a profound effect upon the way in which nature was viewed, creating as they did a number of linguistically discrete object-areas for inquiry, distinct from cosmological ethics.

At the same time, however, one should note that the words and characters he employed came directly out of the Confucian tradition. *Shitsu*, for example, had long been used by Chinese scholars to refer to the material quality of things, as a secondary manifestation of more fundamental metaphysical principles. It was a *kanji* embedded in the philosophical and religious doctrines of Chinese learning. Another example is the very word for "science" (natural philosophy) during this early period, used by scholars (including Baien) from the end of the 17th century until the mid-19th century. This was *kyuu-ri*, meaning the exhaustion or penetration of *ri*, a primal principle set down by Chu Hsi. Such was how the Dutch word *Natuurkunde* was translated, or rather filtered, for use by Japanese scientists. Later on, during the final decades of the Edo era and the beginning of the Meiji period (late 1860s), it was replaced by *kagaku* (literally, "course," or "branch," of "study"), a fairly literal rendering of the German *Wissenschaft*. *Kyuu-ri*, however, was not even abandoned at this late date but instead survived for a time as the term for "physics," under the auspices of the belief, also directly imported from the West, that this field lay at the base of all the other sciences and thus remained in league with traditional concepts of *ri*. Indeed, this connection has never wholly been given up, for it is retained today in the standardized word for "physics," *butsu-ri*—the *ri* of material "things." Such complex blending of Chinese and Western sensibilities was an inevitable part of Japanese science in its early phases.

Thinkers such as Miura Baien reveal that a deep-seated tension had grown up within the *ri–ki* dialectic, one that increasingly divided virtue and knowledge (Craig, 1965). By this time, books in Dutch were being allowed into the country on a regular basis and the government had supported the formation of a community of translation scholars, *rangaku-sha* ("Dutch studies experts"), partly under the idea that the knowledge involved could be of important practical benefit. Visits by Russian and American warships during the early 1790s greatly added to this belief. Conservative scholars, on the other hand, felt that spiritual and moral power, the essence of Japan, would be enough to repel any invaders and that nearly all contact with Western knowledge was contaminating. The effect of these two camps of thought was effectively a freeing of "science" from "ethics" on the one hand, and "virtue" from "science" on the other. Western science came to be viewed more and more as a series of practical techniques, whose province was real-world power in a material sense.[6] This coinage of words, ideas, and outlook proved critical to the founding of a true Japanese science.

V

These brief comments on the intellectual origins of Japanese science return us to a central point: the history of early modern science in Japan is largely, if not entirely, a history of translation. With regard to scientific knowledge, Japan's isolation was then, and has always been, far more a matter of language than of geographical or political seclusion. The fundamental reasons for this are not difficult to state. Few, if any, European scholars were able to speak or read Japanese prior to the late 19th century—and even then, among those (a precious few) who did learn it, the great majority were not scientists or engineers but instead those with interests in artistic, commercial, religious, and political matters. Indeed, this has remained true up to the present. The requirement for Japanese scientists, amateur or professional, to learn foreign languages in order to pursue their field has been a constant element of training for two centuries (a sizeable number of technical journals in Japan are today published in English, for instance). Well into the 20th century, "science" in Japan meant translation. It was modern, up to date, contemporary, to the degree both that it involved "research" and retained its multilingual component, its mixture of the imported and the nativized.

At the time when Western science was first beginning in Japan, the language of use was still Chinese. This ensured a heavy Chinese/Confucian influence over the terminology coined at this time. It also meant that any break with the past, any shift to a more modern view, was bound to be protracted and directly tied to a shift in linguistic fidelity. To modernize, Japanese science had to discover a new language of knowledge.

Beginning in the late 18th century, this new language was Dutch. It is important to note that the term *rangaku,* which originally meant "study of the Dutch *language,"* came to be used to denote the learning of *all* things Western, science in particular. By the early 1800s, works in Dutch had found a relatively secure place in the scholarly community and were supplanting those in Chinese for scientific subjects. There was a desire to get in more direct touch with Western learning and to shed outdated methods and beliefs. The feeling that European learning was "mere technique" (though of considerable practical power) and thus did not oppose or conflict with neo-Confucian precepts allowed Japanese scholars some freedom to choose, linguistically and intellectually, between China and the West. One historian describes the situation in this way:

> Japan had been culturally dominated by China until at least the 18th century, but it was not politically dominated. The Japanese never lost their sense of national identity or their self-confidence. Compared with Chinese political satellites such as Korea and Annam, Japan was in a better position to modernize on its own initiative. The Japanese were free to choose either Western or Chinese science as they pleased. [Moreover] the existence in Japan of a plurality of influential philosophies also provided a sound basis for appreciating new ideas. There had been Buddhism, Confucianism, and Shintoism; why not another approach? (Nakayama, 1969, p. 230)

Such a characterization is too simplistic, of course; the Japanese were never so "free" that they could cast off their entire intellectual heritage in natural philosophy with a wave of the hand. Various continuities were to remain, nearly to the end of the 19th century. Moreover, not all scholars agreed that the West should be so avidly embraced, particularly at the expense of the long-standing loyalty to things Chinese. But despite these complexities, the general attitude that the methods of Western science (often referred to as *gei-jitsu,* or "technique") were significantly superior to Chinese "technique" in a number of areas did grow impressively. Medicine, navigation, and astronomy were three such areas widely recognized by the late 18th century, and military technology followed close behind, particularly after the visits by Russian and American warships. The Tokugawa era, in general, was a period of growing nationalism in Japan, and this too turned many intellectuals away from China. The National Learning movement, which gained steam throughout the late 17th and 18th centuries, adopted a platform to supplant Buddhism, a Chinese import, with Shintoism, a native religion, and to create a Japanese canon of classics for study. Ideas for the founding of schools of national learning, such as those of Kamo Mabuchi and Motoori Norinaga, consistently vilified the "chaos" and "philosophical pretensions" of China, holding instead that it was Japan who governed itself according to the natural laws of heaven and earth. To many Japanese, China signified a teacher who had gone astray, become arrogant, deformed away from the truth. It was, moreover, a symbol of Japanese cultural debt—a debt impossible to make good except through erasure or substitution. As such, China gradually became the whetstone on which Japan's nationalizing consciousness sharpened itself.

This did not mean that Chinese philosophy, especially natural philosophy, was abandoned; on the contrary, it was felt to be part of Japan's own legacy, a necessary cultural ingredient. It was rather that China itself, politically and culturally, had not lived up to the precepts and principles of the ethical cosmos: it had fallen from the "right path" into "false government," "petty rationalizing," and the "addiction to sophistry." Western knowledge, on the other hand, brought with it no direct burdens of morality, no sense of obedience or rebellion to high ethical principles. As "technique," stripped of religious reference, it could be absorbed without compromising loyalties. Even those who regularly abjured all things foreign, such as the influential and ultra-patriotic writer Hirata Atsutane (1776–1843), spoke up on behalf of European science:

> The Dutch have the excellent national characteristic of investigating matters with great patience until they can get to the very bottom. . . . Unlike China, Holland is a splendid country where they do not rely on superficial conjectures. . . . Their findings, which are the result of the efforts of hundreds of people studying scientific problems for a thousand or even two thousand years, have been incorporated in books which have been presented to Japan. (Tsunoda et al., 1958, pp. 41–42).

Here was a perception of Western science conceived in the terms familiar to Japanese scholars; that is, as a literary canon, compiled and handed down through the ages. Such "wisdom," it was believed, could easily be used by Japan in its effort to grow strong and autonomous. Indeed, whether an author was for or against European learning, every blow struck against China was a boost in the long run for Western science. Every step backward for the former proved an advance for the latter. Put differently, Western scientific learning became a cultural-political lever that helped Japan to feel that it was freeing itself of Chinese influence. Even at this early date, it was a tool that aided Japan in its formative efforts to become an independent nation-state.

Growing religious debate, unrest, and philosophical conflict led the conservative Shogunate to pass another ban on "heterodox teachings" in 1790. Yet the overall effects of this attempt at censorship were small. Books from Holland continued to enter the country at an ever increasing rate. Political and cultural orthodoxy remained highly porous and changeable, and it was through this medium that Western learning continued to seep or pour. By the 1780s and 90s, a growing number of physicians and interested samurai[7] had taken up the study of the Dutch language. This had been greatly accelerated by the release to the general public of the first book translated from Dutch into Japanese, a medical work entitled *Tabulae Anatomicae*, originally written in 1731 by a German physician. The resulting translation, the *Kaitai Shinso* (also called "Tafel Anatomia," an apparent half-rendering into Dutch of the Latin title), was published in 1774 and "started a great wave of interest in Dutch learning of every description," but mainly scientific (Keene, 1968, p. 24). Though fears of the "barbarian menace" continued in many quarters, the new wave of interest helped ensure that a large number of works be allowed in. Equally important, neo-Confucian doctrine helped urge officials to perceive such works as adding to the national stock of practical knowledge. In 1811, the government set up the Office for Translation of Foreign Books (Bansho Wage Goyo), an attempt on the one hand to gain control over the content and spread of this new knowledge, but also an obvious acknowledgment of its growing importance. This office had a somewhat checkered career: originally, it was mainly an administrative bureau; in the 1840s, it became a censoring board; finally, in 1855, after Commodore Perry's visit, it was changed into an official training institute that not only taught Dutch, but English, French, and German, as well as courses in most of the sciences. No doubt its most significant influence in the early years was its publication of the first Dutch–Japanese dictionaries. These were no doubt critical to the spread of *rangaku* and to its work in founding Japanese science through the medium of translation.

By the first decade of the 19th century, Dutch had become the language of Western science. It remained so until the beginning of the Meiji era in 1867. By that time, Japanese scholars, translators, and officials had come to realize that their earlier gateway to Europe had in fact been something of a side entrance. English and American naval power, French chemistry and astronomy, and German medicine excited Japanese fear, fascination, and envy. But before a shift of linguistic interest could occur, a greater degree of openness to outside influence had to take place.

During the 1830s and 40s, things went in exactly the opposite direction. As a result of renewed visits by foreign warships and of China's crushing defeat by Britain during the Opium Wars, anxiety in official Japanese circles gained the upper hand. A new round of censorship occurred; European medicine was banned and those who had been promulgating a "Western technique, Eastern ethics," that is, an "open nation" (*kaikoku*) approach, came under increasing suspicion and were forced to give ground before those expounding an "expel the barbarians" (*jo-i*) philosophy. Between these groups, which had existed since early in the century, a battle over terminology ensued, involving above all the very word for "science" itself. *Jo-i* writers and officials saw use of the neo-Confucian term, *kyuu-ri*, as a dangerous appropriation, a sign of degradation. Western science did not deserve this sacred word; it was not concerned with the moral essences of *ri*—essences which, if kept intact, would help keep the Japanese capable of repelling any outside colonial power. *Kaikoku* thinkers, on the other hand, tried to claim compatibility between native and European knowledge. One of the most influential of their number, Sakuma Shozan, went so far as to say that "western learning and techniques are not something alien but a branch *of our own learning*" (quoted in Blacker, 1967, p. 23, italics added). Science and technology, in the Occidental sense, were still knowledge, and therefore, at some level, were in concert with "the Way of the Sages." Both groups, however, saw this knowledge as lacking in any fundamental spiritual or ethical content. Shozan's view eventually prevailed: in order to be true and useful, Western science had to comport with the moral/metaphysical principles ruling the universe. Otherwise, it would be nonsense, or worse, sophistry. At this fundamental level too, therefore, *jo-i* and *kaikoku* thinkers were in agreement.

The existence of these two groups, and of others yet more avidly pro- or anti-Western, indicates the lack of any practical orthodoxy, any agreed-upon set of expectations, regarding Western science. This is linked to the fact that prior to the Meiji Revolution (1867), when the shogunate was overthrown, Japanese science remained largely noninstitutional. *Rangaku* was primarily an activity of individuals, even if they worked under government auspices. Most seem to have had mixed feelings regarding China, yet they were nonetheless loyal neo-Confucian scholars, and their writings and translations from the late 18th century onward very commonly indicate a desire like Shozan's to absorb Western science within the framework of the moral/metaphysical cosmos bequeathed from Chinese natural philosophy. Even as late as the 1830s, it was common practice for *rangaku* scholars to make translations of Dutch works into classical Chinese. Moreover, even the most xenophobic of critics against China or advocates for the new National Learning (e.g., Kamo Mabuchi), would never have advocated something so radical as a purge of Chinese thought and its complete replacement by Western philosophy. Chinese thought, like the Chinese language itself, had too long been part of Japan to disavow entirely. Something akin to its denial would only come later, when science entered its first major phase of institutional growth.

Regarding this phase, it would be too simple to maintain, as many historians have, that modern science began with the Meiji era in 1867, as if by some divine surge of intellectual wind. While it is certainly true that the government took matters firmly in hand at this point, viewing scientific knowledge as the key to modernizing the country, exporting students overseas and importing foreign teachers, this did not mark a complete break from the past. The translation of Western works was an activity long performed with government support and under its supervision. The bureau of translators in Nagasaki comprised a kind of institute for the introduction and transmission of Western learning, and, together with the book repository in Edo, to which many translations were sent, was part of the larger governmental effort to take institutional control over science. So too was the *Kaiseisho*, or "Institute for Western Studies," that the shogunate set up in 1855. These efforts were the brighter side to a long-standing attitude that had its darker moments in such events as the Siebold Incident (1828) and the *Bansha no Goku* affair ("Jailing of the Office of Barbarian Studies"), in which important *rangaku* scholars were arrested, tortured, and forced into making confessions that led either to execution or ritual suicide.[8] In these cases, the government's vicious reaction was a result of its perception that its strict policies of information control had been undermined. It was a sign, once again, of the great power invested in Western knowledge, whether this power was thought to be contaminating or enlightening. The official Meiji program, therefore, was merely an extension of this greater belief. If the logic of dealing with the West had shifted, from denying contact to welcoming limited contact, the government's self-appointed role as doorkeeper had not changed. Books, teachers, and ideas could have good or bad impacts on the Japanese character. It was the government's responsibility to maintain command over the dissemination of foreign ideas. One measure of the long-term success is exactly the point I have been making; namely, that "science" in Japan was for so long equivalent to "translation."[9]

Programs for importing foreign teachers (the *o-yatoi*) and exporting talented students (*ryu-gakusei*) were begun during the 1870s and 80s. The motive to Westernize now came to prominence (though not without some local resistance), especially in technical, literary, and governmental circles (see Gluck, 1985). The 1870s in particular were the heyday of the Japanese Enlightenment (*keimo*), spearheaded by thinkers and writers such as Fukuzawa Yukichi, who helped make it commonplace to vilify the old feudal system and proclaim the adoption of Western ways in all areas of life, from clothing to poetry. Watchwords of the day were "civilization" and "progress." It has been common to interpret this period as one in which Japan sought to "re-found itself on the basis of European culture" (Kobori, 1964, p. 3). Yet the ideology of the moment was not so simple; there was no wiping the slate clean. As Carol Gluck has written, "In those years the apostles of civilization often initiated their exhortations with some variety of the phrase, 'for the sake of national strength and expanding national power'" (1985, p. 254). The impulse toward nationalism, which had long viewed Occidental science as a needed element, never wavered. The move to Westernize only increased this impulse, providing it with a new set of practical agenda and methods. This is why "civilization" and "enlightenment" could go hand in hand with a renewed official idolatory of the emperor, who stood as the personified form of an ancient, even timeless national unity—a unity that was seen as wholly unique to Japan, even an essence of its Asianness, and that would one day (soon) come to justify the country's own colonial "adventures." Nationalism, in other words, meant increasing the internal strength of the country by incorporating the best that other nations could provide, those that had already proved their worth in various areas of modern power.

According to Bartholomew (1989), a total of only about 8,000 teachers from Europe and the United States were admitted to Japan between 1870 and 1900. In contrast, tens of thousands of *ryu-gakusei* were sent overseas. This reflects an important policy of the government, which was to reduce, as quickly as possible, any direct dependence on foreign sources of knowledge and know-how. Pupils of the *o-yatoi*, it was hoped, would replace them, and in greater numbers. In 1872, the government initiated a plan to divide the country into eight academic districts, each with an institute or university of higher learning headed by a European professor. The idea was to create a series of nodal centers, from which qualified students would radiate outward, permeating the body of the nation with the new knowledge (only three such schools, however, in Tokyo, Osaka, and Nagasaki, were actually established). Students returning from overseas, meanwhile, would both help in this process and expand well beyond it, staffing the government, the military, the new universities, and commercial enterprises with a wealth of trained expertise.

This dual effort to nativize Western science, both from within and without, must be understood as an overall rational plan, one based, profoundly, on the powers of translation. It was through translation—into Chinese, into Dutch, and into Japanese—that Western science had always been known. Now this tradition would extend deeper into the true centers where this science had been built, involving a host of new languages such as English, French, German, Russian, Latin, and Greek. The Japanese experience of Western science now became polylingual, and therefore multisocial. Writers and officials may have spoken of it as a single entity: *rangaku* ("Dutch studies") or *bansho* ("books of the barbarians") early on, and later as *seiyo-gaku* ("studies from the Western seas"). But science in Japan was more than ever, both linguistically and practically, an extremely heterogeneous thing.[10]

This matched, as well, a larger linguistic reality I spoke of earlier. By the late 19th century, the situation of written Japanese was enormously plural, confused, a melange of formal and informal, ancient and modern, communal and personal forms. The 1870s and 80s, in particular, were a time of great dissension and debate in official circles about what "to do" with the Japanese language. Many, such as Fukuzawa, viewed it as clumsy, medieval, and inefficient, wholly inadequate to the learning of Western ways, especially science. Proposals were made to do away with *kanji* altogether and to romanize the language (i.e., use Western phonetic symbols). Conservatives, meanwhile, argued instead that an effort must be made to "purify" the mother tongue, either by transforming all Western and Chinese words into newer Japanese ones, or else by doing away with the "foreign" altogether. In the end, the conflict was reduced to a matter of emphasis. There was, after all, nearly complete agreement that Western scientific knowledge held a key to the country's future. The question was how best to nativize this source—what the proper balance should be between adoption and adaptation, between the "Westernization" of Japan and the "Japanization" of Western knowledge and ways. The argument about language was an argument both about practical use and political/cultural symbolism. The "West" however—and therefore "science"—was never a single entity.

While selecting foreign teachers or even *ryu-gakusei* was often done haphazardly (see Bartholomew, 1989), the choosing of certain "target" nations was not. Indeed, the preferences decided upon by the Japanese government present a fairly accurate portrait of the state of Western science at the time. Nakayama (1977), among others, has delineated these preferences. They include:

Germany: for physics, astronomy, geology, chemistry, zoology, botany, medicine, pharmacology, the educational system, political science, economics;

France: for zoology, botany, astronomy, mathematics, physics, chemistry, architecture, diplomacy, public welfare;

Great Britain: for machinery, geology and mining, chemistry, steel making, architecture, shipbuilding, cattle farming;

United States: for mathematics, chemistry, general science, civil engineering, industrial law, agriculture, cattle farming, mining;

Holland: for irrigation, architecture, shipbuilding, political science, economics

Japan recognized that Germany and France both were the leaders in basic scientific research. In the first part of the 19th century, French chemistry (Lavoisier) and natural history (Buffon, Cuvier) had formed the basis for these disciplines in Japan. During the early Meiji years, chemistry and physics were taught in three major languages: German, French, and English. The tilt toward France, however, declined after the Franco-Prussian War. British and Scottish technicians were prized for their expertise in mining, railways, and roads. It was the Germans, however, who came to be seen, especially from the 1880s onward, as the essential model in nearly all areas (medicine particularly), and for university teaching in basic science. The number of German *o-yatoi* was greater than that of all other foreign nationals combined. During the 1880s, German models were used by the Japanese government to reform the entire system of higher education and many political institutions too. In the 1870s only 27% of the *ryu-gakusei* chose Germany as their country for study, but this number climbed to 69% by the 1890s and to 74% by the first decade of the 20th century (Bartholomew, 1989, p. 71).

As a whole, therefore, Japanese science was largely conducted in the languages of several Western countries. Japan's new scientists taught, wrote papers and monographs, and sometimes even conducted conferences in German, English, or French, depending on which nation had been adopted for a particular field. Again, this did not represent a new development, but merely continued an ancient tradition that had begun with Chinese and later moved on to include Dutch during the years of *rangaku*. Now, however, it meant that different disciplines—or different groups within single disciplines—quite literally did not talk to each other. There was an often radical absence of interplay between fields. It was not until the turn of the 20th century that Japanese itself begin to finally replace foreign languages as the medium for scientific learning and inquiry.

Such an important change might have required even more time, if not for an important development. This was the formation of scientific societies, like the Tokyo Chemical Society and the Physico-Mathematical Society, which undertook conscious efforts to compile dictionaries and to standardize important terms. These societies took such linguistic tasks as their first duty. The large number of newly translated works, including textbooks, often applied different nomenclatural systems to the same phenomena (e.g., names of the elements), and this created confusion. Titles derived from French, for example, with regard to chemistry, differed strongly from those taken from the German, which had far fewer words cognate with Latin. Such confusion therefore, in a sense partook of the larger absence of linguistic standards in society. Different nomenclatures in science paralleled the perplexing number of writing systems in literature; just as some Japanese researchers might prefer German or Dutch terms (transliterated into Japanese), or else a mixture of German and French, so did some novelists of the day choose a classical style of writing, others a colloquial style, still others a merging of the two. Science, indeed, was an integral part of the larger literary moment. Standardization began, in effect, when professional groups formed and began to impose order by using the ambient chaos as a reason to promulgate a particular linguistic program. In 1888, for example, under the direction of physicist Yamagawa Kenjiro, a dictionary of physics terminology in Japanese, English, French, and German was published (see Watanabe, 1990), providing Japanese physicists with an unprecedented document for unifying their field and developing pedagogic standards. Such developments in other fields as well made it finally possible for Japanese scientists to lecture and write consistently in their native tongue. Moreover, this took place against a background of increasing demands for "Japanization" in the middle years of the Meiji period, when social and political ideology brought a reactionary wave against the zealous Westernization current of the "Enlightenment," a wave that moved forward on the call for national self-confidence, independence, and cohesion (see, e.g., Gluck, 1985).[11] The new technical power of the state was proven, in no uncertain terms, by Japanese wartime victories over both China (1894–1895) and Russia (1904–1905). Both were largely a result of Japan's modern navy, built and equipped with much help from Britain (Japan's major ally in Europe at the turn of the century).

This, however, did not mean the end of translation as a central part of Japanese science. Japanese students were no less required to study German or English than in prior decades. Models for research, involving new governmental and industrial laboratories, continued to come from Europe (and by now America). The first phase of linguistic standardization had ended by about 1900 and, despite noble efforts, had not been inclusive, in part due to lingering internal factionalism. The pace of research in the West, meanwhile, brought an ever increasing number of new terms which had to be absorbed. Indeed, during the 1910s and 1920s, when the National Research Council of Japan set up a large number of new journals in fields such as astronomy, physics, chemistry, geology, zoology, and botany, it was decided that such linguistic problems would be best avoided and international readership enhanced if all were to be published in English (as I mentioned earlier, they remain so today). Thus was born the *Japanese Journal of Chemistry* (1919), the *Japanese Journal of Geology and Geography* (1922), the *Japanese Journal of Astronomy and Geophysics* (1920), and so forth.

German was not chosen for this effort. During and after World War I, Germany lost its "favored nation" status in science. The war dramatically interrupted intellectual and commercial exchanges between the two countries. In the prewar years, for example, Japan had imported a great deal of its pharmaceuticals, medicines, and industrial chemicals from Germany, but had to search elsewhere for such supplies during World War I. The war revealed a depth of dependence that did not sit well with official hopes for Japanese self-sufficiency. Japan, in fact, actually entered the war against Germany, allying itself with England in the Pacific and using the opportunity to take possession over previous German colonial areas in Asia. The war also helped make it clear to the Japanese government and Japanese scientists and engineers that a larger portion of the industrialized world now spoke English than any other language, and that America in particular had rapidly modernized itself into an industrial and military power of ever expanding influence. In the postwar 1920s and 1930s, scientific research in Japan was carried out according to military and colonial goals: the government, the army, and private industry all worked together to effect the "Manchurian Incident" of 1931, which resulted in a munitions boom that staved off the effects of the Great Depression and caused an estimated 50% of the country's engineers to be employed in military research, with a substantial portion of the remainder, and many basic scientists as well, involved indirectly (Nakayama, 1977; Hiroshige, 1973). It was during this period when "big science" became central to Japanese capitalism and Japanese imperialism, and began to truly achieve, if selectively and for troubling goals, a level commensurate with scientific work in Europe and America. Needless to say, this continued during the years of mobilization leading up to World War II. Throughout the period, America was perceived as both a looming nemesis and an international power to be envied.

After Japan's defeat in World War II, scientific effort shifted from the military sector to private industry and the universities. Science now turned its full attention to America, as a source both of inspiration and appropriation. The details of this shift and its institutional effects, however, are less important here than is the fact that this allegiance, in general terms, has remained intact for a period of 50 years, a length of time nearly equivalent to the whole of the Meiji era. During this period, the United States might be said to have become in the late 20th century what Holland was in the 18th century. Certainly *rangaku* has found its correlative in a variety of species of *bei-gaku* (*bei-*, or "rice," being the designation for America).

From World War I onward, then, a progressive changeover occurred in the model language of Western science. Even by the mid-1920s, German had given place to English, a pattern that continued into World War II, despite hostilities, and that has culminated in the postwar era. This turn toward English must be seen, too, against the larger backdrop of postwar preferences granted the United States and the English language within Japanese society generally—preferences which have arisen from a complex range of motives. Among scientists, the turn to English has involved a psychology of purpose linked to intellectual need as well as to professional and personal status. Since a sizeable majority of articles in international journals are now published in English, since technical conferences are more often held in this language than any other, and since the closest diplomatic, industrial, and academic ties Japan has had in the postwar era have been with the United States, it is inevitable that some command of English would be viewed as an almost absolute requirement for doing science. Added to this is the larger social status that English has acquired in Japan, a status associated with being urbane, cosmopolitan, "in touch," unconfined by traditions viewed as valuable and unique but also as provincial and overly "Japanese."

My point is this: that Japanese scientists today, no less than a century ago, are very often engaged in efforts of translation—reading, writing, or listening to a foreign language or else consuming information that has been transferred into their native tongue from elsewhere. While these activities no longer comprise the whole or the major part of Japanese "science," they are still central to it. For their effects, we must turn again to the realm of scientific language itself.

VI

Early Beginnings of Western Translation

Between 1770 and 1850, a great deal of Western science entered Japan via the Dutch language. Copernican and Newtonian theory, the biology of Linnaeus, Lavoisier's chemistry, the astronomy of Laplace and Lalande, much of Western medicine—all these and more were introduced and taken up, mainly by physicians and scholarly samurai. Dutch, of course, was not the language of origin for most of this knowledge. To many Japanese at the time, however, habituated to thinking of Holland as the epitome of Europe, works written in Dutch were often seen as primary sources. To read these works in the "original" was tantamount to being in contact with scientific knowledge at its generative base. The reality, needless to say, was that many of the books imported from Holland were translations of translations (etc.). Moreover, due to the tenuous political climate with regard to Western ideas, *rangaku* scholars often felt it expedient (for their own survival) to make certain changes of their own.

These characteristics are well shown by an early work that helped introduce the heliocentric view to Japan. This text, by one of the most famous translators of the early period, Motoki Ryoei (1735–1794), was titled *Oranda Chikyu Zusetsu* (An illustrated explanation of the Dutch view of the Earth, 1772). Ryoei's source text was a Dutch translation (dated 1745) of a French original, *Atlas de la navigation et du commerce qui se fait dans toutes les parties du monde*, written in 1715 by Louis Renard.[12] But whereas the Dutch work—titled *Atlas van Zeevaert en Koophandel door de Geheele Weereldt* (Atlas of navigation and commerce for the entire world)—had included a number of important maps as well as a seaman's guide to their use, these very central parts of the work were omitted by Motoki. Not only was the result incorrectly (if revealingly) titled the "Dutch View of the Earth," it also lacked any "Illustrated" dimension whatsoever. It seems more than likely that Motoki left out the maps in part due to fears of government censorship and reprisal. Their absence was a direct expression, one could say, of the official desire for the continued isolation of the country—for the effective deletion of Japan from maps of this type.

The term Motoki selected for "heliocentric" was entirely expressive of the time and, one might say, the general situation of knowledge. This term, *taiyo kyuu-ri*, was wholly in keeping with neo-Confucian natural philosophy, denoting "the exhaustion of *ri* with regard to the sun" ("sun" itself being the ancient Chinese combination of the characters for "thick" and for "yang principle," i.e., positive, male, daytime, etc.). A new theory was thus introduced through largely traditional linguistic means, linked most of all to the Chinese language. This was revealed in another way too. According to Nakayama (1992), Motoki's manuscripts show that, prior to publication, he used the indigenous Japanese phonetic alphabet, *katakana*, to write the names of Greek and European astronomers mentioned in the original text, and of geographical placenames in Europe, Africa, and the Middle East. When it came time to publish his work, however, he felt obligated to change all of these into phonetic *kanji*, in obedience to convention. Scholars of the day were still scholars of Chinese before they were scholars of Dutch.[13]

Rangaku translators such as Motoki began the process of rendering Western terms and concepts by using existing Chinese words or by borrowing them from Chinese dictionaries and lexicons, often of ancient vintage (the Sung dynasty of the 10th–13th centuries, viewed as a high point of mainland civilization, was a favorite reference point). With time and confidence, however, this method declined in favor of another. Rather quickly, scholars found that even the total Chinese vocabulary was inadequate to the task before them. They therefore began inventing their own character combinations. The tradition for this was set by one of Motoki's students, Shizuki Tadao (1760–1806), one of the most important *rangaku* figures in the entire history of modern Japanese science.

Newton and the Language of Physics

It was Shizuki who brought Newtonian theory and vocabulary to Japan, in effect founding the language and therefore the concepts of modern physical science. This he did in an unparalleled work of translation titled *Rekishō Shinsho* (New writings on calendrical phenomena), which appeared in three volumes between 1798 and 1802. Shizuki spent two decades laboring over the translation from which these treatises were derived. His original was a Dutch text, *Inleidinge tot de Waare Natuuren Sterrekunde* (Introduction to the true natural philosophy and astronomy, 1741) by Johan Lulofs, who had translated his text from the British writer John Keill's *Introductiones ad veram physicam et veram astronomiam* (Introduction to the true physics and true astronomy, 1739).[14] In addition, Shizuki made use of Newton in Dutch translation, as well as other scientific writings (Ohmari 1964; Keene, 1968). This indicates that significant collections of European scientific works existed in Japan by this time. Lulofs's rendition, meanwhile, like most Dutch translations, was highly faithful to its source, except where it included a number of clarifying notes or explanations. Shizuki's version, however, could not be so exacting. Many of the terms Newton used were so utterly foreign as to have no easy counterpart in Japanese. They not merely had to be invented, they had to be conceived. They were inserted into a larger linguistic frame that depended as much on traditional terminology as on Western terms; thus the product represents an immediate historical mixture of old and new.

Perhaps as a precaution, Shizuki chose a title that invoked the ancient Chinese term for "calendar making" to designate astronomy in its most general sense (*rekishō*). Other, more specialist terms derived from Chinese classical literature existed at this time too: *tenmon* (literally, "documents on the heavens"), which usually referred to the study of existing writings and was also employed for astrology; *tengaku* ("study of the heavens"), using direct observation; *seisho* (calendar writing); *kenkon* ("heaven and earth," but also "emperor and empress"), a metaphysical term signifying the cosmos as a creation of opposites. Shizuki used all of these. The first two volumes of the *Rekishō Shinsho*, in fact, were written in classical Chinese. But in the last volume, and in his final rendering of the entire work, Shizuki wrote in Japanese, and a certain change in emphasis occurred, generally speaking, away from *tenmon* toward *tengaku*, that is, from "astronomy" as literary study of the Chinese classics and the Chinese calendar, to "astronomy" as a more direct examination of celestial objects themselves. He further took up the term *seigaku*, "study of the stars," which came to be used during the 19th century to refer to the academic discipline of astronomical research. Finally, the word *tensetsu* ("explanation of the heavens"), derived from an earlier work, *Oranda Tensetsu* (1796) by Shiba Kōkan, was also in use when the *Rekishō Shinso* was written. This term, and most of those listed just above, passed out of usage before 1900. By this time, astronomy had come to be called by its now-standardized name, *tenmongaku:* a compromise struck between ancients and moderns.

In his use of such terms, Shizuki shows himself a kind of battle-ground between these two poles. As an official interpreter in the government's employ, he felt his translation had to obey certain conventions—indeed, as a loyal neo-Confucian *rangaku-shi,* he believed strongly in the value of Chinese thought and natural philosophy and attempted, in a number of places, to reconcile Western ideas with this philosophy by tying Newtonian principles back to the *I-ching (Book of Changes).* For example, he attempted to justify the heliocentric view on the basis of Confucian ethical principles:

> There always exists a governing center in everything. For an individual, the heart; for a household, the father; for a province, the government; for the whole country, the imperial court; and for the whole universe, the sun. Therefore, to conduct oneself well, to practice filial piety . . . to serve one's lord well, and to respond to the immensurable order of heaven; [these] are the ways to tune one's heart to the heart of the sun. (quoted in Nakayama, 1969, p. 185)

Thus did Newton make his entrance into Japan in the service of Confucius.

But Shizuki was a modern, too, and quite consciously so. In his day, there were no Japanese or Chinese equivalents for terms such as "force," "gravity," "velocity," "elasticity," or "attraction." Shizuki avoided simple phonetic translation into Chinese sound characters (*kango-yaku*). Instead, he invented words whose *kanji* combinations always attempted to offer a clear, precise meaning. His purpose thus seems to have been a didactic one in part: to provide a usable, practical base for understanding the new science of Newtonian physics. It is obvious, in any case, that he labored to no small degree over his choice of terms and thereby came up with an intelligent and organized system for doing so.

All of the terms just mentioned—"force," "gravity," "velocity," "elasticity," "attraction"—and many others, Shizuki coined according to a simple formula. This involved taking the respective characters for movement, weight, speed, stretch, and pull, and combining them with the character for strength or power (*chikara*). Thus "force" became "movement-power" (power to create movement), "gravity" became "weight-power," and so forth.[15] By doing this, Shizuki gave each term a concrete, easily understood meaning. He created a direct link to everyday experience that the English and Latin originals lacked. Newton himself once described "force" as mysterious and inexplicable, and several of his other concepts, such as "mass" and "velocity," as mathematical abstractions. In a sense, Shizuki made the Newtonian vocabulary more Baconian. He brought it more within the realm of perception—"weight-power" and "pulling-power," for example, are far more visible concepts than "gravity" or "attraction."[16] Some of his other coinages bear this out as well. "Vacuum," for example, he rendered with the ideograms for "true emptiness" (*shinkū*). The focus of an "ellipse" he denoted with the character for "navel" (*heso*). For ellipse itself, he rejected the older Chinese-derived term used by Japanese mathematicians, *soku-en* ("circle on its side"), and replaced it with *da-en* ("oval-circle"). On the other hand, he accepted the traditional Chinese word for "eclipse," with its allegorical, poetic combination of the characters for "day" or "moon" with that for "eat." "Corpuscle he wrote as *bunshi* (literally, "part-small"), from the characters for "segment" or "division" and "child" or "offspring."[17]

Finally, Shizuki coined more difficult words as well, such as "centrifugal force" and "centripetal force." Following his own lead, he designated "force" in these cases with the character *chikara* ("strength/power") by itself, which therefore became a suffix. This, one might note, along with the use of *-shi* for "particle," set patterns that have been followed ever since in coining any related terms (e.g., electromagnetic force, electron, proton, etc.). To render "centripetal" and "centrifugal," Shizuki seems to have advanced into the realm of the poetic. The former he wrote with the ideograms *kyūshin* (literally, "want" or "request" and "heart" or "center"), meaning "seeking the center," and the latter with *enshin* ("recede" and "heart"), meaning "moving away from the center." Shizuki could have easily found more simple and unsuggestive ways to write these same words, as he did for other portions of the Newtonian vocabulary, but he chose instead to be more "Chinese" in his inventions here. This method may relate to his own larger attempt to interpret Newtonian ideas according to the metaphysics of *ki*. As indicated by the brief quotation above, notions of "the center," "inwardness," and "outwardness" were critical to the neo-Confucian cosmos within which Shizuki conceived his own theories of reconciliation. In any case, all of these terms remain unchanged today as the basic vocabulary of physics. Shizuki coined for posterity, together with Miura Baien, the most fundamental language of physical science for Japanese society.

Scholars who have written about Shizuki Tadao—and there have been many (see, e.g., Ohmori, 1964a,b; Nakayama, 1969; Kiyoha, 1975)—commonly relegate him to a "premodern" category, according to the long-standing bias that modern science did not begin in Japan until the Meiji era (see Low [1989] for an excellent discussion of this problem). That such an idea is overly narrow can be gathered immediately from my discussion so far. Despite allegiances to traditional Chinese thought and language, Shizuki clearly inaugurated a critical segment of modern scientific discourse—as any adequate translator of Newton would inevitably have done—laying down vocabulary and patterns of linguistic formation that have remained intact ever since. With Shizuki Tadao, the Japanese trend in scientific language begins to depart strongly from the Chinese tradition, while still including it. It inaugurates a type of quiet "revolution" whereby the past, in terms of discourse, becomes less the single dominating influence than an ingredient in a complex evolution of nativization that would soon involve other elements as well. The language of Newton in Japanese therefore presents a condensation of history at a moment of transition, when fundamental loyalties regarding culture and thought were shifting. To the knowing eye and the tuned ear, the daily use of this language today rehearses this history like an unending ballet. The dance of its movements, its conflicts, may have been reduced to a residue of shadows. But they are there nonetheless, if only behind the thickening veil of "standardization."

Darwin, "Evolution," and "Survival"

I mentioned earlier that few of the *o-yatoi* ever lectured or wrote in Japanese. Linguistically, on a professional level, they were never allowed to "leave home." In the case of the American zoologist Edward Morse, his writings were handled by one of his more gifted students, Ishikawa Chiyomatsu, who, in 1883, helped introduce Darwin to Japan by collecting a series of his teacher's lectures into a book he titled *Dō butsu Shinkaron* (Theory on the evolution of animals). The term for "evolution," *shinka*, Ishikawa either coined himself or else adopted from one of several earlier works published in the 1870s on Darwinian ideas, most notably Izawa Shuji's 1879 translation of Thomas Huxley's *Lectures on the Origin of Species* (1862). In either case, it did not exist in Japanese scientific discourse before the 1870s.

It was, however, an excellent choice: comprised of the characters for "advancement" and "change," it had come to enjoy a wide currency outside science prior to its Darwinian adoption, denoting "progress," as in the "forward movement of society," especially under the auspices of Western science. It was a term that could be said to have embodied two sensibilities at once: Darwin's own Victorian view of evolution as a process of continual improvement and, more immediately, the ideology of the Japanese Enlightenment, with its call for a civilizing nationalism. The political side to Darwin's own language in English, therefore, was aptly retained, to serve the purposes of Japanese ideology and self-imagery. Such nationalism, however dictated that Darwin himself, as an author, would be the last to speak for his own ideas.[18] Instead, despite Ishikawa's efforts, it was Herbert Spencer who came to be adopted as the apostle of evolution theory. By the time of the first Japanese version of *Origin of Species* (1889), no fewer than 20 translations of Spencer's work were already in circulation and widely quoted in writings of the day. Moreover, Darwin's book was translated not by a biologist, geologist, or natural historian, but instead by a literary scholar, Tachibana Sensaburo (see Watanabe, 1990, pp. 68–69). Before this, only four works on the science of evolutionary theory had been published (two from books by Huxley). Even within scientific circles, the little of Darwin that did get through during this period was doubly "translated," being imported through the writings of his own English interpreters.

As Watanabe (1990) indicates, Spencer had been introduced to Japan through American contacts.[19] Japanese authors took up Spencerian ideas and disseminated them broadly, copiously, at times with government support. During the 1880s and 90s, as the intellectual and political atmosphere of Japan grew increasingly conservative and nationalistic, many thinkers, officials, and students found themselves drawn to the concept of a struggle between nations, with "higher" species eventually winning out over "lower" ones (Nagazumi, 1983). Before 1900, the theory of evolution tended to operate ideologically both within and without science. Indeed, before the century was out the language of evolution took on a more striking cast. Terms such as *skinka* reflect the era of their origin, the early Meiji period of hope and "progress." But by the late 1880s and early 90s, the pitch of nationalism had altered to embrace more reactionary concerns about moral standards, loyalty among the people, and national destiny in terms of empire (*teikoku*). Western nations were being viewed more in oppositional terms, as colonial aggressors, and as destructive models for Japanese "character" and "virtue." There was a strong resurgence of Confucian ethics, especially evident in the Imperial Rescript on Education (1890), which linked "virtue" directly with such things as obedience to authority, national sacrifice, and belief in the emperor's divine status.

In this atmosphere, Spencer himself was translated in sometimes hyper-Spencerian terms. Kato Hiroyuki, in his *Jinken Shinsetsu* (New doctrine of human rights, 1882), provides one of the best and, at the time, most influential examples. Not satisfied with a literal rendering of Spencer's famous phrase "survival of the fittest," so central to his philosophy, Kato felt compelled to evoke what he perceived to be its deeper significance and meaning. He wrote it thus: *yūshō reppai,* "victory of the superior and defeat of the inferior." This, he asserted, was "the law of heaven," governing the world of plants and animals as well as that of human beings and the cultures they build (see Watanabe, 1990, pp. 71–74). For a brief time, Kato's "victory of the superior . . ." was actually adopted into biological discourse. Though largely abandoned before the second decade of the 20th century, by which time it had been replaced by a much milder alternative, *tekisha seizon* ("survival of the most suitable"), it was nonetheless revived during the era of militarism in the 1920s and 30s, when eugenics was introduced into Japan. Today, it is no longer used on any sort of regular basis and has been virtually driven from the language of evolutionary biology. Yet it has not totally disappeared. It is still listed in dictionaries, without comment, as an equivalent to Spencer's seemingly immortal phrase.[20]

Names of the Elements: A Complex Litmus of History

As I indicated near the beginning of this chapter, the nomenclature of the elements presents an almost unparalleled example of the collision of cultural and political influences in the realm of Japanese scientific language. It is an example that both includes and goes well beyond the cases of Newtonian and Darwinian language. This is because the relevant nomenclature evolved over a much longer period of time and involved a much greater array of sources. These included a native tradition of names, in part derived from China: elemental substances such as iron, gold, silver, and the like, had long been known by their Chinese or Japanese titles. Then, when the element concept was introduced, these substances were effectively segregated into a new category and their traditional names cast into question. Added to this "native" group was an increasing number of new elemental substances from the West, imported under a range of names, that through the course of the 19th and early 20th centuries reveal a change in loyalty to Dutch, Latin, German, and finally English sources. Reconciling past, present, and future proved to be a task that encouraged no end of new and competing nomenclatural systems. How Western, or Japanese, or Chinese, or mixed, a system might be depended on the intentions of its creator, intentions that nearly always were grounded in some ideological motive.

Western chemistry was first introduced into Japan in the early 19th century through native physicians interested in pharmacology. These men were all trained in traditional forms of Chinese herbal-based medicine and were therefore all scholars of Chinese literature. Their interest in Western medicine was partly a direct outgrowth of this background. Books on European plants and pharmacology led them, rather quickly and with excellent logic, to botany and then to the importance of chemical understanding for the analyzing and making of medicines. The most important early works were published between 1820 and 1850. These have been discussed in some detail by a number of scholars (see, e.g., Sugawara and Kiyonobu, 1990a, b; Tanaka, 1964, 1965, 1967, 1976; Doke, 1973). They include, in particular, three books by *rangaku* scholars: *Ensei Ihō Meibutsuko* (A reference on the products of Far Western medicine, 1822) by Udagawa Genshin and his son-in-law Yoan; *Kikai Kanran* (a poetic Chinese title, meaning literally "A gaze over the broad waves of air and sea," more often translated as "A study of nature," 1826) by Rinso Aochi; and Udagawa Yoan's own *Seimi Kaisō* (Principles of chemistry," in 21 volumes, 1837–1847). The most influential of these was the *Seimi Kaisō*, widely acknowledged to be one of the foundational works of modern Japanese science, not the least with regard to chemical language. A leading historian of Japanese chemistry, Tanaka Minoru, describes its importance as follows:

> The most impressive characteristic of this work is that its author, through largely independent effort, succeeded in translating into Japanese nearly all of [Lavoisier's] chemical nomenclature and terminology. No less was he successful in rendering into easily understandable speech and clear expression the precise description of scientific conclusions, chemical analyses and experiments, the characteristics of substances, etc. Present-day chemical nomenclature in Japan largely owes its existence to the work of Udagawa. Until the fall of the feudal government in 1867, the *Seimi Kaisō* played a decisive role in the spread of scientific chemical knowledge in Japan. (1976, p. 97)

Udagawa Yoan was a complex and brilliant man, a wide-ranging polymath of enormous curiosity and intellectual vitality. He not only wrote on chemistry, but produced works about Western botany, zoology, history, geography, music, linguistics, and mathematics. While still very young, he was adopted into the Udagawa family, well known both as hereditary Chinese-school physicians to the Tsuyama clan and as outstanding *rangaku* scholars and translators. His adoptive father, who had translated as many as 30 books on Western anatomy, urged Yoan to first study Chinese philosophy and composition (*kogaku-ha*, or "school of ancient studies"), which the young man eagerly did. "If you lack the ability to compose a Chinese sentence," he was told, "you cannot achieve medical learning either. . . . Do not forget that translation is important work and worthy for a man to sacrifice his whole life to" (quoted in Doke, 1973, p. 104). One should note, again, that no contradiction existed here: being a physician meant mastering the sacred works of Chinese medicine, but at this point in history it had also come to mean studying and adding to the increasing number of translations of Western medical texts, too, many of which were only available in Chinese. For those such as Udagawa Genshin, the "ability to compose a Chinese sentence" seemed to have had a double meaning. It meant, on the one hand, the old rhetorical view of knowledge, based on repetition and imitation (so similar to the rhetoric-based education in the West), but it also meant an ability to add to the stock of knowledge itself, by translating and interpreting works of Western science into the language of Japanese scholarship.

Whether or not Yoan understood such advice in this way, he came to practice it, and very soon. Before he was even 20 he had largely mastered the Dutch language and had begun writing technical articles on various botanical and chemical substances. One of these, a small booklet entitled *Seisetsu Botanika Kyō* ("Sutra of Western botany," 1822), shows an extraordinary mingling of Eastern and Western influences and more than demonstrates the unique capability of Japanese scholars to employ both without any sense of inherent contradiction. The book, in fact, is known as a work that played a pivotal role in introducing Linnaean botany into Japan. It does so, however, in the literary form of a traditional Buddhist sutra, employing a total of only 75 lines, each made up of 17 *kanji*. Thus did the basic late-18th-century concepts of Western botanical theory enter Japan via a scheme of rhythmic chanting dating back to the Sung dynasty and possibly earlier. Where Shizuki Tadao had placed Newton in the arms of Confucius, here was a still more intriguing example of Western thought in Eastern form.[21]

The *Seimi Kaisō*, Yoan's masterwork, is a compendium of translations from various Dutch books, with commentaries, and the author's own experiments. In his preface, Yoan states that the *Ensei Ihō Meibutsuko*, which he helped write with his adopted father, was for specialists. It was "too profound and difficult for beginners, so I decided to write my own book *Seimi Kaisō* in more plain and accessible terms using in full the elucidative descriptions in simpler Western chemistry books [as well as] my own chemical knowledge accumulated through various experiments" (quoted in Doke, 1973, p. 113). His intention, clearly, was a pedagogic one: to make Western science more available to educated Japanese. The books he chose to translate from were mainly introductory in nature (see Tanaka [1976] for a discussion of these, of which there appear to be at least 20). The most important seems to have been a Dutch translation of a German version (*Chemie für Dilettanten*, 1803) of William Henry's *Epitome of Chemistry* (1803), itself largely a simplified rendering of Lavoisier's *Traité Élémentaire de chimie* (1789). These works, aimed at lay audiences and beginning university students, had all been popular in Europe. Yoan's choice was therefore well considered. Moreover, it is crucial that he wrote the *Seimi Kaisō* not in Chinese, as was previously the rule, but in his native tongue.

Udagawa set out a clear definition of "element" as a "primitive indivisible substance" and proposed a system of names. Part of this system, and the name for "element" itself, he borrowed from the *Ensei Ihō*, on which he collaborated with his father-in-law.[22] The word for "element" was *genso*, employing the characters for "origin"/"beginning" and "simple"/"essence." The latter of these ideograms, *-so*, he used as a suffix for a number of other elements for which no names yet existed, for example, hydrogen, which was written as *suiso*, or "water-essence," carbon as *tanso*, or "coal-essence."

The first volume of the *Seimi Kaisō*, devoted to a discussion of element theory, lists 58 names, in dictionary order by their first sound.[23] A fair number of these, if well-known metals or other substances, were allowed to keep their traditional titles. These were often originally Chinese in terms of the characters used to write them, but had acquired Japanese pronunciations. Other elements, meanwhile, were taken directly from Latin or Dutch and translated phonetically via the *on-* (i.e., Chinese sound) system. As was the case with other scholars, Udagawa seems to have first written these names in the native phonetic alphabet, *katakana*, in his handwritten manuscripts, but then changed to the conventional *kan-on* system for publication, adding miniature *katakana* alongside them to indicate proper pronunciation (Sugawara and Kiyonobu, 1989). At times, this led to interesting "errors." For example, since Chinese had no hard "k" sound, a name like "chlorine" could only be approximated, as "su-ro-rin." In contrast, three elements in particular, platinum, bismuth, and molybedinum, the author translated from the Dutch on the basis of literal meaning (which had a basis in the vernacular alchemical tradition). Platinum, *witgout* in Dutch (literally, "white-gold"), was rendered by Udagawa with the two appropriate characters (as *hakkin*). Molybdenum, as *waterlood*, in Dutch (literally, "water-lead") became *sui-en*.

It is essential to note that Udagawa's term for "element," *genso*, and his use of the—*so* suffix—which later became standardized for Japanese chemical nomenclature—are also modeled directly on Dutch words, which, in turn, were derived from German. In particular, the Dutch/German ending -*stoff*, was the direct inspiration for Yoan's scheme of translation. This ending is used to write "element," as *Grondstoff* (Dutch) or *Urstoff* (German), and appears at the end of many element names (e.g., *Koolstof/Kohlenstoff*, "carbon"). This is significant to the whole of Japanese chemical discourse not simply because of its influence but because of its larger meaning in terms of cultural inheritance. Udagawa's scheme represents, in effect, a historical choice between two competing linguistic traditions in Europe. One of these was that of Lavoisier himself, who had revolutionized chemical nomenclature by creating a system of naming roughly patterned on Linnaeus. In this system, the critical activity of each elemental substance was embodied in its title, via Greek or Latin roots. The gas whose central principle was to combine with other substances in order to make acids (the "dephlogisticated air" of Priestley or "vital air" of others) could thus be termed *oxy-gene*, meaning "acid-producer" (*oxy-* being from the Greek, *oxus*, for "sharp," "bitter"). To Lavoisier, as to many other thinkers of the French Enlightenment, language was an analytical instrument that could be used to both contain and employ the natural order for human use and understanding. His systems of names, for the elements, acids, and bases alike, was seen to have great power and was adopted before long both in France and England and subsequently in America. Though Dalton's atomic theory later amended many of Lavoisier's concepts, the latter's nomenclature remained in use and has obviously carried down to the present.

In contrast to this scheme, however, was the German chemical and mining tradition, which had its own names for substances that later came to be known as elements. This tradition had buried within it an older, mainly alchemical view of matter as containing various "essences." Unlike Lavoisier's concept of *substance chimique*, which focused on a capacity for interaction, the German view tended to see the chemical universe in terms of "pure" and "impure" matter constantly engaged in revealing itself, its primal truth, through various principles of activity. To Lavoisier's *oxy-gene* or *hydro-gene* ("water-producer"), 18th-century German chemists coined the terms *Sauerstoff* ("sour-essence") and

Wasserstoff ("water-essence"). No primal suffix, defining of the final substance of "element" itself, ever existed in the French scheme. A concept such as *-stoff* had no place in a nomenclature built out of qualities and exchanges. In the early 19th century, moreover, as the Lavoisieran system gained currency in many parts of Europe, German chemists refused to alter their nomenclature. No doubt this refusal was linked, in some part, to the climate of strong negative feeling toward France, which, under Napoleon, had defeated and humiliated both Austria and Prussia and had helped bring about nationalizing reforms in all areas of the intellectual life of these countries. In any case, the long-term result was that the Germans, the Dutch, and other northern European countries never abandoned their own native systems of chemical nomenclature.

As a *rangaku* scholar, Udagawa inherited the Dutch/German tradition. This had been ordained in the early 17th century by the shogunate's choice of Holland as Japan's only major trading partner with the West. In coming to Japan, European science was forced to pass through the filter of Dutch linguistic realities. In the case of chemistry, this meant that a quiet yet momentous acceptance of Lavoisier adapted by a Teutonic onomastic scheme took place.

According to Sugawara and Itakura (1989), a total of 23 elements in the *Seimi Kaisō* were translated directly into native Japanese, that is, not into the Chinese sound system. These included: (1) traditional Japanese names with one character (in most cases originally derived from Chinese); (2) traditional names with two or more characters (some of Chinese derivation, some not); (3) literal translations from the Dutch (e.g., platinum, as discussed above); and (4) translations using the *-so* suffix. The remaining 35 were given in *kanon-yaku*, being taken phonetically either from Latin or, less often, Dutch. At its earliest stage of formation, therefore, in terms of sheer quantity, the nomenclature of the elements seems most obedient to Chinese influence, which at this time, the early 19th century, still commanded the language of scholarship. But a crucial change had occurred: the most basic term in all of the new chemistry, *genso*, had been rendered in a manner loyal to Western terminology. This was the historical signal of the direction the coming vernacular revolution would follow within Japanese science.

To better understand this revolution, it is important to point out that Udagawa's system of names was in competition with another scheme. Rinso's *Kikai Kanran*, published in 1826, was a popularizing attempt at explaining basic Western concepts of physical and chemical phenomena. Written in Chinese, but with a lucid style that reveals solid understanding of the science involved, the book became a standard reference for

rangaku scholars, just as Udagawa's did (Tanaka, 1967). It was basically a translation of an elementary Dutch textbook, *Natuurkundig
Schoolboek* (Textbook in natural philosophy), written by Johannes Buijs
for use in upper-level grammar schools of the day, and was itself later
translated into Japanese and expanded with a commentary by Kawamoto
Komin (*Kikai Kanran Kōgi*, 1851). With regard to his naming system
for the elements, Rinso adopted the German/Dutch -*stoff* tradition but
translated it differently, rendering the term "element" itself with the characters *gen-* and *shitsu*, meaning "original" and "matter"/"quality" and
thus employing -*shitsu* as a suffix equivalent to Udagawa's -*so* (e.g., hydrogen became *suishitsu*, or "water-quality"). As I noted above, this
term had a long history of use within Confucian metaphysical philosophy, indicating something close to a merger between material and spiritual natures. In using it, the author was proposing a nomenclatural system that adopted Western terminology into traditional Chinese forms.

Despite the popularity of *Kikai Kanran* and its support by influential writers such as Kawamoto, Rinso's system did not flourish. Indeed,
Udagawa's was not only preferred, but by the 1860s had been extended
to include a number of traditional, single-character Japanese names; these
were now being written with the -*so* suffix too. During the 1870s, another important translator and writer of chemical works, Ichikawa
Morisaburo, proposed that all nonmetallic elements be put in this form.
Ichikawa's proposal, however, was not adopted. In fact, it proved to be
merely one in a veritable explosion of new naming schemes that were
put forward in the politically chaotic years between 1860 and 1880,
when Japanese society as a whole and the scholarly community in particular was rocked by another of the periodic debates regarding the place
of, and the proper balance between, traditional and modern (Western)
ways. It is striking to see just how closely the rival nomenclatural systems within chemistry reflect the patterns of larger ideological issues
during this time. Each system suggests a particular type of vision for
Japanese chemistry, a canonizing of fidelities to certain features of the
past or hopes for the future. When seen from this viewpoint, these systems divide into five basic categories, characterized by: (1) Chinese influence; (2) traditional Japanese influence; (3) Chinese and Japanese influences combined; (4) Western influence; and (5) Western and Chinese
influences combined.

The Chinese influence can be discerned in two systems proposed
during the 1860s and 70s.[24] One of these, created by Mizaki Shosuke,
made use of the suffix *ch'i* (-*ki* in Japanese)—the same character so central to yin–yang natural philosophy—and also the suffix -*sho*, meaning

"spirit," "energy," or "purity," another metaphysical association. Another scheme, offered by Kiyohara Michio in his translation of a Chinese work (*Kagaku Shodan*, 1873), stated that each element should be given its own single ideogram, which would often involve creating new characters by combining two existing *kanji* according to a prearranged system of prefixes to indicate "solid," "nonsolid," or "metal."[25] Both systems had their "standard" Chinese ingredients. Where the former adopted a conventional terminology (with its many traditional associations), the other took advantage of a creative logic inherent in the writing system itself, something that had been often done through the centuries by Chinese authors. Either way, both proposals sought to filter Western ideas to Japan through the Chinese tradition. Both systems revealed a combination of loyalties, yet granted China via language.[26]

Fidelity to Japan's own literary tradition, meanwhile, came in the form of a curious and entirely unique scheme offered by Kiyomizu Usaburo in 1874, in his work *Mono-wari no Hashigo* (Steps to a division of things), written in *hiragana* and meaning "analytical procedures." Kiyomizu's idea was to replace nearly all element names with new titles written in classical Japanese—the Japanese of the *Manyōshu* and other ancient poetry and prose collections. "Element" itself would be written as *ō-ne*, with the characters for "great" and "root." Individual elements would use *-ne* as a suffix and would employ classical words, written in *hiragana*, as a prefix. Thus, oxygen became *sui-ne* (*sui* being equivalent to *suppai*), "bitter-root," or more poetically, "the root of all bitterness." Some elements would keep their traditional titles, and the previously unknown rare earth metals would retain their Western sounds. But all would be written in the native syllabary of *hiragana*, and thus the "feel" of the classical tradition would be maintained as a central sentiment.

The mixing of Chinese and Japanese influences, on the other hand, came in a predictable form. In the 1860s, for example, just prior to the Meiji Restoration, Ueno Hikoma published a work in Chinese entitled *Seimi Kyoku Hikkei* (Official manual of chemistry, 1862) in which he suggested that all substances with long-standing Japanese names should retain them and all others should be rendered from Latin into *kan-on* (the Chinese sound system). This made for more than a few clumsy titles: even simple names such as "chromium" had to be written with as many as five or six characters, that is, with a character each for the sounds *ku-ro-mi-u-ma*. Rendered in this way, the Western origins of these names were effectively erased. To read them, moreover, meant that one had to be a Chinese scholar, still a reality for most chemists in the 1860s but increasingly less so thereafter. Finally, despite its title, Ueno's book ignored the Udagawa nomenclature, which had already become the most commonly used by that time. Ueno's ambition was to impose a new "official" scheme, more in line with traditional scholarship.

The most complex and varied category of nomenclatural proposals, however, were those that looked Westward. This was perhaps to be expected, but the diversity of ideas is still striking. Some of these ideas were directly consequent to the new and sometimes radical proposals for linguistic reform. Kawano Tadashi's *Seimi Benran* (Handbook of chemistry, 1856), for example, proposed writing all the names of the elements in the Roman alphabet. This was tantamount to swallowing the European/American system whole, in its own native state.[27] Ichikawa Morisaburo, on the other hand, suggested extending Udagawa's -*so* nomenclature to include a much larger number of substances, in effect filling Japanese chemical language with the Teutonic notion of "essence." From the late 1870s on, a growing fidelity to German science inspired the convention of changing any older Dutch terms into German ones and replacing certain Latin terms with their German equivalents as well. The main conflict, at least in the early stages, was whether to express the chosen titles in *kan-on* or in *katakana*. The latter had been used sporadically for a few elements since the 1820s. It was Ichikawa, following the example of Mizaki Shōsuke, who finally helped set the pattern in the 1870s (Sugawara, 1984).

Mizaki, in fact, who died prematurely in 1874 (at age 36), was extremely active as a translator of chemical works in the several years before his death. In his books one sees the old and the new visions of chemistry as embodied in nomenclature combined to a remarkable degree. Mizaki had studied under a well-known Dutch science teacher in Nagasaki, F. Gratama, whose lectures he translated into several books that were widely used. His most influential work, however, was published after he had left Nagasaki and had been recruited by the government to take up a professorship of physical science at Tokyo University Medical School. This work was a translation of a Dutch version of C. R. Fresenius's *Anleitung zür Qualitativen Chemischen Analyse* (1869), which had become the standard introductory textbook on chemical analysis throughout Europe. In transferring this work to Japanese, Mizaki helped introduce a number of important new concepts, including Dalton's atomic theory and Avogadro's molecular hypothesis. But it was his handling of chemical nomenclature that caught the attention of his contemporaries. Mizaki divided the elements into two basic groups, those with their traditional Japanese names intact (e.g., the common metals) and those imported from Europe, which he translated from Latin by means of *katakana*. Next to this list, he gave equivalents in English, also in *kana*, and in *kan-on*, for which he chose the Chinese sound character combinations most often used by chemists. He had used this system of both juxtaposing and intermingling Eastern and Western names before, in his translations of Gratama's lectures (especially *Rika Shinsetsu* [New treatise on physical science], 1870), a work which Ichikawa apparently drew inspiration from for his own *Rika Nikki* (Diary of physical science, 1872). Mizaki's list became a standard part of chemical textbooks, where it served like a multilingual dictionary. Moreover, Mizaki's translation of Fresnius's textbook was widely adopted as a basic work throughout Japan and helped spark debate over the term "chemistry" itself (Mizaki and Ichikawa had both ignored the older term *seimi* entirely for the newer term, *kagaku*).

Mizaki and Ichikawa represented a newer generation of translator-scientists. They, too, like Udagawa and other early chemists, were from families in which the father had traditionally been an official physician of some sort. Yet, as young men who had come of age in the years of the Meiji Enlightenment, their careers differed from those of their predecessors in both fact and in spirit. They were recruited into government service, not merely as translators but as teachers too, as replacements for the *o-yatoi* in the new scientific institutes of Osaka and Tokyo. They were regularly exposed to a greater range of linguistic materials than were Udagawa and Rinso, and their fidelity both to European science and to Japanese comprehension of it—at the expense of Chinese traditions, shown by their adoption of Western nomenclature in the form of *katakana*—was a temporal mark of their higher "civilization." Those who opposed them, and who remained adherents of the *kan-on* system even into the 1880s, were generally older men who had stayed in the medical profession, and who, in addition to employing Western therapeutic techniques, continued the traditions of herbal treatment and pharmacology linked both practically and philosophically to China. This division within the chemical community was a reflection of similar conflict in the power relations of Japanese scholarship as a whole.

From roughly this point on, the story of chemical language in Japan begins to shift. The use of *kan-on* declines rapidly, while more and more German and English words enter the language, signaling the larger change of political and cultural loyalties regarding knowledge. The turn away from Chinese was true throughout Japanese society as the result of the nationalist movement that promoted a Japanese identity based on things Japanese, especially the Japanese language itself.

The government had assumed the job of building public schools and instituting curricula that encouraged the study of the Japanese classics and, in terms of "things foreign," Western languages, especially German and English. By this time, Germany and England were viewed as the most powerful European nations, both intellectually and militarily. The final humiliation of China, in Japanese eyes, was its quick defeat to Japan's own military in the 1894–5 war. Thus, in the period from roughly 1880 to about 1900, China came to lose nearly all prestige as a crucial cultural reference point. The full vernacularization of Japanese intellectual life, like the introduction of European science, now took place under the canopy of freedom from the "old" and, in some measure, continued servitude to the "new."

Beginning in the 1880s, Japanese chemists, like society generally, struggled to gain control over their own discourse. This struggle is reflected in three major position papers by the nation's leading professional chemical organization, the Tokyo Chemical Society (*Tokyo Kagaku Kai*), presented between 1886 and 1900. In these papers, the Chinese sound system is essentially abandoned as a method for writing any element names. Those titles expressed in *kanji* are reserved for Japanese names only, including Udagawa's system utilizing the suffix -*so*, which was extended to the halogen family as a whole (Sugawara and Itakura, 1990a). All other names are written in *katakana* and are derived from one of three sources: European titles common to Latin, German, and English; names used in English only; and names in German only. Reviewing the three standardizing proposals in some detail, Sugawara and Itakura (1990a) noted changes in the emphasis given these various categories over time. These changes are detailed in the table below.

		Totals	
Element Name Categories	1886	1891	1900
Japanese	16	23	23
Common to Latin, English, German	24	26	31
English only	13	12	0
German only	11	13	18

(Data from Sugawara and Itakura, 1990a)

The changes shown are a mirror of the larger political circumstance of Japanese chemistry at the time. Under the *o-yatoi* system, two groups of chemists had been established in Japan, one under English-speaking teachers from England and America, the other instructed by those from German-speaking countries. Because of this social reality, the Society felt it had to forge a compromise. Yet the most striking aspect to the numbers given in the table is the disappearance of names in English and the increase of those in German. More generally in chemical discourse, one sees at this time the first use of the Roman alphabet to express chemical equations and also the adoption of the German system for naming common chemicals: for example, "silicate" being written *keisan-en* (literally, "jade-bitter-salt" from the German *kieselsaures Salz* (originally, "*pebble-bitter-salt*") or "sodium bicarbonate," *jû tansan-natrium*, being translated literally from *doppelkohlensaures Natron* ("double-carbon-bitter sodium"). The growing favoritism granted to Germany in the basic sciences, physics and chemistry most of all, is here in evidence. Meiji nationalism had found the perfect model in Bismarck's Prussia, with its centralized system of education, its use of intellectuals as civil servants, its unparalleled chemical industry, and its imperial discourse of spiritual force and national destiny. Germany was the country Japan wanted to emulate. Germany had been defeated and humiliated by Napoleon at the beginning of the 19th century, but in 1871 it defeated and humiliated France, becoming the new military, intellectual, and cultural power in Europe. This story could not but draw Japanese admiration and serve as a background narrative for official hopes of Japan's transformation into a world power.

The German influence did not last. Germany's participation in the "Triple Intervention" along with Russia and France, by which Japan was ordered to return some of its conquered territory to China after the 1894–1895 war, left a bitter political aftertaste. This, however, did not culminate until World War I, when Japan sided with England and (in a small way) helped destroy German influence in the Pacific. From that point on England and, somewhat later, the United States became Japan's models, and English began to be perceived as the more important language with regard to international science. English thus progressively replaced German in the Periodic Table of *katakana* names. Though some noises were made during the xenophobic era of the 1930s to eliminate *bango* ("barbarian speech") from all areas of Japanese scholarship, this was never taken wholly seriously by the government or the scientific community. In the early post–World War II period, linguistic reform, which sought to simplify and standardize the entire language for a host of aims, did away with many *kanji* that were either complex to write or that had very limited use. In the sciences, this meant the replacement of certain *kanji* with their sound equivalents in *katakana*. This process, along with the great new interest in the United States, led some chemists to advocate replacing all traditional Japanese names with English ones, but this suggestion was not adopted. In 1949 and again in 1955, the Ministry of Education issued a series of dictionaries and other publications intended to standardize scientific language as it was taught in the schools. In the sections on chemistry, several element names (fluorine, arsenic, and sulfur) were changed: in 1949 they were first stripped of their original characters and given simpler ones (those included in the *Tōyō Kanji*), then, in 1955 they were rendered into *kana* (Sugawara and Itakura, 1990b). Dictionaries of that time, particularly in specialist areas, often speak in their prefaces about a need to "streamline" or "advance" education and knowledge at all levels in order to "contribute toward a progress in national strength vis-à-vis Europe and America" (Mamiya, 1952, p. v).

Conclusion

Just as science is plural with regard to its disciplines, so are there "sciences" between languages, and nationalities. A laser may seem to work exactly the same in the United States or France as it does in Japan, but the people who build and operate it and the discourse they use to do so are by no means identical—nor, at a basic cognitive level, are the observations and recordings they might make of its effects in a particular experiment. All these things, of course, are part of science too. The objective truth of the laser's operation, and the principles by which it is manufactured are philosophically profound and undeniable. But in the real world of human activity and knowledge building, these principles become misleading if taken purely in these terms, for they always occur in localized form. They are not "universal" (the same everywhere) as much as they are "shared" (made intrinsic to each new cultural site)— or, rather, their "universality" resides in this very fact of their being renativized time and again. The necessity of translation reveals that they have no final, independent existence apart from language.

To say that translation has been important to the spread and exchange of scientific knowledge—and thus to research, technology, and a thousand forms of modernization throughout the world—is therefore to be guilty of understatement. As an act, translation has made the world we call "modern" possible, nothing less, and in the most practical sense.

Translation on a massive scale has enabled the physical transfer of knowledge and know-how between nations and cultures; it has been the process by which this knowledge could be given away, sold, or stolen, and by which it could then be nationalized and rendered into forms that have acted as forces for stabilization or change. Technical translation is the basis for the dissemination of nearly all technology. Before machines, there must be words; before science can exist, in ideas and institutions, there must be language for its "truth." Translation has thus been essential to contemporary politics, to trade, to the history of capitalism as a whole. However invisible, however ignored, technical translation has proved itself a thousand times over as a vast and tireless worker in the spread of Western culture elsewhere in the world, and to Japan perhaps most of all.

The larger issues that the subject of science and translation raises regarding the nature of human knowledge, differences in perception and thought between cultures, and the like, are questions that exceed the limits of what I have tried to do here. They are questions that demand cognitive study, anthropological analysis, and so on. Those who study the history of scientific language in a particular country, I feel, should cast a wary eye on the claim that such language expresses fixed national "essences." It would be more profitable to avoid this ambition of finality and to look instead for the intimate cultural sources and realities contained within a particular science. The point would be not to defeat "universalism," that aging bugbear of left-wing criticism, but to trace the role of international exchange, the shifting patterns of global influence, in short the politics of scientific knowledge. The example of Japan, as I have tried to show on a simple level, presents an intriguing test case for this type of analysis. The intertwining of political, literary, philosophical, and scientific developments, and their changing relationship to deliberately chosen Western influences, are all particularly evident in the Japanese case. The surface of one of these factors is often the refracting mirror for others.

Indeed, I have found this to be true today, in an unexpected form. I mean that the past history of Japanese science, its international dimension, is to be found in the very reality of how this history is now being written, by the Japanese themselves. Anyone interested in this area of study will be forced to read articles and books *by Japanese authors* written in four or five different languages: Japanese, English, German, Dutch, and French. Each of these languages, moreover, tends to be divided along traditional lines regarding the field(s) where they have had the most influence. The many essays and articles written by Tanaka

Minoru on the history of Japanese chemistry, for example, are nearly all in German; those by Nakayama Shigeru on the history of astronomy are in English.

Where, in the West, might this sort of thing happen? Where might one find the correlative to a well-known Japanese author writing in German on the development of Chinese natural philosophy (Saigusa, 1962)? Or surveying in French the diffusion of European science in the Orient (Kobori, 1964)? Where in America or in Europe is it felt that to reach an international audience one must adopt a voice utterly different from one's own? In more recent years, I have noted a tendency for most Japanese historical journals to publish fewer articles in French and German and an increasing number in English. This, too, obviously enough, matches the larger and more long-term tendency within Japanese science as a whole to adopt this language as its international model. Such is the state of Japanese scholarship, which therefore wears its history on its sleeve. In seeking to enter this realm, then, one must also don this garment and be swept up in its colors, patterns, and fashions.

Translation, in one form or another, remains at the heart of the content and the experience of Japanese science, whether for those who practice it or those who study it. It is my hope here to have shown something of the truth of this, and to suggest its importance for science in general. Cicero once remarked that the same sun shines on people everywhere. But, as science itself would tell us, the light we receive is never quite the same from one place to another.

Notes to Chapter 5

1. According to historians, moreover, the battle against Newton involved an effort to save Descartes, mainly by mathematics. This effort, which ultimately failed, must nonetheless have been substantial, absorbing much labor. For it seems to have pressed itself deeply upon L'Académie, becoming something of a ritual or unified sensibility, such that "French experimental philosophers became expert mathematicians and mathematical manipulation of data rather than creative experiment became the peculiar characteristics of French physical science" (Brockliss, 1992, p. 79). According to Brockliss, the suppression of Newton also acted to pave the way for an "explosion of scientific creativity" (in physics and astronomy) during the later Revolutionary and Napoleonic periods, when thinkers such as Laplace and Lalande began to pursue in lush detail the consequences of Newtonian theory. See also Guerlac (1981) for a more extended discussion.

2. Most international journals today include an abstract written in the native language (that of the country of publication), followed immediately

by an English translation. Both versions are assumed to offer a precisely equivalent "content," thus their juxtaposition provides an excellent "control" opportunity to directly examine changes in stylistic standards. As an instance, here is the final sentence of an abstract in the *Bulletin of the French Geological Society* dealing with the structural geology of Tunisia: "En effet, quel que soit le site initial des séries les plus compétentes, celles-ci ont finalement tendance à occuper une position structurale haute en fin de déformation" (Turki et al., 1988, p. 399) (Therefore, whatever the initial setting of the most competent strata, these have a tendency, finally, to occupy a high structural position at the end of deformation). In the English translation of the abstract, this comes out as: "Therefore, despite the initial configuration, the most competent strata will reach the highest structural position at the end of deformation." This is not only more compact than the French original, but involves changes in meaning.

The French has a certain open quality to it, but more than this, it is far more qualified than the translated English: "ont finalement tendance à occuper" is rendered, simply, as "will reach"; "une position structurale haute" becomes "the highest structural position" (something of a mistranslation). The English tends toward direct statement, a disinterested and epigrammatic tone. The science within it appears much more the province of factuality, or attempted factualism—the stating, outlining, suggesting, or predicting of a knowledge with few literary roundings and little time for nuanced equivocation.

3. It is, perhaps, an irony of history that the greatest work of Heian literature, the famous *Tale of Genji* (*Genji Monogatari*), was written by a woman, Murasaki Shikibu, in *hiragana*. At this time, such a form of Japanese allowed for a much greater range of expression and a more detailed and convincing portrait of the details of the age than almost anything a male author might have been permitted to conceive in literary Chinese.

4. This reading aid took the form of *furigana* (literally, "attached" *kana*), which are small *hiragana* or *katakana* placed immediately next to a given character to indicate its reading. Use of this notational system had existed previously but became especially prevalent during the late Edo period. It is interesting, historically speaking, to recall in this connection the origin of *katakana*, as an aid to the translation of Chinese texts. My point is that written Japanese had become, in its later stages of development, something so complex as to be occasionally "foreign" to its own users. *Furigana*, in a sense, were a means of translating the now-native language to its own users, a sign that the millennial process of nativizing Chinese (characters) was still incomplete.

5. In the postwar era literary authors have continued at times to employ older styles of writing, especially those in fashion during the later Edo and Meiji eras (the famous writer Mishima Yukio provides one of the most well-known and interesting examples). This is sometimes done as a con-

scious archaic construction, appropriate for effect in certain situations, but it also approaches a wholesale linguistic policy for some writers (Mishima among them). Either way, it adds yet another dimension of history to the visual surface of Japanese.

6. In fact, two schools of thought existed with regard to this view. One maintained that science was a means to material power, nothing more. It was useful and should be pursued for that reason alone; it was not a path to the deeper moral and ontological essence of things. The other school of thought, however, saw in Western science the embodiment of a kind of rational inquiry that could actually bring one closer to *ri*, to the essential principle governing all things, and thus to virtue as a state of higher understanding.

Those who tended to follow the latter, the so-called Eastern spirit, Western techniques school of thought, were, in a manner of speaking, classicists. They were political and intellectual conservatives, for they believed that true wisdom, the final aim of all inquiry, could be found only in the writings of the ancient Chinese sages, whose guidance must be applied to everything that came afterward. This view actually comported better with the shogunate, which was always of two minds about Western learning, one admiring, the other fearful.

7. Samurai represented the upper class, the top of the social hierarchy, in Tokugawa Japan and were educated as such. Those of more prominent families were taught Chinese and the Chinese classics, much as upper-class children in Europe were schooled in Greek and Latin. Possession of knowledge was viewed as a necessary responsibility for samurai, not merely a means to enforce and maintain their social position but a required element of the civilized man or (much less often) woman. It was therefore, in part, to be expected that this class would one day come to dominate the sciences, once these became invested. Such an interest gained greatly from the later 18th century onward, when European prowess in military weaponry, shipbuilding, navigation, and other practical effects became evermore clear and undeniable. A mixture of national and provincial loyalty (to Japan and to one's own lord), to private ambition, and to curiosity, prompted some samurai to take up the study of Western science. The more profound event, however, was the Meiji Restoration of 1868, which dissolved the old feudal order and divested the samurai class as a whole from its earlier (largely hereditary) guarantees of employment. Thus divested, yet possessing both education and the idea of its worth, samurai were more economically and culturally prepared to take an interest in Western science. Since Chinese, linguistically, is structured more similar to Western languages than is Japanese, this class was also in a better position to learn both Dutch and other European tongues when the need arose.

8. The Siebold Affair, in particular, is one that historians have pointed to as an indication of the repressive measures the government could take. Philip Franz von Siebold was one of the first Europeans allowed to enter

Japan and teach there. He arrived in 1823 from Germany, as physician to the Dutch trading post, and was permitted to teach medicine and biology in Nagasaki until 1829. In 1828, the shogunal astronomer, Takahashi Kageyasu, was given the official task of drawing up a map of Sakhalin Island, for which little information existed in Japanese, but much more in Dutch. Takahashi contacted Siebold and made a trade for this information by showing the latter several coastal survey maps of Japan itself. This was highly illegal, and the government reacted viciously. A large number of people were sent to prison; Takahashi himself died there only a few months later, apparently after being tortured. Several interpreters who had transferred messages between Takahashi and Siebold were also imprisoned. Siebold was held for several months and then expelled to Europe. Finally, reflecting the depth of the government's immediate paranoia of "contagion," all of Siebold's students were arrested.

9. It is in this light that one should measure the frequent criticism leveled at Japanese science prior to the 20th century that, with few exceptions, it was too often devoted to translating texts rather than to performing actual observations and experiments. One should understand, however, that translation was not merely an established route to well-paying posts and social status. It was part and parcel of two more far-reaching aspects of Japanese society: the literary tradition itself, which depended in large part on movements from Chinese to Japanese and back again; and, equally important, the Confucian model of organization, which sought to make intellectuals employees of the government.

10. It is important to note, too, that the government's program never included any major provision for teaching foreign teachers Japanese. Those o-yatoi who did learn were strongly discouraged from using the native language in their professional work. Their books, articles, and lectures were translated by their students into Japanese, who therefore carried out the government's plan for knowledge control in a more subtle, political sense by maintaining authority over the development of scientific discourse in Japan. Westerners, that is, almost never added to this discourse directly.

11. For example, during the mid-1880s, it was still possible for someone such as Mori Arinori, who had studied in England and was now minister of public instruction, to follow the lead of Enlightenment pioneers such as Fukuzawa Yukichi and propose abandoning the Japanese language altogether for English. When, a few years afterward, in 1889, Mori was assassinated by an ultranationalist Shinto fanatic, many interpreted the event as a sign of change in the national temper.

12. This example, and that which follows, can be found in Nakayama (1969, pp. 170–187).

13. No doubt, ironically enough, in the conservative atmosphere of the time, Motoki would probably have been "read" himself as subversive, overly influenced by Western ideas, had he used his native language even

that much more than was common. But it is just as likely that he obeyed convention because he was himself conventional, and desired to have his work appreciated by the scholars of his day.

14. MacLean disputes this, saying that in his searches through the documentation on materials shipped to Japan "I have never found record of the importation of Keill. Even if we accept that the Japanese received Keill from an official of the Factory on Deshima, before 1806 they were incapable, in view of their limited knowledge of mathematics, of understanding Newton" (1974, p. 57). The date given is when Lalande's *Astronomy* was imported. No other information, however, is given and thus one is left with MacLean's opinion only. Given the importance of the issue, this does not seem sufficient. We are, after all, in possession of Shizuki's manuscripts, in which passages are present that very closely (at times literally) parallel Keill's work. Moreover, there is the language of modern physics itself, which clearly dates to the *Rekisho Shinsho*. Unfortunately, Nakayama (1969) and Shimao (1972) merely state the derivation but do not provide documentary evidence. The problem thus remains unresolved. I have chosen to follow these latter writers, however, because MacLean seems to be disputing the existence of Shizuki's work itself, and also because the presence or nonpresence of Keill as a source is largely irrelevant to the larger question of terminology.

Accepting that it did serve as Shizuki's text, however, several interesting things might be said. Earlier editions of Keill's work had appeared in his native English, as *Introduction to Natural Philosophy* (1720) and *Introduction to the True Astronomy* (1721), these in fact being his own translations and revisions of original Latin manuscripts written in 1702. Keill was an avid disciple of Newton, and, along with his brother, James, a well-known popularizer of Newtonian ideas (Knight, 1972). He wrote first in Latin to reach a broader and more sophisticated audience of scholars throughout Europe, then in English for the sake of his own countrymen. Lulofs, meantime, was a professor of astronomy and philosophy at the University of Leiden and used Keill's Latin version as his own text. At the time, a number of other excellent popularizations of Newton already existed in English, the most wellknown of which was Henry Pemberton's *A View of Sir Isaac Newton's Philosophy* (1728). Keill, however, in a competitive spirit, sought a broader distribution for his own work (as well as Newton's) outside of England (as mentioned earlier, Newton was being resisted in France). That Lulofs's version eventually found its way to Japan, there to become an essential basis for Japanese physics, is a historically striking confirmation of Keill's ambition in the waning years of Latin's existence as the international language of science, this having continued due to the classical nature of European higher education.

15. In his review of Shizuki's terminology, Nakayama (1992) states his belief that the character *chikara* may have been adopted from well-known works on magic and the supernatural, especially those providing incantations for overcoming one or another form of evil. The implication is that

"force" with regard to the material universe was derived from a similar concept regarding the immaterial cosmos. It is Professor Nakayama's view that Shizuki's originality may have been overstated in the past. Be that as it may, however, the point seems irrelevant in the larger scheme of meaning here. The same type of borrowing, after all, frequently took place in Western science. But more importantly, Shizuki's significance to Japanese science is as a *translator*, a borrower and maker of words simultaneously. It makes little sense to reduce his "originality" by noting that he drew direct inspiration from his *own* culture as well as from others.

16. "Attraction," in fact, had acquired two Chinese equivalents by 1800, according to Nakayama (1992). These also employed the character for "strength/power," but attached it on the one hand to the *kanji* for "suck" and on the other, to that for "absorb" or "adopt." It is not clear whether Shizuki knew of these. In any case, it is likely he would have rejected both. In early drafts of his translation he had chosen for this term not "pull" but instead an ideogram with several meanings: *kyū* (verb form, *motomeru*), meaning "want," "seek out," "pursue." This he later abandoned but then used instead for the more complex designation of "centripetal force" (see following text).

17. This term, with its complex associations, was later taken up for a brief time as the designation for "atom" in the early 19th century, only to be accepted 50 years later, following debate within the Tokyo Chemical Society, as the word for "molecule," which it remains today (see Sugawara et al., 1986).

18. To say that something of Darwin, in a sense, existed *avant la lettre* in Japan, however, would be to turn the truth on its head. It is more that Japanese nationalism, which pervaded intellectual matters in the form of an official and private desire to be "civilized" and "enlightened," was not entirely dissimilar from the species of nationalism then prevalent in Europe, Britain perhaps above all, which had come to see itself as the pinnacle of such "civilization."

19. One of these, early on, had been none other than the famous philosopher and art scholar, Ernest Fenollosa, one of the very few *o-yatoi* who ever published materials in Japanese.

20. Indeed, because it is not labeled "archaic" in even the most up-to-date and widely purchased Japanese–English dictionaries, it still shows up every now and then in translations into Japanese by nonnative speakers.

21. This work also indicates that Japanese scientists were now writing and publishing books of their own, thus seeking a broader audience than had previously existed. Udagawa's term for botany, meanwhile, "Botanika," was obviously a phonetic transliteration into Chinese of the Western Latin word. Only a few years later, however, he stopped using this term and changed it to *shoku-gaku* ("plant studies"), as in his work *Shokugaku Dokugo* (Confessions on Botany, 1825?). This nativization did not become standardized

usage. Instead, it was replaced by the combination *shokubutsu-gaku* (literally, "study of plant-things"), adopted from a Chinese work of 1857 (see Nakayama, 1992). Here was a case, in other words, where terminology evolved *away* from a Western standard.

22. It seems probable the relevant nomenclature was the brainchild of Yoan himself, since it was he more than his father-in-law who felt such a strong interest in chemistry and had studied it in depth.

23. Several of these were nonelements. For example, three had been proposed by Berzelius for this designation but were discarded later in the century. All of these Udagawa named with the *-so* suffix, for example, *onso*, or "warmth-essence."

24. The idea of "influence" here needs to be qualified. The element concept, after all, seems not to have been accepted in China until the 1870s (Sugawara and Itakura, 1990b), substantially after it had become current in Japan. Mizaki, meanwhile, was directly influenced by Chinese translations of certain Western texts.

25. Solid was designated by the character for rock, *ishi*; nonsolid was to be written with the abbreviated radical for water (known as *sanzui* or *mizu-hen*) and would included gasses as well. Metal would be indicated by the character *kane*, meaning "gold" or "metal."

26. It is difficult to distinguish in every case whether Mizaki, Kiyohara, and other chemists who translated Chinese works into Japanese were actually advocating the mentioned nomenclatures, or whether they were merely being faithful to the original texts. Yet in Mizaki's case, as pointed out by Sugawara and Itakura (1990a), there was a conscious replacement in every case of Udagawa's *-so* with *-ki*. And more generally, that no attempt was made to employ existing Japanese usage (e.g., that of Udagawa) indicates, to an important degree, that choices were made to accept the Chinese model and introduce it into Japanese chemical circles. Loyalty to the original source text, in other words, translates itself, in this case, as loyalty to the culture from which it derived.

27. For the reasons described above, this would have necessitated a series of new choices, however, because the elements in English were not exactly the same as those in French, and both were distinctly different from the elements in German.

References

Barnstone, W. (1993). *The Poetics of Translation*. New Haven, CT: Yale University Press.

Bartholomew, J. R. (1989). *The Formation of Science in Japan*. New Haven, CT: Yale University Press.

Bassnett-McGuire, S. (1991). *Translation Studies*. Rev. ed. London: Routledge.

Blacker, C. (1969). *The Japanese Enlightenment*. Cambridge: Cambridge University Press.

Bloomfield, L. (1939). *Linguistic Aspects of Science*. Chicago: University of Chicago Press.

Brockliss, L. W. B. (1992). The Scientific Revolution in France. In R. Porter, and M. Teich (Eds.), *The Scientific Revolution in National Context* (pp. 55–89). Cambridge: Cambridge University Press.

Catford, J. C. (1965). *A Linguistic Theory of Translation*. London: Oxford.

Coyaud, M. (1977). *Études sur le Lexique Japonais de l'Histoire Naturelle et de Biologie*. Paris: Presses Universitaires de France.

Craig, A. M. (1965). Science and Confucianism in Tokugawa Japan. In M. B. Jansen (Ed.), *Changing Japanese Attitudes Toward Modernization* (pp. 133–160). Rutland, VT: Charles E. Tuttle.

Craig, A. M. (1969). Fukuzawa Yukichi: The Philosophical Foundations of Meiji Nationalism. In R. E. Ward (Ed.), *Political Development in Modern Japan* (pp. 22–39). Princeton, NJ: Princeton University Press.

d'Alverny, M.-T. (1982). Translations and Translators. In R. L. Benson and G. Constabe (Eds.), *Renaissance and Renewal in the Twelfth Century* (pp. 421–462). Cambridge, MA: Harvard University Press.

Doke, T. (1973). Yoan Udagawa—A Pioneer Scientist of Early 19th Century Feudalistic Japan. *Japanese Studies in the History of Science*, no. 12, 99–120.

Fabre, J. (1988). Les Series Paleozoiques d'Afrique: Une Approche. *Journal of African Earth Sciences*, 7(1), 1–40.

Fuji, K. (1975). Atomism in Japan, 1868–1888. *Japanese Studies in the History of Science*, no. 14, 141–156.

Gluck, C. (1985). *Japan's Modern Myths: Ideology in the Late Meiji Period*. Princeton, NJ: Princeton University Press.

Guerlac, H. (Ed.). (1981). *Newton on the Continent*. Ithaca, NY: Cornell University Press.

Habein, Y. S. (1984). *The History of the Japanese Written Language*. Tokyo: University of Tokyo Press.

Hiroshige, T. (1973). *Kagaku no Shakai-shi* (A Social History of Japanese Science). Tokyo: Chuoko-ron.

Hirota, K. (1988). *Meiji no Kagakusha* (Japanese Chemists in the Meiji Era). Tokyo: Tokyo Kagaku Dojin.

Hovrani, G. F. (1972). The Medieval Translations from Arabic to Latin Made in Spain. *The Muslim World*, 62, 97–114.

Kamata, H. (1993). Beppu-wan oyobi henshuchi-iki no shinbun chikakozo to sono sei-in (Deep subsurface geologic structure and genesis of Beppu

Bay and adjacent zones). *Chishitsu-gaku Zasshi* [Journal of the Geological Society of Japan], 99(1), 39–46.

Keene, D. (1968). *The Japanese Discovery of Europe, 1720–1830.* Stanford, CA: Stanford University Press.

Kelly, L. G. (1979). *The True Interpreter: A History of Translation Theory and Practice in the West.* Oxford: Basil Blackwell.

Knight, D. (1972). *Natural Science Books in English, 1600–1900.* London: Portman Books.

Kobori, A. (1964). Un Aspect de l'Histoire de la Diffusion des Sciences Européennes au Japan. *Japanese Studies in the History of Science,* no. 3, 1–5.

Lindberg, D. C. (1978). The Transmission of Greek and Arabic Learning to the West. In D. C. Lindberg (Ed.), *Science in the Middle Ages* (pp. 52–90). Chicago: University of Chicago Press.

Lockheimer, R. F., (1969). Development of Science in Japan. In K. H. Silvert (Ed.), *The Social Reality of Scientific Myth* (pp. 154–170). New York: Harcourt Brace.

Low, M. F. (1989). The Butterfly and the Frigate: Social Studies of Science in Japan. *Social Studies of Science,* 19, 313–342.

MacLean, J. (1974). The Introduction of Books and Scientific Instruments into Japan, 1712–1854. *Japanese Studies in the History of Science,* no. 13, 9–68.

Mamiya, F. (1952). *Toshokan Dai Jiten* (A Complete Dictionary of Library Terms). Tokyo: Japan Library Bureau.

Marshall, B. K. (1977). Professors and Politics: The Meiji Academic Elite. *Journal of Japanese Studies,* 3(1), 71–98.

Matsumura, A. (1982). *Kindai no Koku-go, Edo kara Gendai e.* (Modern Japanese from the Edo Period to the Present). Tokyo: Sakurakaede.

Nagazumi, A. (1983). The Diffusion of the Idea of Social Darwinism in East and Southeast Asia. *Historia Scientiarum,* 24, 1–17.

Nakayama, S. (1964). Edo Jidai ni Okeru Jusha no Kagakukan (The Scientific Views of Confucian Scholars during the Edo Period). *Kagakushi Kenkyuu,* no. 72, 157–168.

Nakayama, S. (1969). *A History of Japanese Astronomy: Chinese Background and Western Impact.* Cambridge, MA: Harvard University Press.

Nakayama, S. (1977). *Characteristics of Scientific Development in Japan.* New Delhi, India: Centre for the Study of Science, Technology, and Development.

Nakayama, S. (1978). Japanese Scientific Thought. *Dictionary of Scientific Biography,* 15 (suppl. 1):728–758. New York: Scribner's.

Nakayama, S. (1992). Kindai Seiyō Kagaku Yōgo no Chu-Nichi Taishaku Taisho-yo (A List Comparing Chinese and Japanese Loan Words from

Western Scientific Nomenclature). *Kagakushi Kenkyu*, ser. 2, *31*(181), 1–8.

Nakayama, S., D. L. Swain, and Y. Eri. (1974). *Science and Society in Modern Japan*. Cambridge, MA: MIT Press.

Oettinger, A. G. (1959). Automatic Translation. In *On Translation*, 240–267. Harvard Studies in Comparative Literature, No. 23. Cambridge, MA: Harvard University Press.

Ohmori, M. (1964). A Study on the Rekishō Shinsho, part 1. *Japanese Studies in the History of Science*, no. 2, 18–26.

Ohmori, M. (1964). A Study on the Rekishō Shinsho, part 2. *Japanese Studies in the History of Science*, no. 3, 81–88.

O'Leary, D. (1949). *How Greek Science Passed to the Arabs*. London: Routledge and Kegan Paul.

Porter, R. (1992). Introduction. In R. Porter, and M. Teich (Eds.), *The Scientific Revolution in National Context* (pp. 1–10). Cambridge, UK: Cambridge University Press.

Saigusa, H. (1962). Die Entwicklung der Theorien vom "Ki" (Ch'i) also Grundproblem der Natur-"Philosophie" im alten Japan. *Japanese Studies in the History of Science*, no. 4, 51–56.

Sakaguchi, M. (1964). Studies on the *Seimi Kaisō, part 2:* II—The Original of Translation. *Kagakushi Kenkyu*. no. 72, 145–151.

Sakaguchi, M. (1968). On the Chemical Nomenclature in the *Seimi Kaisō*. *Kagakushi Kenkyu*, no. 85, 10–21.

Shimao, E. (1972). The Reception of Lavoisier's Chemistry in Japan. *Isis*, *63*(218), 309–320.

Sugawara, K. (1984). Mizaki Shosuke no Kagaku-sha to shite no Katsudo (The Chemical Works of Mizaki Shosuke). *Kagakushi Kenkyu*, ser. 2, *23*, 20–27.

Sugawara, K. and K. Itakura. (1989). Bakufu, Meiji Shoki ni okeru Nihongo no Gensomei (I) (Names of the Elements in Japanese during the late Edo and early Meiji periods). *Kagakushi Kenkyu*, ser. 2, *28*(172), 193–202.

Sugawara, K. and K. Itakura. (1990a). Bakufu, Meiji Shoki ni okeru Nihongo no Gensomei (II) (Names of the Elements in Japanese during the late Edo and early Meiji periods). *Kagakushi Kenkyu*, ser. 2, *29*(173), 13–20.

Sugawara, K. and K. Itakura. (1990b). Tokyo Kagaku Ka ni okeru Gensomei no Toitsu Katei (The Process of the Standardization of Japanese Element Names by the Tokyo Chemical Society). *Kagakushi Kenkyu*, ser. 2, *29*(175), 136–149.

Sugawara, K., N. Kunimitsu, and K. Itakura. (1986). Atom no Yakugo no

Keiseikako (The Process of Translating the Term Atom into Japanese). *Kagakushi Kenkyu*, ser. 2, 25, 41–45.

Tanaka, M. (1964). Hundert Jahre der Chemie in Japan, Studien uber den Prozess der Verpflanzung und Selbstandigung der Naturwissenschaften als wesentlicher Teil des Werdegangs modernen Japans (Mitteilung I). *Japanese Studies in the History of Science*, no. 3, 89–107.

Tanaka, M. (1965). Hundert Jahre der Chemie in Japan (Mitteilung II): Die Art und Weise der Selbständigung chemischer Forschungen während der Periode 1901–1930. *Japanese Studies in the History of Science*, no. 4, 162–176.

Tanaka, M. (1967). Einige Probleme der Vorgeschichte der Chemie in Japan. Einführung und Aufnahme der modernen Materienbegriffe. *Japanese Studies in the History of Science*, no. 6, 96–114.

Tanaka, M. (1976) Rezeption chemischer Grundbegriffe bei dem ersten Chemiker Japans, Udagawa Yoan (1798–1846), in seinem Werk, *Seimi Kaiso*. Beiträge zur Geschichte der Chemie in Japan. *Japanese Studies in the History of Science*, no. 15, 97–109.

Tsunoda, R., W. T. de Bary, and D. Keene. (1958). *Sources of Japanese Tradition*, Vol. 2. New York: Columbia University Press.

Turki, M. M., J. Delteil, R. Truillet, and C. Yaich. (1988). Les Inversions Tectoniques de la Tunisie Centro-Septentrionale. *Bulletin de la Societé Géologique en France*, 4(3), 399–406.

Van Houten, F. B. and R. B. Hargraves. (1987). Paleozoic Drift of Gondwana: Paleomagnetic and Stratigraphic Constraints. *Geological Journal*, 22, 341–359.

Watanabe, M. (1971). Darwinism in Japan in the Late Nineteenth Century. *Actes, XIIe Congres International d'Histoire des Sciences*, 11, 149–154.

Watanabe, M. (1990). *The Japanese and Western Science*. Transl. by O. T. Benfey. Philadelphia: University of Pennsylvania Press.

Yoshida, T. (1984). Tenbo: Rangaku-shi (Outlook: The History of Dutch Studies). *Kagakushi Kenkyu*, ser. 2, 23, 73–80.

6

A Case of (Mis)Taken Identity

The Question of Freudian Discourse

᠁᠁᠁᠁᠁᠁᠁

Was einst des Knaben Spiel und Freude war,
Wird nun dem Mann zur Arbeit und Gefahr.

(What once to the boy were games and pleasure,
For the man are toil and danger without measure)

—C. F. Meyer, *Huttens Letzte Tage*

I cannot dreame that there should be at the last day
any such Judiciall proceeding, or calling to the Barre,
as indeed . . . the literall commentators doe conceive:
for unspeakable mysteries in the Scriptures are often
delivered in a vulgar and illustrative way, and being
written unto man, are delivered, not as they truly are,
but as they may bee understood.

—Sir Thomas Browne, *Religio Medici*

360

I

Sigmund Freud, founder of psychoanalysis, reformer of Western concepts of the individual, the family, and their deformations, is quite possibly the most influential writer of the 20th century. No other author has had his ideas permeate public sensibility to such a degree. No other author has himself been the subject of so much study and writing—both academic and popular—as Freud. No other author has been returned to, time and again, like a classic neurosis, for renewed inspiration or calumny, and by writers in an ever-expanding universe of disciplines that have come to include not only medicine, psychology, and philosophy, but also history, sociology, literature, anthropology, film studies, and art—indeed the whole of the humanities and social sciences. Like a fountain of magnificent riddles or a shallow grave of final origins, Freud has been re-discovered and stumbled upon time and again. It is a fate of the present century that we cannot let him rest in peace.

But what kind of writer was Freud? What type of language did he use, and where did it come from, historically speaking? Is his language that of science? Is it literature? Is it, instead, a unique, self-conscious blend of the two, or perhaps an instance of neither? On the other hand, is it ineluctably German? Can it be properly rendered into other languages? How badly has it been handled by translation thus far, and why? Just where does it fit in with the larger realm of our discourses about "truth"?

Such questions have been much debated and never resolved. In one form or another, they tend to return to the clashing rocks of the "scientific" versus "literary" argument, which has a long history both within and around the edges of psychoanalysis, and which has become again, along with "the German question" (i.e., the question of translation), a major concern in the 1980s and 90s, in no small part due to doubts cast upon the integrity and accuracy of the *Standard Edition* of Freud in English by such writers as Bruno Bettelheim (1982), Darius Ornston (1982), and Patrick Mahoney (1982). Bettelheim in particular accused the Strachey translation of being "seriously defective," of delivering to us a fictional Freud who never existed except in the translator's mind.[1]

Whether or not one considers the result "defective," the basic perception is undeniable. The *Standard Edition* is a work that readers today with knowledge of both German and English almost universally agree is more unified, congruent, and internally consistent in tone and terminology than is even suggested by Freud's German originals. Many students of Freud have expressed their surprise at the "discovery" that this version of him is easier to understand than its German source; for some, it even "seems theoretically more correct" (Junker, 1987, p. 317). Yet this should hardly be cause for wonder. Indeed, the *Standard Edition* is a single work, and a single "Freud" emerges from it. This Freud is the creation of a single translator (really, a single group of translators supervised by a general editor), whose purpose was, after all, to produce a coherent *oeuvre*, a grand Author, inevitably fictitious, in full control of his material, setting down the contours for orthodoxy. Strachey, in fact, admits as much from the outset, when he says: "The imaginary model which I have always kept before me is of the writings of some English man of science of wide education born in the middle of the nineteenth century" (*Standard Edition [SE]*, vol. 1, p. xix). A signal term here is "kept." For if we can take Strachey at his own word (and the evidence seems good that we can), it means that not only did he see fit in a sense to exchange Freud the Viennese Jew for Freud as T. H. Huxley (or an equivalent), but that this "English man of science" was itself a kind of fixed, unchanging monument bearing little relationship to the living, struggling author who lived through, and wrote about, some of the greatest and most disturbing changes of the late 19th and early 20th centuries.

There is something critical in this. Talk about whether or not Freud was a scientist, after all, revolves not upon Freudian language but upon ideas of "science," against which Freud's language is then measured (see, for example, Hook, 1960; Fisher and Greenberg, 1977; Grunbaum, 1984). In nearly every discussion, pro or con, a shadowy imitation of Strachey takes place, with "science" invoked as a kind of grand essence, fixed and beyond history. If Freud wrote like a "scientist," he obeyed certain rules that define this category in a final, inalterable sense, rules that apply equally to the 19th century as to the 20th, to 1900 as to 1990. In the same way, the claim that Freud was a "literary artist" has similarly been based on the idea of "art" as a transcendental precinct, one by which any writer of any epoch can be gauged, whatever his intentions, whatever the institutional realities of publishing, whatever the standards then existing for "fiction" and "prose." Thus, where one group of scholars might speak of a "fusion of science and humanism" in Freud's writings (Gedo and Pollock, 1976), while another perceives "a continuing and necessary tension" between these categories (Ticho, 1986), both are really arguing a slightly different set of universal standards.

How, then, does Freudian discourse comport with the science of its *own* time? For if we know this, we are in a far better position to speak about Freud as a living writer rather than as a fixed or embattled institution. Even asking the question helps one confront a critical limit that should never be lost sight of. Freud may well be, quantitatively, the most influential author of the present century. But whether we view him as an "originator of discursive practices" (Foucault) or as an initiator of "new forms of interpretation" (Paul Ricoeur), we need to remember that he has had no impact whatsoever in a vast realm whose acceptance he sometimes seemed to want most of all and which has become the largest and most powerful realm of human knowledge in history—I mean, of course, the natural sciences: chemistry, physics, biology, geology, mathematics, and astronomy. These were fields that Freud drew ripely upon, particularly for his imagery. His frequent use of metaphors taken from fields such as hydraulics, crystallography, stratigraphy, biology, and so on amounted to a claim for legitimacy as well as for expressiveness, and is entirely in keeping with a very common 19th-century mode of prose writing. The employment of such imagery, that is, was prompted by the desire to share in the vast symbolic and literal heroism granted to science in the wake of the Industrial Revolution. In Europe it was typical for historians, politicians, essayists, and journalists from roughly the 1830s onward to employ figures derived from scientific knowledge or to imitate "scientific factualism" in their writing. Among American authors, also, such borrowings are abundant, being found for example in the speeches of Daniel Webster and the essays of Ralph Waldo Emerson (see, e.g., "Great Men," the introductory essay to the latter's *Representative Men*). Freud used the "scientific," often with great skill and imagination, in a kind of standard 19th-century way. But, like Emerson or Leopold von Ranke, he himself had no effect on the fields from which he drew so many of his metaphors. Beyond psychoanalysis itself, Freudian influence has been limited to the humanities and their scientized derivatives, the social sciences. This is important, since it emphasizes the position of this discourse historically, in terms of its legacy. By the first decade of the present era, the natural sciences had already left the realm where a writer like Freud could have had such an effect, where a language like Freud's would have even been acceptable.

II

During the time that Freud was in school, in the late 19th century, scientific discourse in much of Europe and America was in the midst of major change, reflective of larger developments within the practice of scientific work and thought.

Between 1850 and 1900, Western science was transformed from a loose collection of local societies, research institutes, and academic programs into a series of densely professionalized disciplines, highly centralized within each nation, and often underwritten by direct government and commercial support. From a source of fascination and curiosity, science became a recognized origin of nearly unlimited power. This elevated status, borne aloft on the wings of continuous technological advance, drew an increasing number of students to the sciences. In Germany, most budding scientists received a traditional classical education during their *Gymnasium* years, being subjected to rigorous technical training and apprenticeship only thereafter. By the end of the 19th century, however, this had changed, such that science was being taught in the lower grades too, with the effect that lecturing styles became more functional, textbooks more simplified and direct, and the general language by which future professionals were introduced to technical subjects more instrumental. A focus on transmitting the most essential skills, facts, and ideas resulted. Little room was left to offer the type of philosophical discussions (on the nature of life, matter, etc.) that had been common earlier.

At the same time, the mid-to-late 19th century saw a tidal shift in scientists' own view of their work, in Germany most of all. Leadership here was seized early on, in the 1840s, with the founding of the Berlin Physics Society (Berliner Physikalische Gesellschaft) under the leadership of Emil Du Bois-Reymond, Ernst Brücke, and Hermann von Helmholtz. This organization, which came to have an enormous impact on all of German science, was not made up of physicists as one might expect, but instead by a handful of young doctors, physiologists in particular, whose shared philosophy was indicative of a pivotal break then occurring in technical circles. This involved the abandonment of an older, more speculative *Naturphilosophie* (natural philosophy) for *Naturwissenschaft* (natural science), a materialistic-mechanistic view of nature. That the manifesto for this movement came from a group of physicians was no accident: medicine, after all, was perceived to suffer the greatest "gap" between true science, based on rigorous experimentation, and metaphysical guesswork, with its invocations of immaterial, half-spiritual "life forces." The fundamental urge behind the Society was to remake medicine in the image of physics, an "organic physics" so to speak: in Du Bois-Reymond's famous phrasing, "No forces other than the common physical-chemical ones are to be understood as active within the organism" (see Tuchman, 1993). This urge perfectly expresses the trajectory of the group's most famous and influential member, Helmholtz, who began his career as a surgeon, taught physiology, and then moved into a number of cross-disciplinary areas in which he applied the ideas, methods, and mathematical reasoning of physics to the processes of perception, nerve transmission, hearing, color vision, and other cognitive areas. Helmholtz has often been called a "polymath," one of the century's greatest scientists, and it is undoubtedly true that he made major advances in every area just mentioned. His contribution to physics itself, exemplified by his early book *Über die Erhaltung der Kraft* (On the Conservation of Force, 1847), was extremely profound, resulting in a host of new discoveries and hypotheses, and, along with his work in cognitive physiology, earned him a vast reputation while still a young man. The uncompromising view Helmholtz and the Berlin Physics Society held of what constituted "real science" became, before long, a model for German science as a whole. That this view first arose within medicine was perhaps forgotten; but this did not prevent it from having its continued effects here, as elsewhere.

The "Helmholtz School," as it has been called (Bernfeld, 1944), reigned nearly unopposed by the 1860s and 70s. Mechanistic approaches to natural phenomena dominated and frequently had a literal dimension: following Helmholtz's book on the conservation of force, analogies between machines and organisms became common. Helmholtz had argued that organic processes, like mechanical ones, consumed and generated energy (heat) and so had to obey thermodynamic laws, especially the conservation of energy law. Older medical and biological ideas that proposed "vital forces" as directing the action of life ignored this truth, he said; such "forces," having no source of energy, were not forces at all, but mere chimeras ("ghosts" he once called them). As an example of the deductive power of the new philosophy, this writing by the "Young Helmholtz" won its author overnight acclaim and placed the Berlin Physics Society at the recognized forefront of German *Naturwissenschaft*. This position, it should be said, was at least partly due to the public presence of those such as Du Bois-Reymond and Helmholtz himself, both of whom often gave popular lectures, wrote about larger cultural matters, and spoke on the topic of science and philosophy. In fact, their general materialist outlook had come to exist in other fields as well. The renowned chemist Justus von Liebig, in particular, seems a possible forerunner (or earlier correlative) to Helmholtz in having applied, even by the 1830s, the experimental and quantitative methods of his discipline to the study of life. The Helmholtz School was a sign as much as a stimulus for the larger drift in 19th-century German science.

Insisting that this science had to move into the laboratory, take up mathematical analysis, and speak the language of "energy" or "quantity" had great appeal for a younger generation of German scientists imbued with the idea of progress yet frustrated by the apparent lack of it under *Naturphilosophie*. These young men were eager to stake out new territory by following the examples of great experimentalists like Helmholtz and Liebig, whose laboratories had become meccas of the new science for researchers throughout the Western world. The standards they adopted reveal a passion that verges on religious asceticism: absolute rigor and impersonal logic were demanded in everything; no meditative reasoning was allowed; mathematical formulations were everywhere required or recommended; new instruments and laboratory techniques were constantly implemented; strict causal explanations, backed by exact data, were the final goal. In his essay "Über das Ziel und der Fortschritte der Naturwissenschaft" (On the aim and progress of natural science, 1869), Helmholtz explicitly announced that the era of "book learning" and "reflection" was over; the necessity now was for extreme fidelity to detail and implacable discipline: "Only when the observer clings obstinately to his subject, such that all his thoughts and all his interest are fixated upon it and he cannot break free from it over the course of weeks, months, even years . . . until he has utter command over all details and feels certain of every relevant result that might be obtained in the present—only then does he produce a capable and valuable work" (1896, vol. 1, p. 370). Clearly, this could only be delivered in a prose equally tuned to the prose of anonymous truth. Only in a language that approached being a form of mathematics could science attain such rigor and be truly purged of its amateur past.

By the 1870s, such thinking prevailed in the larger medical community as well, where resistance had been strong, due to depth-rooted traditions. A "somatic" view of illness came to be accepted and standardized, such that every disease, whether affecting the organs or the emotions, was assigned a physical origin (Coleman and Holmes, 1988). As a result, older semi- or nonscientific areas were transformed into technical disciplines—in Germany, this included not only physiology but also neurology, psychology, and psychiatry. In all these fields, a break with former attitudes and "theories" was proclaimed essential for progress. The new physiology and psychology were to be rigorously experimental in nature, as developed by Ernst Brücke, Gustav Fechner (founder of psychophysics), and Wilhelm Wundt (founder of physiological psychology), while the new psychiatry, as expressed by one of Freud's teachers, Theodor Meynert, would no longer look to "treatment of the soul" but instead seek its basis in cerebral and neural anatomy.

New fields, new instruments, new findings, new methods, and new scales of investigation all meant a wholesale expansion in terminology and jargon, as each specialty sought to distinguish itself from all others. Most technical vocabularies exploded during the last quarter of the 19th century (technical dictionaries first began to appear regularly at this time). Because of linguistic conventions within science, tied to the classical tradition in education, the great majority of new terms were coined from Greek and, especially, Latin roots. This loading of scientific discourse with insider terminology, combined with the asceticism of the Helmholtz School, was another encouragement for an impersonal, functional style of writing. Scientists were now led in the direction of writing narrowly focused studies of single experiments; journals defined the market for their work. No longer, as in the days of Darwin or even the "Young Helmholtz" himself, did they turn regularly to the composition of essays or books intended for a general audience. If they did write "books" it was in the form of a research monograph, a collection of articles or lectures, or a classroom text. Moreover, even in their public lectures, delivered as early as mid-century, the Helmholtzians reveal a strong tendency toward instrumental uses of language, under the dictates of "factualism." The following, for example, offers a clear instance of this from an 1855 lecture Helmholtz gave on the subject of human vision:

> The eye is an optical instrument formed by nature, a natural camera obscura. The only essential difference from the instrument employed by photographers is that instead of the frosted glass square or light-sensitive plate, there lies at the back of the eye the sensitive retina, in which light generates a response that is transmitted via bundled nerve fibres in the optic nerve to the brain, the physical organ of consciousness. In its external form, however, the natural camera obscura certainly differs from the artificial one. Instead of a square wooden case, we find the round eyeball. The black color, with which the case of the camera obscura is lined is replaced in the eye with a second, thin, brown-black layer, the chorioid membrane. (1896, vol. 1, p. 90)

By 1900, this type of straightforward, unnuanced expression (far more laden with technical terms) was the rule in most of German science. The scientific community now seldom wrote for the public any longer, at least with regard to their professional work—unless it was to "translate" this work into more ordinary language. This, however, did not prevent some of the more prominent members of this community from speaking out on other topics. Indeed, the growing belief that "scientific thinking" could solve nearly any problem of human understanding and behavior spurred men such as Du Bois-Reymond to write extensively on politics, culture, and the German state. His proclamation, made in the 1870s, that "natural science is the absolute organ of culture, and the history of natural science is the only proper history of humankind" dovetailed neatly with a newly unified Germany that viewed itself, especially in the wake of its victory over France, as the political and scientific leader of the modern world (Gradmann, 1993, p. 5).

⑤⑤

The language of science was therefore transformed according to the dictates of the new philosophy, the contours of labor, and the institutional requirements for a more sober, straightforward style of expression. With each passing decade, nearly every field, new or old, experienced a notable increase in linguistic efficiency and a correlative decline in literary technique. The overall magnitude of change was perhaps less dramatic in Germany than elsewhere—not because Victorian-type eloquence continued to be valued, but because, on the contrary, a large portion of German scientific prose had *already* abandoned such eloquence. This was especially true in chemistry, a field that Freud studied. Here, a focus on procedure had long been paramount: one was required to report the exact manner in which an experiment or therapy was performed, the materials and equipment it utilized, and the quantitative data it produced. This focus had developed as a standard in the 18th century and before long it spilled over into the study of natural processes, beginning with plants and later advancing into human biology and medicine, where phenomena (such as illnesses) came to be seen as if they were "experiments of nature." Justus von Liebig was the first to fully combine the two fields of chemistry and biology, treating plants in terms of their chemical makeup; his language shows that the "literary" dimension to scientific writing was already beginning to ebb in some areas by the middle of the 19th century. The following, for example, is from his famous work *Die organische Chemie in ihrer Anwendung auf Agrikultur und Physiologie* (Organic chemistry in its application to agriculture and physiology, 1840):

> The chief components of vegetation, whose relative quantity renders other components negligibly small, include carbon and the elements of water; in total, less oxygen is included than carbonic acid.

It is therefore certain that plants, which assimilate carbon from carbonic acid, must possess the capability to decompose carbonic acid;
the makeup of the chief components of plant life assumes the separating of carbon from oxygen; the latter, during the living processes
of the plant, must be returned to the atmosphere while carbon is
being combined with water or its elements. (p. 3)

In Freud's own student days (the 1870s) meanwhile, the standard
for the most advanced scientific style was being newly set by Emil Fischer,
a highly successful and productive organic chemist, whose strict laboratory methods and formal demeanor were more than matched by the sobriety and restraint of his writing. Fischer had been the first to unravel
the structure of complex biochemical substances, such as glucose and
other sugars (work for which he received the Nobel Prize in chemistry in
1902). He did this by a unique method of breaking these substances
down into their basic constituent parts and then building them back up
again, via a step-by-step process of chemical synthesis. In his writing,
one notes a very similar overall procedure: an initial laying out of the
problem, followed by discussion of existing ideas and facts, then an in-
cremental construction, through logical elimination and deduction, of a
series of new results of increasing scale. The style in which this is rendered seems almost contemporary to the late-20th century. It is highly
"technological," with barely a hint of literary expressiveness, even in its
opening and closing phrases. The "I," as narrator is either missing altogether or reduced to a rhetorical nod ("The results are not sufficiently
certain. Therefore I will take up this study myself.").

Such a style became prevalent among German chemists by the 1890s,
but was not employed as a direct model for writing in every field. Indeed, Fischer was even attacked by some fellow scientists for his "uncompromising asceticism." Yet he nonetheless stands clear as a revelation of the direction in which things were moving. His was by no means,
at this time, the final word; but that word, when it eventually came, had
the ink dry on it long before.

Medical work, meanwhile, and therefore medical discourse, was
being made to obey the new standards for science under the dominance
of "experimental physiology." Led by researchers such as Ernst Brücke,
one of the founding members of the Berlin Physics Society, medicine
began to look away from fields like anatomy and comparative zoology,
with their emphasis on form and structure, and toward chemistry and
physics, whose methods of analysis seemed better suited to evince the
deeper functions and material processes of life and therefore disease.
The desire was to reduce organic reality to its most reductive, elemen-

tal level. Older notions that held "vital forces" to be the seat of life's activity were viewed as not merely speculative but mystical, tinged with religious spiritualism. German medicine had been associated with the universities since early in the century, and it too was now swept up by the general trend within the larger academic community toward *Wissenschaft* (experimentation and theory in the sciences; scholarly research in the humanities). Within this system, medical discourse in German took on a predictable professional cast. Unlike France, where writers of the first rank such as Louis Pasteur and Claude Bernard wrote very much as individuals, aiming much of their work at a wider readership and employing forms of eloquence not unlike those prevalent in Victorian England (see Chapter 2), German practitioners of the new experimental medicine (and related fields) wrote a prose consciously meant to embody the tenets of "scientific factualism" on which their new disciplines were based.

As early as the 1870s, the vast bulk of medical writing took the form of periodical articles intended solely for other professionals. The most widely known medical periodical in the German-speaking world, the *Deutsche Medicinische Wochenschrift* (German medical weekly), supposedly offered articles intended for the general public, but in reality few laymen could read it without difficulty. Even front-page matter was often highly technical, as the following example from an 1874 article by Robert Koch suggests:

> A variety of microorganisms have been found in infected surgical lesions. Yet these findings do not by themselves justify the hypothesis that surgical infection is merely a result of the penetration and proliferation of microorganisms in the body—in a word, a parasitic illness. For it can be justly maintained against this evidence that microorganisms are often lacking in cases of definite surgical infection, and in other cases are present in insufficient numbers to explain the observed symptoms of illness or their fatal result.[2]

The writing here, as a sample of medical discourse at the time, is neither "tortured" nor "labored," as Freudian scholars (interested in cursory comparisons) have often claimed. It is obviously formal, reserved, stuffed with content, not especially congenial to metaphor or other figurative kinds of language. Koch's German, like that of his medical colleagues, was dense and difficult, but it was not, one should note, as impenetrable and clipped as that of the chemists. At this point, in fact, most medical writing still retained the general syntax of German literary and scholarly language, with its multiple phrases, its complex levels of internal association, and its tendency toward a one-sentence/one-complete-thought containment.[3]

It was in physiology, a subject Freud himself studied in depth under the eminent Ernst Brücke, that the new scientific standard was set. As the most influential physiologist of his era, Brücke favored a distinctly straightforward, fact-driven manner of discourse that before long became the model for a great deal of medical writing. Freud worked at Brücke's Institute of Physiology in Vienna for six years (1876–1882) and no doubt knew his teacher's lectures well, having later spoken of their renown in German scientific circles. These lectures were issued in a revised form during Freud's own tenure at the Institute as *Vorlesungen über physiologie* (Lectures on physiology, 1881; 3d ed.). They reveal a type of writing that, far more than Koch's, imitates the flattened simplicity of Liebig:

> A distinction of the following type has been proposed: that organisms are active on an internal basis, while mechanical devices are active *per accidens*. In other words, an organism undergoes action within and for itself without anything else being done to it, whereas, in order to be activated, a steam-engine must be heated, a timer must be set, etc. When viewed more closely, however, this distinction is untenable: for the organism must be heated—that is, the animal must take nourishment in order that its activity be maintained, and we will see later that nourishment in the case of animals serves an entirely analogous role to that of coal, with which steam-engines are heated. (p. 1)

This is roughly the limit of metaphorical flourish that one finds in Brücke, a limit bound to a materialist-mechanistic vision. In places, especially in his detailed discussions of the human body and its functions, his writing approaches the information delivery of a late-20th-century textbook:

> Hair consists of three substances. In the hair stem one first distinguishes an outer layer of thin, platelike cells whose edges overlap each other like roof tiles and which comprise the cuticula of the hair. If one adds a drop of sulfuric acid to the hair, this cuticula detaches itself in scalelike masses from the hair's surface. Underneath lies the *substantia propria*. This consists of cells that extend to a point at both ends and of spindles which are flattened against each other and thereby appear square. They compose by far the chief material of the entire hair. (p. 318)

In his introductory chapters, where he discusses such fundamental topics as "The Organism," "Matter," and "Force"—thus revealing the

Helmholtzian perspective (human biology as handmaiden to physics)—
Brücke might call on Descartes or offer a phrase in the original Greek.
As shown in both passages above, he also sprinkled his writing with
Latin words. All of this reflects the classical training then still required
of scientists. Connections to the nonscientific, literature-based tradition
of education, deep as they were, could not be erased so easily—nor did
the image of the scientist in Germany turn completely away from this
tradition, which, after all, remained a base for even the most arcane
areas of expertise. Bernfeld (1949) meanwhile, notes that Brücke actually wrote a number of essays and books on topics such as poetry, art,
education, and music,[4] seemingly in contradiction to the strict demands
of his own new science. Yet every one of these works has a type of missionary purpose for this very science. Instead of trying to reach beyond
its bounds, they seek to draw their various nontechnical subjects deep
into the scientific metropolis, discussing art or poetry or music almost
purely in terms of Helmholtzian physiology, as problems of physics-laden
psychology. Under Brücke's care, the arts and education become special
cases of cognitive, brain-motor process. In the same way that his colleague Du Bois-Reymond subordinated all human history and culture to
the effects of scientific advance, so does Brücke reduce human creativity
to the "conservation of forces" theory, thus to science in content form.
Needless to say, Brücke's style and tone are the same here, whether he is
discussing art and teaching or blood, nerves, and hair.

The degree to which the young Freud might have accorded with
this general attitude is not wholly clear. It seems doubtful that he ever
did so completely, based on his own published writing during this same
time period. Prior to the 1890s, while at Brücke's Institute, Freud published several articles (his very first) on the neural anatomy of primitive
fishes. His teacher had given him certain problems on the histology of
the nervous system to work out, and he began by investigating the spinal cord of these lower animals. The titles of these articles (here translated into English) give some indication of the general style in which
they were written: "Concerning the Origin of the Posterior Nerve Roots
in the Spinal Cord of Ammoncoetes (*Petromyzon planeri*)" (1877); "Notice Regarding a Method for the Anatomical Preparation of the Nervous
System" (1879); "On the Structure of Nerve Fibers and Nerve Cells of
the Freshwater Crayfish" (1882). It would seem clear that Freud wrote,
at this early stage, as a scientist of the reigning persuasion—but not
entirely. On closer look, his language even in these apprentice works is
not without a formal elegance and drama—certainly a tone and flow
that greatly exceeds that of Brücke in its sophistication:

> In the course of my studies on the structure and development of the medulla oblongata I succeeded in working out the following method, which will be found a powerful aid in tracing the course of fibres in the central nervous system of the adult and the embryo. . . . This method will never fail (as all methods of staining by chloride of gold will do) if the specimen be not overhardened and brittle, as it is sure to be if kept in the hardening fluid an inordinate length of time. . . . By this method the fibres are made to show in a pink, deep purple, blue or even black colour, and are brought distinctly into view, while the grey substance, vessels and neuroglia, lost in the slightly tinged background, are not obtruded upon the attention of the observer. . . . In the adult . . . the ensemble of fibres is much too complicated for analysis; in the new-born and the embryo, the nerve-fibres alone are strikingly brought out, and those bundles, which are already possessed of a medullary sheath, are distinguished by darker colouring from the others. (1884, pp. 86–88)

Though written in English, this article is indicative of Freud's early, scientific style in German as well. It is direct, explanatory, but not without literary effects. A degree of literary energy is added to what might otherwise be an utterly flat description of technique. Even at this early stage, that is, Freud cannot help but write against the grain of the "scientific" even while encased by its requirements.

Freud's other early teacher, Theodor Meynert, was a somewhat different sort of writer from Brücke. He seems a kind of half-step to the type of discourse Freud would one day come to compose—that is, he retains a number of elements from scientific discourse of older generations but seems also to have moved no small distance toward the new standard of instrumentality and factualism. Freud left Brücke's Institute in 1882 (most biographers describe this as a result of Freud's decision to obtain clinical experience for a possible career as a doctor, since the pursuit of pure science in an incoming-earning academic post was closed to Jews at the time). He had chosen to join the Vienna General Hospital, where Meynert, one of Europe's foremost psychiatrists and brain anatomists, worked and lectured. Meynert was decidedly a relative of Brücke's in terms of his general scientific outlook, but he also believed strongly in the importance of anatomy and structure as a means to understand the nature of illness. He believed that mental disorders were neurological in a general physiological sense; they were the result of anatomical or physical deformities, injuries, lesions, or the like. Consciousness was not purely a matter of physicochemical action but a result of complex stimulations to the cerebral cortex, the "cortical I" (das cortical Ich), he called it. Meynert was thus committed to the somatic view of mental reality, but drew also from older notions of "excitability" and "irritation" prevalent in the early part of the century (see Chapter 3). Freud remained with Meynert until 1886, when he returned from a brief yet pivotal 6 month period of study under Jean Martin Charcot, France's eminent neurologist. It was during this period that Freud shifted away from the somatic interpretation of all mental illness, siding in a sense with an older realm of explanation (as I shall point out below), and thus broke with his intellectual patron. This break, as so many writers have shown, was critical to Freud's future thought and career, indeed to the founding of psychoanalysis.

While at Meynert's Institute of Cerebral Anatomy, however, Freud had been granted a particular honor. The old professor, having recognized on the basis of Freud's technical publications his strong verbal abilities, offered his student instructor's position (*Dozent*) and even promised to bequeath his own extensive (and profitable) lecturing work upon his retirement. It is interesting therefore to observe Meynert's own writing style, with a view to its possible affinities for Freud. The following, for example, taken from "Clinical Lectures on Psychiatry, Given a Scientific Basis" (*Klinische Vorlesungen über Psychiatrie, auf Wissenschaftlichen Grundlagen*), appeared in published form in 1890 but derived from a series of lectures that Meynert had been giving over the course of the previous decade, during the period of Freud's residency:

> Mania and melancholia count among these illness of the forebrain, whose anatomical basis has not yet been determined as a process. Without excluding anatomical processes, I place them today, reservedly, among that group of diseases resulting from a disruption of nourishment to the forebrain. Clearly this interpretation is insufficient, for mania and melancholia lead to idiocy, in which it is certain that more fine and coarse anatomical transformations occur. Immediately following the development of melancholia there takes place an extensive atrophy of the brain. Yet its symptomatic form conceals itself beneath an illness-mechanism lacking all anatomical transformations of the forebrain, and this mechanism must in every circumstance be included within the process of the disease. (1890, p. 5)

Meynert was a scientist more conscious of the uses of language than Brücke. Here and elsewhere his sentences display qualities of pacing, logic, and neologism that are largely absent in writers who slavishly followed the new standards of scientific style. Though not seen in this passage, Meynert is also much more fond than Brücke of making reference to historical, literary, and philosophical figures, quoting such fig-

ures as Byron and Schopenhauer in a discussion on melancholia for instance. The subtitle of his book—"for students and physicians, lawyers
and psychologists" (*für Studirende und Ärtzte, Juristen und Psycholo-
gen*)—seems to reveal the author's eagerness to attract a broad and well-
educated audience (a professional one mainly, though "studirende" really means "those who are studying," thus possibly including anyone
who might have an interest in the subject). Yet again, much of this is
offset by other aspects of Meynert's style. Such things, that is, hardly
compensate for his arch formality, his constant repetition of technical
terms, his refusal to employ metaphor and other literary techniques, his
general dullness of tone and effect. Meynert is in the same mold as Koch—
not quite so technological in his language as Brücke, but, in the end, no
less sober and concerned with efficiency.

It was in France that Freud encountered a very different set of linguistic models. French contemporaries of Koch, Brücke, and Meynert
followed a separate medical tradition from their German counterparts.
This was the clinical tradition, focusing not on mechanical explanations
of cause but on careful descriptive analyses that gave priority to observation (usually of symptoms), which then became the basis for diagnosis
and, afterward, for hypothesizing about mechanisms and origins of disease. While they too had accepted the idea of physical (somatic) causes,
because the French relied far more on narratives of description, holding
back from high-minded theory, their writing was often more literary,
more involved with the telling of stories, more characterized by interesting asides. Unlike many German research doctors, who were specialists,
French medical authors were often highly urbane individuals with active
interests in philosophy, literature, and art; they were dedicated to
"progress" in its largest sense, and to embodying the French ideal of
l'homme civilisé, "the cultured man." Formal and measured, French
medical writing is also learned, varied, full of diverse rhetorical elements,
aimed more often at both informing and entertaining its readers. French
researchers were also prone to reshaping their university lectures into
public addresses, given before local societies and lay audiences of various types, and were much more likely to deliver these in elegant language, spiced with anecdotes, humor, artistic or literary allusions, and
the like. Where German physicians, by the 1880s and 90s, were generally
more interested in advancing knowledge, and sharing it locally among professionals, French doctors appear to have been concerned with disseminat-
ing medical understanding more broadly, thus advancing their field directly
as a national enterprise and their own national fame as part of this.

Charcot, in some ways, provides one of the best examples of this

French medical "scholar." Reserved, even diffident in person, he was nonetheless known as a *savant* of his day, a sort of national treasure, who held regular *soirées* at which well-known writers, artists, musicians, politicians, and scientists were in attendance. His uncompromising sense of ethics and honor went hand in hand with verbal eloquence, a combination that made him a figure of profound influence (Guillain, 1959).[5] Charcot applied rigor and great precision to his observation, diagnoses, and writing. He worked very hard on each of his lectures, cultivating a style that would charm and educate his audience, that would embody the higher, unifying knowledge and understanding of which he believed medicine capable. A great many of his talks were written as essays, delving deeply into philosophic matters, the nature of scientific knowledge, or the history of medicine; they draw on his readings in English, German, and Italian literature, and call on the likes of Vico, Herder, and Taine as well as Hippocrates, Vesalius, and Harvey. In one of his more well-known works, entitled *Maladies des viellards* (Illnesses of the aged, 1881), Charcot begins with an extended discussion of "empirical medicine versus scientific medicine":

> You already know the object of these studies; we should be able to embark immediately upon the material, as we are of the opinion of Condillac in thinking that general considerations are better placed at the end of a course than at the beginning. However, there exists a tradition, known as the classical, which creates in the professor an obligation to first expound in a manner more or less categorical on certain fundamental questions, and to offer in some degree a declaration of scientific faith. . . . There does exist a simple means, let us say natural, by which one might approach these larger questions: namely, to examine, through the progressive development of scientific culture, how they themselves arise and how they have been resolved. History therefore becomes an agent of critique. By following this path, moreover, one soon recognizes that in no epoch whatsoever did pure observation become possible without supreme efforts being exercised to gain mastery over the will to hypothesize. (*Oeuvres Complètes*, hereafter *OE*, vol. 7, pp. iii–iv)

The subtitle for this introductory essay, one might note, is "Parallele entre les anciens et les modernes" (parallels between the ancients and the moderns). Medicine for Charcot was often coterminal with the philosophical and literary tradition, and also with its battles: an intimate familiarity with the entire history of medical writing and writing on medicine, since Hippocrates, was entirely necessary. According to Guillain (1959), Charcot's library was full of books on subjects in the humanities, including philosophy, literature, psychology, art, and history. He was very well read in both ancient and modern literature, claimed Shakespeare as his favorite author, and apparently often read authors in the original Greek and Latin. In his intellectual habits and possessions, and in his related pleasures, he therefore offers the only clear model for the mature Freud among any of his teachers (see, e.g., Timms, 1988).

Charcot employed a variety of specific narrative forms in his writing. He wrote not only lectures and clinical reports, but case histories, anatomical analyses, historical surveys, reflective summaries, discussions of art, and critical reviews. The structure of many of his texts is interesting to contrast with that of Emile Fischer. Instead of simply stating the problem at hand, breaking it down into analyzable fragments, and then reassembling the whole to create a conclusion, Charcot commonly begins with an embrasive statement of the topic, followed by a description of the events and thoughts that led him to study it. Next, he offers evidence in the form of a series of both clinical and personal observations, upon whose meaning he then reflects, asks questions of, and finally draws tentative conclusions. Scattered throughout are discussions of methodology, expressions of doubt, and statements on the often-contingent nature of medical knowledge. Certain conventions of scientific discourse commonly found in German are also obvious: in his later papers on hysteria and hypnotism, for example, Charcot's writing becomes jargon-heavy and refers to its author in the third person ("Les recherches entreprises . . . par M. Charcot et, sous sa direction, par plusieurs de ses élèves . . . " [The research here undertaken . . . by M. Charcot and, under his direction, several of his students . . .]; *OE*, vol. 9, p. 297). Yet in this same collection, he also ventures into meditations on the power and systems of naming (in a section notably titled "Essai d'une distinction nosographique"), and still more significantly, into art.

Charcot was himself an amateur artist, a skilled drawer, painter, and caricaturist. It is a testament to both his expansive style of thought and the openness of his discourse that he employed his sensitivity in this area toward a highly important discovery of clinical fact, in a manner that did not reduce the original subjects themselves to mere "data" (as had Brücke). The visual dimension to Charcot's teaching was famous at the time: he was the first medical scholar in France to make use of projection equipment; he kept a large photograph library; and he drew illustrations on the blackboard continuously while he was lecturing. In all this, he carried on a tradition begun by Vesalius, in which seeing, drawing, and knowing—learning to read symptoms, to draw their appearance, and to translate them into words—were merged. From the

1870s onward, moreover, Charcot began to take note of, and to lecture on, the appearance of hysterical symptoms evident in artworks displaying the human figure in the throes of demonic possession. As a frequent traveler to European museums, he kept notebooks on such works, wrote up his analyses of them and eventually, along with his colleague Paul Richer, produced two books, *Les demoniaques dans l'art* (The demonic in art, 1887) and *Les difformes et les malades dans l'art* (The deformed and the sick in art, 1889). These books are remarkable in the history of late-19th-century medicine. Within them, artworks from various eras are discussed as actual case studies. Charcot's philosophical questionings, personal remarks, his sense of connoisseurship, as well as historical information on each of the works themselves are also included. The total effect is truly a unique blending of subjects and styles, all in the name of "science." On the basis of what he found in these artworks, moreover, Charcot drew two conclusions wholly startling to his colleagues, yet extremely important to Freud's own work: first, that hysteria was by no means a 19th century illness, brought on by the conditions of "modern society"; and second, that it was an illness not limited to women. History—here, art history—was indeed a powerful "agent of critique" for Charcot. In comparison with Brücke, for whom art was tantamount to a release of "excess cognition," Charcot presents one of the few cases in the history of science where a contemporary technical discovery was made from "ancient" artistic material, and where the discourse of such discovery remained, in some part, true to humanistic appreciation.

Some flavor of Charcot's writing on this subject can be seen in an early lecture entitled "Épisodes nouveaux de l'hystero-epilepsie" (New episodes in hysterical epilepsy), published in 1878 (reprinted in *OE*, vol. 9, pp. 289–296). This lecture includes a number of sections on the clinical appearance and diagnosis of hysterical convulsions, hallucinations, and paralysis. It speaks of induced catalepsy (a condition in which the limbs hold any position they are placed in), both in humans and animals, citing references as old as 1636, and ends in a brief section of several paragraphs titled "De l'hysterie dans l'art," where the following lines occur:

> We project onto this screen an exact copy of a fresco that struck my attention in a church in Florence, during a recent voyage to Italy. It represents a miracle of St. Philippe of Neri, an ascetic regarded as having the power to recognize the chaste by means of their special scent.
> He is shown here curing a possessed individual (1595). Note how he focuses his attention on the sick person before him. Examine

the posture of the latter; it is that of an hysterique in a hypnotic or somnambulistic state. Look at the disposition of the limbs: it is an exact copy from nature. Without knowing it or desiring it, the saint has produced the effect that we are engaged in studying. (*OE*, vol. 9, p. 295)

Charcot here offers us a portrait of his own classroom methods, his own enthusiasm, indeed his very gestures. The atmosphere of the moment is entirely palpable; we are sitting in a late-19th-century lecture hall, darkened for a few minutes so that we may observe a very different sort of "specimen." In but a few lines, art, history, personal experience, and humor are all present. Indeed, they are what place us in that moment, what makes possible our imaginative journey back more than a century in time. They are elements that do not merely expand the boundaries of "science" but that allow this "science" to re-create its very context of discovery and insight. This is all the more true when one considers that the subject itself, hysteria, was not yet viewed by the medical profession as an illness. (At the time most physicians regarded hysteria as a form of malingering, and therefore as a topic of no medical interest; see, e.g., Gay, 1988.)

No doubt Charcot's more expressive science was not lost on Freud during his tutelage under the great neurologist during the winter of 1885–1886, when the latter was at the height of his powers and fame. Given what we have seen of Freud's own writing at the time, it is not difficult to imagine the appeal Charcot must have had for him. During his visit, Freud had been exposed to lectures of exactly this type. One measure of their lasting effect upon him is a paper he wrote nearly four decades later, "A Devil-Neurosis in the 17th Century" ("Eine Teufelsneurose im Siebzehnten Jahrhundert," 1923), where he follows Charcot's example directly by analyzing a work of art for its portrayal of neurotic symptoms (he does much more than this too). The most direct form of homage he paid to Charcot, however, can be found in the eloquent eulogy Freud wrote shortly after his mentor's death ("Charcot"; 1893), in which one finds in nascent form the very kernel of the Freudian project:

No one can object that the theory of a splitting of consciousness, as a solution to the enigma of hysteria, is so improbable that it must be forced upon the impartial and untrained observer. In fact, the middle ages had chosen this very solution, in explaining the origin of hysterical phenomena as the result of demonic possession; it would merely have been a question of replacing the religious

> terminology of that dark and superstitious age with a modern scientific one.
>
>   Charcot did not follow this path to the clarification of hysteria, yet he nonetheless made rich use out of existing accounts of witch trials and possessions, in order to show that the appearance of neurosis at that time was the same as today. (*Gesammelte Werke*, hereafter *GW*, vol. 1, pp. 31–32)

It is easy to imagine Freud, as he wrote this, recalling to mind the very class in which Charcot spoke of the painting in Florence. To the young Viennese student, the French master "was not a brooding speculator, not a thinker, but instead an artistically gifted nature—as he once said of himself, a *visuel*, a seer" (p. 22). Indeed, Freud has much to say about his power as a lecturer, stating that "as a teacher, Charcot was wholly captivating, each of his lectures being a small work of art in its structure and composition, complete in form and in manner so penetrating that for the rest of the day one could not get the words that he had said out of one's ear, or the object of his demonstrations out of one's mind" (p. 28). This quality of retention is strong stuff, especially coming from one who would later write about the power of teachers as father figures.[6]

    It is, of course, common in traditional analysis, literary or otherwise, to pinpoint origins and influences. Freud has been many times reassembled as a series of tracings left by those who influenced him. Yet it seems clear from his own essay that the young neuropathologist was deeply moved and altered by his experience with Charcot, and at a critical point in his career. That winter was surely one of discontent, born of Freud's separation from his fiancée, his confusion regarding his future career, and his growing dissatisfaction with Meynert's somatic provincialism. Charcot's mixture of eminence and magnaminity, which enabled the great man to treat his students as something more than mere intellectual patients, no doubt had its impact on Freud. One can hear it in Freud's, descriptions of Charcot that *this* was the type of "science" he wanted to practice. It becomes all the more clear in a brief anecdote he could not resist offering in the early pages of his eulogy:

> Charcot never tired of defending the correctness of seeing and classifying in pure clinical work against the encroachments of theoretical medicine. One day a small troop of us foreigners, brought up in German school–physiology, gathered to annoy him with an objection to his clinical innovations: "But that can not be," challenged one of us at some particular point, "it indeed contradicts the theory of young Helmholtz." He did not reply: "So much the worse for the

theory; facts that arise from clinical work take precedence," or the like, but he said to us instead something that made a great impression upon us: "*La théorie, c'est bon, mais ça n'empêche pas d'exister*" [Theory is fine, but it does not prevent what exists from existing]. (*GW*, vol. 1, pp. 23–24)

Here, in a nutshell, is both reason and expression of Freud's own conversion, his turning away from Meynert and "German school–physiology." Perhaps if, despite his Jewishness, he had been allowed to hold a high-level academic position in science, things for Freud might have been different; he might have tried to reclaim, upon his return to Germany, a place within orthodox science. But this door was closed, and the constant echo of its closing must have made Charcot's words and encouragement ring all the more with appeal. Freud's essay is a profound repayment of debt.

Finally, it is critical to remember that the first book Freud ever wrote was a translation of Charcot's own *Leçons sur les maladies du système Nerveux* (*OE*, vol. 3, 1879), rendered as *Neue Vorlesungen über die Krankheiten des Nervensystems, insbesondere über Hysterie* (1886).[7] Here was the 29-year-old Freud, in other words, tracing again Charcot's pointer in the lecture hall, his steps through the Salpêtrière (an asylum in Paris), yet this time in a form that nativized the master both to the German language and to Freud's own touch. Here he was, going over every line, every word, every expression of his teacher's, tracing out everything in his own hand (literally and figuratively) that made the latter an author, a creator of discourse. In his foreword to the book, Freud says he added the phrase "especially about hysteria" to the title, with Charcot's blessing, for the simple reason that it was here, above all, that the author had inspired a "new epoch in the worth of a little known and therefore ignorantly denied form of neurosis" (p. vi). The symbolism seems acute: it was in this area that Freud would carry on Charcot's work. By specifically adding "hysteria" to the title, he announces that he himself will inherit the master's wand. Indeed, in the translation itself Freud seems to have changed Charcot's exact wording fairly often to suit his own purposes—he is not a bad translator, that is, so much as he is an appropriating one. But the one thing he never sacrifices entirely is Charcot's complexity of style, its various elements and forms of expression, which, in fact, the English translation by G. Sigerson does remove, performing the same type of scientizing of language that would one day be visited upon Freud too (see Charcot, 1877–1889).

Freud thus became an author, a writer of books, by putting himself in Charcot's place, by taking on his style, by becoming his copyist. This simple fact seems to have been ignored or passed over by the vast majority of Freudian scholars. Yet to any writer, as writer, its significance must be apparent. On the occasion of the centenary of Charcot's birth in 1925, Freud was asked about his thoughts regarding his teacher's influence upon him, particularly regarding the "discovery" of psychoanalysis (Codet and Laforgue, 1925). Now as famous as Charcot had been nearly a half-century earlier, Freud gave a simple reply: nothing, to his mind, had changed since the day his eulogy had appeared in print.

III

Strachey, no doubt, would have opposed making Freud a Frenchman, even of the most cultivated sort. One wonders what he made of another famous German author, who, in fact, was writing at the time Freud was a student and whose career sought to combine the styles of new and old science, *Naturphilosophie* and Helmholtzian experimentalism. This was Wilhelm Wundt, founder of "physiological psychology," who first coined the phrase *experimentelle Psychologie* in the 1870s, who later opposed Freudian ideas as "unscientific" and "metaphysical," yet whose own work from the 1890s onward was the object of similar establishment criticism (see Decker, 1977; Danziger, 1990).

Wundt held a chair in philosophy at the University of Leipzig, but he had been trained in medicine and had worked as an assistant to Helmholtz at Heidelberg. His early books were on human physiology, sense perception, and animal psychology, whose ideas he then sought to unite under one roof in a work called *Grundzüge der physiologischen Psychologie* (Basic characteristics of physiological psychology, 1874). In this book, Wundt stated his intent to "give psychology a mediating position between the natural sciences and human sciences" (1893, p. 3). This meant, on the one hand, setting up an Institute of Experimental Psychology and publishing the results of laboratory work in the realm of stimulus–response behaviors; on the other hand, however, it also meant a more speculative type of writing in the area of social or cultural psychology (*Völkerpsychologie* was Wundt's term). Without going into detail, it can be said that the general movement of Wundt's thought was away from the more positivist, Helmholtzian aspects of his theory and toward the sociocultural, even folkloric ones, that is, away from "science" and toward "philosophy." Rather than straddling the fence between these two regions, or breaking it down, Wundt seems to have crossed or skidded over from one side to the other. Unfortunately, his action was exactly counter to the larger movement of the age. The infant social sciences and the new generation of psychologists had "neither sympathy nor understanding for a position that seemed to put in doubt psychology's claim to be counted among the natural sciences. What Wundt had tried to hold together now drifted apart" (Danziger, 1990, p. 403). By the 1920s, as Freud rose in reputation, Wundt fell further and further into disrepute, especially in America and Britain, where the "scientific" reigned supreme. Unlike Freud, whose ideas had achieved that label, Wundt was seen as a "failed mystic," a relic of the very age whose end he had earlier proclaimed. Like Freud, he moved in his career from the positivist side of things to a more metascientific position. But he went too in far this direction, and in a manner that permitted no way to cross back.

What of Wundt's narrative style? What type of prose did he write? If, in his attempt to blend science with other contents, he presents a certain correlative to Charcot (and Freud), how does this appear in his actual texts? A sample, taken from his *Grundzüge* (4th ed.), offers some idea:

> The human mind does not favor the collecting of experiences without at the same time weaving into them its own speculation. The first result of such natural reflection is the conceptual system of language. In all domains of human experience, therefore, there are certain concepts that science, before applying itself to its own work, finds ready-made, as the results of that primal reflection which has left behind its abiding deposit in the conceptual symbols of language. Thus are heat and light concepts from the domain of external experience, which arise from direct sensory response. (1893, p. 10)

The American translator of Wundt's book, E. B. Titchener (a professor of psychology at Cornell University) states in his preface that "Wundt's style has often, of late years, been termed diffuse and obscure. . . . It has, perhaps, in a somewhat unusual degree, the typical characteristics of scientific German; the carelessness of verbal repetitions, the long and involved sentences, the lapses into colloquialism, and what not" (p. x). The above excerpt would appear to bear this criticism out to some degree—except, of course, that it is anything but *typical* scientific German. It is instead something else, a kind of half-metaphysical philosophizing. It is not Charcot, yet it is not Koch either, nor Liebig, Brücke, or Meynert. The key to Titchener's comment comes in his later claim that "a special difficulty in Wundt's style . . . is his increasing tendency to clothe his ideas in conceptual garb" (p. xi). The problem, in other words, was that Wundt simply wasn't technical enough. He had to be made more "concrete," more truly scientific-sounding. Comparison of the above passage with Titchener's own version leaves little doubt of this:

> The human mind is so constituted, that it cannot gather experiences without at the same time supplying an admixture of its own speculation. The first result of this naive reflection is the system of concepts which language embodies. Hence, in all departments of human experience, there are certain concepts that science finds ready made, before it proceeds upon its own proper business,—results of that primitive reflection which has left its permanent record in the concept-system of language. "Heat" and "light," e.g., are concepts from the world of external experience, which had their immediate origin in sense-perception. (1904, pp. 16–17)

The scientizing aspects of this rendition are striking. Moreover, Titchener's prose is better than Wundt's, meaning it is more flowing, consistent in tone, and clear in sense. It makes Wundt appear a better writer than he actually is—better, exactly, by being more "scientific." It is predictable, then, that the translator is most deforming of the original where Wundt launches into an extended philosophical discussion of "soul" and "mind/ spirit" (*Seele* and *Geist*), where such authors as Plato, Aristotle, Kant, and Leibniz are invoked. Here it is apparent that Titchener is unequal to the task: he struggles to bring the text into line with his conception of German philosophical discourse, inserting Latin terms, Greek words, and parenthetical explanations, all absent from the original— in the name of a hopelessly pedantic "reflection" (later in his career, Titchener was to come under the sway of Ernst Mach's positivism and become a proponent of reductionist views in the United States). In this case, Wundt is the more cogent, writing as he is in a long-established tradition, inherited from Hegel onward. But Titchener *is* faithful, despite his own best efforts, to the irreconcilability of science and idealistic philosophy at this point in history. No doubt he was entirely representative of his time (in America most of all): a German "science" had to be as scientific-sounding as possible; German philosophy, on the other hand, had to be something else altogether, something obscure, classical, indulgent, but definitely not "science." His failure to make Wundt a continued success in the United States was both a result of the latter's own shortcomings and of history itself. Wundt's laboratory, after all, had been an enormously popular destination for American psychologists in the 1870s,

80s, and even 90s: to the "new psychology" it was the same type of mecca that Liebig's Institute had been to the "new chemistry" decades before, a site to which such influential Americans as G. Stanley Hall and William James had journeyed in search of a model. But "science" soon passed Wundt (and Hall) by, and the language he employed no less. Rather than a midway point or an anachronism, he represents a kind of anomaly, a half-vestigialism from the early days of *Wissenschaft*.

IV

Wundt's style was emblematic of his greater failure as a "scientist." The translator seems to have done his best to "scientize" the original, but could only do so up to a particular point. Beyond that, too much was needed. One suspects that just as Wundt's ideas and expressive forms were far less evocative and original than Freud's, so was Titchener less than the equal of James Strachey and, more importantly, Ernest Jones. It was Jones, after all, who was truly Freud's "bulldog" for the English-speaking world. It was he who took control, both behind the scenes and in plain sight, over the translation of Freud's works, with a clear view to medicalizing his language in order to help ensure his acceptance by the scientific and medical establishments.

This has been shown in some detail, for example, by Riccardo Steiner (1987, 1991), who indicates that Jones knew very well what "cultural policy" he was pursuing in taking on the burden—nearly commensurate with his own life's work—of making sure that psychoanalysis was perceived as, and consistently called, a "science" in England and America. A doctor himself, member of a profession that had only recently wrestled itself free of a prescientific past, Jones well knew what this appellation meant in terms of the politics of knowledge at the time. He knew the reverberations of calling Freud the "Darwin of the mind, who has replaced the metaphysical or poetical phrases of Schopenhauer, Nietzsche, Bergson, Shaw, by a scientific and biological one." He knew the importance of going still further than this, and stating that

[Freud's] shifting of the emphasis in psychology away from the intellectual to the instinctive, and his derivation of the higher and more complex mental activities from lowlier forms more nearly akin to those characteristic of animals, represent a momentous progress in scientific thought; for the reduction of the mental to biological terms

> ... seems the only satisfactory way of bringing psychology into line
> with the organic sciences, and of extablishing a harmonious rela-
> tionship between it and physiology. (quoted in Steiner, 1987, p. 53)

This was almost a direct swipe at Wundt. When Jones wrote these words
in 1913, the demise of Wundt's "physiological psychology" was becoming apparent (Freud himself took it to task in the same year, in *Totem
und Tabu*). A gap therefore existed in the realm of a truly medical theory
of mind, with possibilities for deep diagnosis and therapy. Freud—or
rather Jones—was there to fill this gap. Jones's "policy," in fact, was
comprehensive. It involved: compiling a code of technical Freudian terms
(begun as early as 1914); conceiving a standard edition in English;
grooming Strachey to lead the project and devaluing other early translators (notably A. J. Brill); and founding the *International Journal of
Psycho-Analysis* (1919–1920). These were all part of "Jones's attempts
to achieve an hegemonic control of the translation of Freud" (Steiner,
1991, p. 389).

Freud himself remained, for the most part, removed from the process of English translation. This may seem odd, given that he had been a
translator himself, of a number of works besides that of Charcot (Watson,
1958). Yet, due to his feelings about America and Britain, the former of
which he considered anti-intellectual, his attitude was ambiguous about
the rendering of his own work into English. According to Peter Gay
(1988), he largely, if not entirely, agreed with the idea that in "technological" America a more scientized version of his thought was required
to implant psychoanalysis. But the view he bore toward his own language, more generally, was always somewhat undecided. He seems to
have been torn between hopes that his field would eventually be counted
equivalent to one of the natural sciences (thus, in effect, agreeing with
Jones's own ambitions) and far more humble confusions regarding the
often figurative, speculative character of his writings. In any case, the
medicalizing of his writing proceeded without any major opposition from
Freud himself.

To review all the changes Strachey made and the material he added
would require a book of its own. As Laplanche, Cotet, and Bourguignon
(1992) have pointed out, Freudian language is built upon literally thousands of compound German words that Freud himself coined, but that
were frequently transformed into Greco-Latin terminology by Strachey—
that is, into a vocabulary inaccessible to the average 20th-century reader.
In so doing, Strachey, under the close-eyed direction of Jones, effectively
created the alienated "layman" at the same time that he created the Freud-

ian *medicus*. And it can hardly be denied that this result often strips Freud, unceremoniously, of his humanism, his Viennese culture, his classical education, his understanding of French, his preference for Greek words (over Latin), his very Jewishness. All this, moreover, occurs in addition to the noted loss or suppression of humor, wordplay, verbal subtlety, occasional lyricism, and frequent irony—elements that radically separate Freud's discourse from what was becoming standard in contemporary German science.

The result seems blameworthy, and indeed both Jones and Strachey have been much blamed of late. Bettelheim (1982) in fact, represents only an early form of this genre of contemporary accusation. Together Jones and Strachey have been many times impugned for dastardly deeds and of conduct unbecoming on the fields of language. These accusations, coming mainly today from within the analytic community,[8] have a productive violence; they impel us to examine Freud carefully. Yet their quality of censure, often tinged with scholarly contempt, seems, in the eye of history, a blurry view. Certainly it is true that Jones's "policy" was a conscious one. His machinations, which even included misrepresentation and possibly fraud, are well documented in Steiner's extended study (see also the excellent essay by Paskauskas [1988]). At times, Strachey himself grew exasperated with Jones's insistence on using technical terms for ordinary German words, including the primal *das Es*.[9] But Jones was insistent, and Strachey's patrician reserve and his general unwillingness to go to war over terms seems to have sealed Freud's fate in English.

What this reflects is Jones's commitment to "Freud" (the institution), even more than to "science." In effect, he was doing exactly what the period required of someone with his grandiose ambitions. There is more than a little evidence that he consciously compared himself to T. H. Huxley, the great (though selective) champion of Darwin. Indeed, by calling Freud the "Darwin of the mind," Jones placed his own efforts to support Freud in direct parallel to Huxley's defense of Darwinian ideas. Yet, beyond the obvious hucksterism here, one should not dismiss the parallel out of hand. Huxley, after all, possessed exactly the type of eloquence, cultivated learning, heroic bearing, and gift for debate that late Victorian England demanded of its most successful intellectuals. Jones, on the other hand, more the shrewd zealot than eloquent public defender, nonetheless understood at some level what was needed to implant Freud firmly in the drier soil of mid- to late-20th-century Anglo-American scientific culture. Medicine in general and psychology and psychiatry in particular were highly positivist in their outlook and self-image by this time (see, e.g., Hale, 1971). Psychology in particular, especially after 1900, was coming under the sway of men such as Edward Thorndike, for whom measurement, quantification, and experimentation were paramount. The mitigating influence of William James—who himself had tried to draw attention to Freud as early as the mid-1890s—was no longer central, his writings by this time having moved on to education, religion, and philosophy. Jones was no doubt aware, too, of the American and British attitude toward German science, what John Dewey, on the occasion of William James's death in 1910, called an "unreasoned admiration of men and things German" (quoted in Meyers, 1986, p. 1). This prestige cannot be exaggerated.[10] By the 1880s, German universities were considered to be the pinnacle of scientific culture in the world, populated by the greatest minds and producing the greatest discoveries of the age. Consequently, it was this idealized, by-then traditional image and expectation of American intellectual culture that Jones drew on in his reengineering of Freudian discourse.

For these and other reasons, a more literary Freud would have had far less chance of being embraced in America, under almost any circumstances. The oft-noted fact that his works could be read and understood by any German-speaking individual of decent education would likely have been his downfall in America or England, where "science" had come to mean a particular type of writing and thought, an insider's game employing "theory," "hypothesis," and a language of distant, technological character. Certainly the thinness of Freud's obedience to the norms of "science" would have weakened acceptance of his theoretical metapsychology; but it would also have had its impact on reception of his clinical ideas, on his vision of psychotherapy. The fact that Freud himself spoke in medical terms, of "mental illness" and "treatment," also meant that he required accommodation within the frame of contemporary (scientific) medicine and its stylistic demands. In America, these demands were far closer to the German model than the French one. Moreover, given the density of sexual imagery and the centrality of sexual concepts in Freud's writings, scientizing his language meant sanitizing it too, that is, making it more acceptable to a scholarly (and, later on, popular) audience still thoroughly Victorian in its morality.

It thus seems too simple to call the Strachey translations "bad," or "impoverished," or simply "wrong." On the contrary, in a sense they greatly "improve" and "enrich" Freud, giving his works a power and influence that they would not have had. The problem in fact comes back to certain ideas of translation. The demand that Strachey (or any other translator) should have "remained absolutely true" to the original Freud is really a demand that translation should follow the ideal of producing an equally timeless original, and that the translator should hold himself apart from the influences and standards of his or her time in order to fulfill this higher purpose. Such a belief comes down hard on one side of the oldest and most insoluble debate in translation, that between paraphrase and faithfulness. It also ignores the reality that "faithfulness," no matter how strictly defined, has never been a stable idea through the centuries, but has itself been endlessly reinterpreted and redefined. The exacting replica produced in one era is not that of the next; the Homer of Chapman is not that of Lattimore. Indeed, it is a common saying among translators that great works should be rerendered at least once every half-century or so. Languages do not remain stationary, after all, but change over time due to many influences. The idea of retaining Viennese culture or Freud's "Jewishness" (or even locating it) would probably never have occurred to an exacting translator of the 1920s (when Strachey began his work), nor perhaps even to one of the 1950s. Yet these things gain great poignancy today, both because of what has happened since Freud wrote, politically and intellectually, and because of the passage of time that has enabled us to reflect on such things.

Freud was at least partly scientized by the very age in which he wrote. The example of Titchener and Wundt also makes this clear. The standard of the day for translating texts denoted as scientific was to make them technical in the English idiom. This was fidelity. Jones was acting as cupbearer for larger realities as well as his own; without his "bulldoggery," Freudian ideas may never have achieved the influence they did. His placement of Freud alongside Darwin prepared the way for Strachey's image (his "English man of science"), and, taken at its word, would have put Freud back in the late 19th century or even before. Yet the truth of the Strachey translation is that it presents us with a much later Freud, historically suited to the early- to mid-20th century, the time in which Strachey himself wrote.

This truth gains in meaning when compared to the reception Freud received in his own culture and language. As pointed out by Decker (1977), the German response was far more complicated than either Freud or Jones ever admitted. Both men felt it important to proclaim the total denial and neglect of Freudian ideas early on, thus all the more elevating subsequent acceptance to the level of a near-religious conversion. From the beginning, however, Freud's books were reviewed in both medical and literary journals in Germany. They attracted mixed response in both quarters; some reviewers praised them, some called them relatively conventional, some rejected them outright. As a whole they gained little if any important support within the medical and psychological communities and thus had almost no effect on these disciplines. In part this was because—unlike their form in English—they could not benefit from Strachey's efforts. Freudian discourse stood naked before its German readers. Its "nonscientific" character, its suppositions, leaps of faith, indulgent speculations, and highly metaphorical character could not be disguised. Language, certainly, was not the whole problem, but it was undoubtedly an important factor. In Decker's words,

> [Freud] eschewed experimentation and was thus condemned as unscientific by physicians . . . Freud dared to give universal explanations, so his broad concept of sexuality was attacked. . . . Freud stressed the etiological significance of hidden forces and was denounced for his belief in the unconscious by virtually every experimental psychologist. . . . Freud's writing style cast discredit upon his conclusions because he did not limit himself to the dispassionate presentation of facts. (1977, p. 323)

Freud was thus generally seen as harking back to the prelaboratory period of *Naturphilosophie*. His use of such terms as "soul" (*Seele*) ran him fast aground on contemporary Germany psychology. Wundt himself had helped set the tone by stating that one ought to "reject immediately the meaning that naive, everyday linguistic usage attaches to "soul." . . . What is found here . . . is a metaphysical hypothesis" (1893, p. 10).[11]

Thus the professional response. But Decker and other writers are entirely wrong to say that because of these reactions, "psychoanalytic ideas did not become an intimate part of German culture," that they enjoyed, at most (before World War II), a brief and self-consuming flare of interest among "certain intellectuals" (1977, p. 322). Psychoanalysis proved to be an influence of wide scope among German artists, poets, and writers during the era between roughly 1917 and 1940. Its importance to expressionism, both in poetry and painting, is well documented. Among the more famous writers touched or scarred by Freudian ideas, meanwhile, were Arthur Schnitzler (a physician himself, who favorably reviewed Freud's translation of Charcot and whom Freud himself once referred to as his own "double"), Thomas Mann (who actually wrote an eloquent essay on Freud's essential place in German intellectual life), Hugo von Hofmannsthal, Carl Sternheim, Hermann Hesse, Franz Kafka, and Carl Zuckmayer—some of the greatest names in German literature of the early 20th century. Names, in short, that can hardly be considered to have resided on the "fringes" of *Deutscher Kultur*. Indeed, in Germany most of all, it was clear from the beginning that "the intellectual attractions of psychoanalysis [can be] quite independent of its therapeutic [medical] claims" (Robinson, 1993, p. 226).

Rejected by the science and philosophy of his native language, Freud was nonetheless embraced by its art, by its literary figures most of all. What Strachey helped achieved in English, perhaps, was to "save" Freud (the phenomenon) from this very same fate in Britain and America. By deleting the discourse of "soul" (which he translated as "mind" or "psyche"), by imposing a Greco-Latin lexicon, by evening out and standardizing the rest, Strachey broke many of Freud's more obvious ties to the prescientific era. One is inevitably led to wonder: what might have been the effect in Germany itself had it been possible in 1925 to translate Freud *back* into German from the *Standard Edition*?

And what of the search today for Freud's original meaning—*der Ur-Freud*, so to speak? This, too, should be given its own historical context. Hovering in much of the recent debate seems to be a sense that psychoanalysis needs to be saved, that it has weakened or failed in some essential way. As a possible truth of the human subject, it must be returned to its origin, which means Freud's original German. The suggestion is therefore that its truth (if it exists) in effect *belongs* to the German language. On the one hand, this begs a cognitive question; on the other, however, it poses the problem of motive. No doubt, for example, there is at this point in the 20th century an urging to revitalize something deep within psychoanalysis, to rediscover it at its wellspring: the comments of those such as Bettelheim, Ornston, Mahoney, Gilman, and others who call for a new translation are often as much directed at a dissatisfaction with the practice of the field as with Freud's representation in English. Those who attack this position in turn, meanwhile, claim that its "anti-scientific revisionism" is really an attempt to "threaten the very basis of psychoanalysis" (Wilson, 1987, p. 299). The stakes are thus very high, the highest in fact. At some level, one is arguing whether the field has any justification for being, whether, or to what degree, it is a fraud. It is well known that psychoanalysis has suffered a serious decline in prestige over the past 2 decades; it no longer enjoys its former eminence in psychiatry, whether academic or professional. The reasons for this are many and complex: new drug treatments; a vast number of new popular therapies; the advent of a more empirical, neurological and genetic psychiatry; and the general separation of institutional psychoanalysis from institutional medicine have all played their part (Marcus, 1984, pp. 256–264). These, admittedly, may be symptoms more than causes. But it is clear that in an era when the general level of madness, violence, abuse, and depression in society appears to be rising on all sides, and when there has been an increasing tendency both among the public and within the medical community to hope for biomechanistic solutions (e.g., drugs, gene therapies, biotechnology), the image of psychoanalysis as a fundamental personal and social benefactor has largely collapsed.

The search for an Ur-Freud must be seen against this background. Its deeper hunt for a post-Freudian beginning, a new point of brightness in a land of shadows, comes back to language as the last and final possession. Today, this is what it means to return to the "master." There is the sense that one can go no further, and must go this far. Whether it is an act of burial or of archaeology cannot yet be said.

But as an outsider looking in, I cannot help but feel both the required vanity of this effort and, in some part, its in-vain character. For the question of language has another context, too. Whether conducted in English or German, whether hopeful of excavating a pristine truth or of creating a new one, this search for the Ur-Freud is now an impossibility, for reasons I have already hinted at. This is simply (but not so simply, in fact) due to the reality that English and German have evolved since the time Freud wrote—and one of the powerful elements in this evolution has been the growth of the rich connotative universe that has come from Freudian language itself. The meaning of words such as *ego*, *transference*, and *personality* are not the same today as they were in Freud's day. Both within and without the psychoanalytic field, they have acquired a range of new associations, suggestions, theoretical glosses, and so forth that obviate any final semantic exhuming that might be applied to them. Just as one cannot return the word *repression* to its prefascist frame, one cannot eliminate a century's worth of use of Freudian discourse anymore than one can back-strip a word like *evolution* to the meaning it had in Darwin's day. Such an attempt would only pile on a new set of associations, not remove those already in place. Thus to try and delete *ego* and replace it with *I*, to effectively dissociate Freud from his most famous trademark, would be, ironically enough, to go back on the standard of obeying Freud's own use of everyday language (for what is more common today than this word?). Indeed, at the time Freud wrote, use of *das Ich* had a long tradition in German, as everyone knows, just as "ego" does today. Replacing it with "I" might be truer to one idea of "history" (as a source of origins), but would be false to another, as a lived and spoken reality.

This brings us back, full circle, to the issue of Freud's "scientific" versus "literary" essence. Determining such an essence, after all, or its heroic mixture, is currently also part of the search for the Ur-Freud. But here the desire to peel away history really involves putting a claim on Freud in terms of these two categories which, though divided today, were not so aggressively divergent at the turn of the century, as an author like Charcot reveals. Those who have recently spoken of Freud as uniting the resources of "scholar," "artist," and "scientist" (see, e.g., Laplanche, Cotet, and Bourguignon, 1992, p. 157) apply to him categories that had not yet completely come apart as individuated areas of discourse, in the days when Freud was a student. Less than a generation earlier, the field of authorship had been a much larger, encompassing landscape. In Darwin or Liebig's day, scientists, professors, and novelists often still exchanged stylistic urges and devices freely. Their triptych was still long in coming.[12] As we have seen, the new medical researchers such as Koch often employed an academic, semi-literary style of writing, while even the most ardent followers of the Helmholtz School regularly included the words of poets and philosophers as part of their science. Scientific discourse, as a whole, had charted its island course but had not yet arrived there in full. The issue of Freud's language, Freud "as a writer," is therefore much less what type of author Freud might be compared with today—meaning, a "today" projected backward—than what his correlatives might have been like in his own time, or even earlier. This, I would suggest, is more accurate as a type of historical search for origins and placements.

Immediately, for example, it makes us reconsider the standard Freud in terms of institutional preservation. It is very significant that the *Standard Edition* (1955–1974) in English, and both the *Gesammelte Werke* (1940–1968) and *Gesammelte Schriften* (1924–1934) in German, do not contain any of the early, neurological writings. Missing entirely is everything written prior to the early mid-1890s, the initial articles on hypnosis, hysteria, and neurosis. Only in the first obvious, traceable, steps toward psychoanalysis does Sigmund Freud appear, in standard form.[13] A full 15 years of work and writing, containing as many as 65 different, mainly technical pieces (Gray, 1948), are effectively censored due to their noninclusion. Even apart from the issue of translation, therefore, the official "Freud" is a fiction: Freud, the writer, does not publicly exist today prior to having become "Freud," the founder of psychoanalysis.[14] The problem is different from that of Freud's letters or personal papers: access to these has been made difficult (as is so often the case) by family matters, concerning issues of privacy and the like. Freud's neuroscientific works, however, were published and are in the public domain. Their deletion from the corpus therefore qualifies as the result of a deliberate "policy," at first that of Jones, but continued thereafter by the profession as a whole—making it seem as if "psychoanalysis arose without any antecedents, full-blown, as it were, from the forehead of Freud" (Gilman, 1991, p. 340). If this policy is itself, perhaps, linked to the long-standing Western tradition of turning important authors into mythic *summae auctores*—those whose "genius" allows or drives them to create great world-shaking systems unfettered by historical influences—it makes all the more pressing the question of how Freudian discourse, as a whole, evolved from the "science" of one era into a more uncategorizable type of writing thereafter. The question, moreover, is altered for this "Freud" in comparison to the "Newton" or "Darwin" or "Einstein" of past imagining, since Freudian influence has been so strong in exactly this area, the use and interpretation of language itself.

Recent interest in the problem of Freudian discourse seems to have brought this problem to the fore. Finally, an English edition of *The Complete Neuroscientific Works of Sigmund Freud* is being assembled, to appear in four volumes and to be published as a supplement to the *Standard Edition*. Along with the writings themselves, the editor (Mark Solms) will also reportedly be offering "extensive editorial commentary on the biographical context of the individual works, their neuroscientific merit, their historical importance, and the implications for psychoanalysis" (Ornston, 1992c, p. 100; for an example of such commentary, see Solms and Saling, 1986). Thus, it appears, the gap will soon be filled, but not without retaining the original absence. The title—*Complete Neuroscientific Works*—codifies the notion that these works form a separate whole from the "Freud" we know (and love). As an author, Freud will remain broken between institutional pieces that satisfy certain existing power relations and cultural images. Realistically, of course, this cannot be helped: the *Standard Edition*, as a document, is itself a completed work and cannot be simply opened up and added to (another reason, one might say, for its redoing). In the end, this problem is also part of the historical dimension of Freudian discourse. There is a distinct irony here: a later, controversial "science" (psychoanalysis), largely created by translation, has relegated an earlier and more straightforward "science" (Freud's neuroanatomical work) to obscurity. Indeed, the irony deepens when one considers that this later "science" calls back styles of discourse that predate the earlier "science." It is one of the abiding achievements of Freud's writing that it began well within the confines of institutional science and only later moved toward its edges, there to erect a discourse of far more ambiguous qualities.

V

To understand the historical position and importance of these qualities, we need to define them. We've seen how Freud's early style in the neurological works contained some literary and dramatic touches, yet remained wholly acceptable to the scientific community of the 1870s and 1880s. At this date, even in Germany, such elements, though now greatly reduced, were still a common, even expected part of technical style. By the 1920s, this was no longer the case, even in fields such as geology, where literary ingredients lingered somewhat more than elsewhere, due to the exploratory, reconnaissance nature of much field study (see Chapter 2). Freud's language in particular works, meanwhile, has been examined by many writers, especially of late (e.g., Mahoney, 1989; Laplanche, Cotet, and Bourguignon, 1992). My intent is to take up a text that, despite its great importance, has been overlooked by most of this analysis and that, through its theoretical strivings and its many eager calls on scientific imagery, provides one of the very best opportunities to compare Freud's writing to that of the science of his time.

This work is *The Ego and the Id* (*Das Ich und das Es*, 1923), in which the author first outlines his famous theory of the ego, the id, and the superego.[15] Strachey says in his introduction to the book that it "is the last of Freud's major theoretical works. It offers a description of the mind and its workings which is at first sight new and even revolutionary; and indeed all psycho-analytic writings that date from after its publication bear the unmistakable imprint of its effects" (1962, p. 4). In brief, the book represents an attempt at founding a scientific metapsychology by offering a series of fundamental principles for the explication of human psychic process and composition. The effort is not to be scholarly or artistic, but to outline a structural—that is, technical—portrait of the inner workings of the individual. Thus it is an important work, indeed, one of *the* most important if we consider its general impact on Western ideas of "mind." It is the critical paper in which Freud finally comes up with a "structural theory" that would thereafter serve as a justification for psychoanalytic work and thought. My analysis of it will be fairly simple and straightforward. I am less interested in such things as mimesis, narratology, or poetics than in the progress of Freud's images, his stand as an author, the relationship he seems to have to his own writing, and the scientific or unscientific qualities of his language, perceived in historical terms. I will not attempt to define all of Freud "as a writer," only what this one work appears to reveal along the lines just mentioned. A work such as *Das Ich und das Es,* which combines the most "technical" side of the mature Freud with the highest level of theoretical significance, can only be viewed as emblematic in some form. Nonetheless, the verbal pyrotechnics and inverted universal claims of deconstructive readings and the like I leave to those more interested in "Freud," the eternal phenomenon, than in Freud, a man who wrote a particular type of discourse, comprised of particular claims, at a particular time in Western culture.

To begin, the basic structure of the book seems logical, well thought out. In its first two chapters, *The Ego and the Id* summarizes existing ideas on consciousness, unconsciousness, and the ego (the "I"), identifies the need for new thinking on the matter, and then introduces the id ("It"); the third chapter is devoted to how the ego and superego develop (the latter having been postulated in earlier works by Freud), how a division in what was previously called the ego is required; the fourth chapter attempts to reconcile the new tripartite structure with "the two classes of instinct," Eros and Thanatos; and the fifth and final section examines the relations between ego, id, and superego. With the possible exception of reversing the last two sections, the whole would appear to offer a rational, straightforward development of the theory. Such, as we will soon see, was far from the case.

It is important to note that a revised and abbreviated version of the book appeared 10 years later in lecture form. This was included in Freud's *New Series of Introductory Lectures in Psychoanalysis (Neue Folge der Vorlesungen zur Einführung in die Psychoanalyse,* 1933), a collection aimed at a wide, though well-educated, audience. The book was patterned on the earlier *Introductory Lectures,* published in 1918 and highly successful in terms of sales. The *Neue Folge* was apparently intended to help raise money for the International Psychoanalytic Press, which had fallen on hard times. Yet its larger goal was obviously more than this. In seven, relatively brief essays, it gives an unparalleled overview of Freudian ideas in exceptionally concise, well-argued form. It is, in my opinion, among the most eloquent, tightly reasoned, well-structured, and persuasive texts that Freud ever composed, even including *Totem und Taboo* (which the author himself prized for its style).[16] These essays, that is, were not written hastily or haphazardly; their considerable grace argues that Freud hoped for them to have as great a missionary effect as his earlier *Introductory Lectures.* Writing popularized versions of his own work was, moreover, a way to help convince the public that psychoanalysis was indeed a "science": the 1920s and 1930s were the time when science writing came fully into its own as an expository genre, when many famous scientists, including Einstein, began composing books to explain their own ideas to the public. By following this path—which, in essence, reflects a particular stage in the growing inaccessibility of technical knowledge and discourse—Freud was himself "claiming" via genre that his work was scientific, too technical for a popular audience to comprehend without mediation.

Comparison between *The Ego and the Id* and its condensed version, *The Dissection of the Psychic Personality (Die Zerlegung der psychischen Persönlichkeit)* shows the latter to be so superior as an example of writing, so much more organized and controlled in terms of its argument, the order of its elements, its narrative logic, and the tactical variety of tone, that the earlier, longer work seems almost a hazy, often wordy, and unfocused type of raw material. Indeed, even the titles here are reflective of the difference: whereas *Die Zerlegung* perfectly expresses the aim of the theory, along with what might be termed its medical pretensions, *The Ego and the Id* leaves out altogether this goal of psychic dissection, as well as the critical third member of the trinity itself, the *Über-Ich* (superego). No doubt Freud was in better command of his material a decade later. Yet Walter Schönau's (1968) statement that, stylistically and intellectually, Freud was above all a lecturer, seems apt and persuasive. Perhaps, as with *Die Zerlegung*, it is when he is following in the direct literary footsteps of his teachers, Brücke, Meynert, and above all, Charcot, that Freud is more at ease and at home in his eloquence. Indeed, he never really abandons the lecture as a mode of speaking, one that was so central to 19th-century science. Even in a work such as *Das Ich und das Es*, its traces are incontrovertible, if not immediate.

Freud opens his text with an interesting warning: "In this introductory chapter there is nothing new to be said, and it cannot be avoided repeating what has often been said before" (*GW*, vol. 13, p. 239). This is not merely a disclaimer. It tells us that what follows should be considered established knowledge, firm enough so that the experienced reader can even skip to the next chapter, where something "new" will appear. It is an expression of authorial confidence, a promise that no surprises or problems will arise. Freud thus begins as a writer more than a speaker (lecturer). But things change immediately thereafter:

> If I could suppose that all those interested in psychology would read this work, I would also be prepared to accept that a portion of my readers would have stopped already at this point and would accompany me no further, for here is the first shibboleth of psychoanalysis. To most of those schooled in philosophy, the idea of something

psychical that is also not conscious is so incomprehensible that it must appear to them absurd and refutable simply on the basis of logic. I believe this arises only because they have never studied the related phenomena of hypnosis and dreams, which—except in cases of pathology—compel such an interpretation. Their psychology of consciousness, however, is incapable of solving the problems of dreams and hypnosis. (p. 239)

The reader is included here directly, as part of Freud's own insertion of himself into the narrative. There is a wink at the faithful, an admonishment to the yet-unbelieving. Several different possible readers make their appearance. A little further on Freud even engages in a bit of rhetorical dialogue with his philosophical opponents (a favorite device of his), putting words in their mouths ("Philosophers would indeed object: No, the term 'unconscious' is not applicable here . . . "; p. 240). He is therefore laying out and dividing up (dissecting?) the body of interested parties, giving them each a voice, assigning rational and irrational labels, and attempting to draw them all into the welcoming theater of his thought. He does this gently, not harshly, seeking to convince every reader of his reason and experience. But the point is that he does it, as a series of narrative moves, at a time in history (the 1920s) when such uses have already long passed out of scientific discourse generally. We have seen that even Freud's teachers, part of an earlier generation of scientists, were divided between using such techniques (Charcot) and not using them (Brücke). Thirty years later, no such division would have occurred; science, in terms of discourse, meant something very different from what it did in 1890. Even at the level of opinion, but especially at the level of "theory," the kind of writing that Freud was engaging in was no longer "scientific."

In the rest of his introductory chapter, Freud traces the reasoning by which psychoanalysis understands the repressed and its bearing on the unconscious. As part of this, he lays out his idea of mental processes as involving three basic "levels": the unconscious ("the repressed, which is dynamically unconscious"); the preconscious (what is "latent, but capable of being made conscious"); and the conscious. There is some confusion as to Freud's use of these forms: in one example, he states that the preconscious can be thought of only in a descriptive sense, not a "dynamic" one, whereas he elsewhere implies that the "dynamic" view of psychoanalysis is exactly what demands such a division of the unconscious into two parts (repressed and latent-consciousness). Indeed, one of Freud's foremost pupils, Sandor Ferenczi, confronted Freud in a letter with this confusion, and the latter, recoiling in humility, responded by stating "Your question . . . has positively horrified me . . . 'descriptive' and 'dynamic' have simply been transposed [in the text]" (see Strachey, 1962, pp. 60–63). Strachey himself, in an appendix to his translation of this work, attempts some fancy literary footwork to explain away the problem of "what Freud really meant." But it seems clear from a second reading that a good part of the difficulty resides in the ambiguity surrounding the term "dynamic," which is, after all, used in a metaphorical way. Strachey identifies it with the repressed only. Freud himself, however, attempts to dodge the issue in another way:

> Despite everything, we have accustomed ourselves rather well to this ambiguity of the unconscious and have managed adequately with it. As far as I can see, it is itself unavoidable; the distinction between conscious and unconscious is finally a question of perception, to be answered with a Yes or a No, and the act of perception itself provides no information regarding the basis on which something will have been perceived or not perceived. No one can complain that the dynamic has only an ambiguous expression in what appears externally. (p. 242)

The author therefore sees the difficulty as resulting from expectations of certainty. The "dynamic" is real, and part of its reality is that it is expressed ambiguously. But the idea of the "dynamic" itself—especially the "dynamic of soul," *die seelische Dynamik*—has been given in an openly metaphoric manner that leaves it imprecise no matter what might follow. "We have arrived at the term or concept of the unconscious by another road, through working with experiences in which the soul's *dynamic* [*seelische Dynamik*] plays a role. We have realized, that is, we have been forced to accept, that there are very powerful processes and representations in the soul—here a quantitative, therefore economic factor arises for the first time—that can contain all the effects for the life of the soul that ordinary representations can . . . only that they themselves do not become conscious" (p. 240).

The "dynamic" therefore resides in the concept of powerful, invisible processes and ideas at work in the "soul" (mind). This, of course, bears some connection with older medical notions of "vital forces" and shares much, as well, with similar ideas in literature, Romantic literature most of all. But here, Freud must in some part divest himself terminologically of this inheritance, viewed as so antithetical to the science of his day, and thus he invokes, simply but figuratively, such words as "quantitative" and "economic," both of which add a further updated, technical gloss to "dynamic" yet leave it no more precisely defined than before (in his translation, Strachey puts the word "economic" in italics). A simple definition would no doubt have gone a long way to clarifying the "ambiguity" problem—indeed, this would have been easy: in physics, the term "dynamics" refers to the causes of motion of separate bodies, and thus deals with the concept of forces and their relationships. But the metaphor stands as a metaphor. Freud refuses any direct call upon Helmholtz. He seems to prefer the grace of evocation.

With regard to style, this chapter, and much of the first half of the book, is restrained, expository, full of other technical images of more than one kind:

> We have formed the idea of a coherent organization of processes in the soul of each person and call this his or her "I." To this "I" is attached one's consciousness; it controls the access to motility: that is, the discharge of excitations to the outside world; it is the same court of the soul [*seelische Instanz*] which exerts control over all its constituent processes and which goes to sleep at night yet never ceases to administer censorship over dreams. (p. 243)

A bit further on, regarding the ego ("I") again, the author shifts back from this blending of biological and juridical imagery to a more topographic/mechanical scheme:

> We have said that consciousness is the *surface* of the soul's apparatus; that is, we have attributed to it a function within a system, which is spatially the first encountered from the external world. . . . How, then, do we account for those inner processes that we might group— roughly and inexactly—together as thought-processes? Do they rise up, from somewhere within the inner apparatus, as displacements of the soul's energy on its path to final action, reaching the surface and thereby giving rise to consciousness? Or does consciousness come to them? We note that this is one of the difficulties which arise when one takes the spatial, topical idea of events in the soul seriously. (p. 247)

In the second and third chapters, meanwhile, Freud persists in talking of consciousness and the ego in terms reminiscent of chemistry, respectively as the "residues" (or remains) of memory and as a "precipitate of object-incorporations [cathexes, *Objektbesetzungen*] that have been given up" (p. 257). But this is not the last of it. The ego is also treated in medical terms, as analogous to anatomical phenomena: the perceptual system is said to form its surface just as the "germinal disc sits upon the ovum" (p. 251); and a bit farther on, the ego is said to resemble the "cortical hommunculus of the anatomists" (p. 254).

For the most part, therefore, the ego is given to us through a heady olio of scientific metaphors—physical, chemical, biological, mechanical, and medical. Its lawyerly aspects, one might say, are more or less an extension of this cosmos of images focusing on material objects and on process. This cosmos is not entirely a stable one, strictly speaking; making the ego both a "precipitate" and a kind of "ovum" is not wholly conducive to technical rigor, even in its imaginative flights of theory. Freud's own hydra of metaphors makes one suspect he is himself, here as elsewhere, on a kind of exploratory, speculating odyssey (this point has been made by many other authors as well, e.g., Mahoney, 1987), whose Ithaca is not yet known. It seems as if he were guiding himself toward a central idea, lurking among the many heads of his images. The ego, that is, seems to be something so material in substance and meaning that it can draw to itself all the sciences of the world.

🖪🖪

The id, meanwhile, is something else again, and calls upon a different array of representations. Indeed, so powerful are its emanations that the ego itself is profoundly altered whenever brought into narrative contact with this "lower" entity. Here is how Freud introduces the id ("It"):

> Now, I think we shall gain great advantage if we follow the suggestion of an author who maintains, in vain and out of personal motives, that he has nothing to do with a rigorous, higher science. I refer to G. Groddeck, who constantly emphasizes that what we call the "I" behaves essentially passively in life, that, in his expression, we are "lived" by unknown, uncontrollable forces. We have all been subject to these same impressions, even if they have not overwhelmed us to the exclusion of others, and we should not despair of designating a place for Groddeck's insight within the framework of science. I propose we account for it by calling that entity [*Wesen*] which arises out of the perceptual system and next becomes preconscious the "I," and the other psychic substance, into which the "I" extends and which behaves as if it were unconscious, the "It," after Groddeck's usage. (p. 251)

Suddenly, as it were, the ego is weakened into re-definition. It becomes passive, worthy of a more minor role in the scheme of the psyche. The references to Georg Groddeck (a physician-writer, sympathetic to psychoanalysis) are entirely literary in style, even conversational. They hardly obey the rituals of clipped citation (name, date, note at bottom or end of text) that had by this time become established for scientific articles. Freud gives place in his narrative to another writer's idea, which he discusses not in the mode of a "rigorous, higher science," but instead with appreciative, even friendly deference. This writer enters and displaces the author for a moment. Freud also draws on us, the audience, to maintain control ("We have all been subject to these same impressions. . . ."). But the result is that his earlier scientific stance, along with his characterizations of the ego, must now be largely forgotten, or at least overprinted. The nonscientific, brought in here with the id, is not so much transformed into science and ego but instead works its own transformation upon these.

This passage is followed by a drawing that attempts to offer, in diagrammatic form, the various positions and relationships between the ego, id, the preconscious, the repressed, and the perceptual system. One might say that this picture is vaguely anatomical, except that all labels are placed inside, whereas in the case of medical drawings they are kept outside for maximal visibility. Freud's sketch, however, is not a technical illustration; its parts seem half like organs, half like design elements, and are thus much too purely schematic and at the same time are not abstract enough to serve as a formalized image for demonstrating mental processes. If portrayed as a chart or flow diagram, perhaps, instead of in this half-representational manner, the figure might approach being such an illustration. But this is not what Freud asked of it. He calls it merely a "drawing" and asks that "no special meaning" be given to it.[17] It is mostly an attempt on his part to try and keep all the pieces of his narrative straight, to collect them in some kind of order, and to do this in an immediate, visual way—all the better to halt or freeze the overweening complexity he has set loose. He lets us know this in effect, with self-deprecating humor (something else that has largely disappeared from scientific discourse), when he notes he has inadvertently drawn the ego as wearing a "hearing cap" ("*Hörkappe*," i.e., this being the acoustic portion of the perceptual system), after the example of cerebral anatomy, except that this cap sits clumsily on the ego with a slant (like a beret?).

The drawing, then, seems a sign of Freud's own sense that things have started to veer away from clarity and control. And indeed, it is in the next few paragraphs that he first begins to call upon a set of new, dramatic, and highly suggestive and also confusing metaphors. These, as we shall see, extend far beyond the technical analogies he has employed to this point for the ego. What eventually ensues by the end of the work, in fact, is a kind of imagistic battle for influence, by which the ego, already bereft of its initial materiality, becomes progressively engulfed in an ever more wild jungle of aggressive figures, nearly all having to do with the id. Some hint of the coming storm can be seen already here, in the second chapter, when Freud writes that "the 'I' is that part of the 'It' which . . . strives to bring the influence of the external world to bear . . . endeavoring to put the reality principle in place of the pleasure principle, which rules without limit in the 'It.' . . . The ego represents what one calls reason and discretion [Strachey has "common sense"!], in contrast to the 'It,' which contains the passions" (p. 252). Following this, a striking image appears, one of Freud's most famous:

> The functional importance of the I is expressed in the fact that in its normal state . . . its relation to the It [is like that of] the rider who needs to rein the superior strength of his horse, with the difference that the rider attempts to do this with his own strength, the I with what is borrowed. The comparison can be carried a bit further. Just as the rider, if he does not want to be separated from the horse, is often left with no choice but to lead it where it wants to go, so is the "I" given to converting into action the "It"'s will, as if this will were its own. (p. 253)

The image is both visual and experiential, not cerebral—one thinks (perhaps) of Charcot, pointer in hand. Yet, what is happening to the ego, narratively speaking? As a metaphoric and a rational center, a core of scientific imagery, the ego is beginning to lose its substance. It is now locked in a struggle. At one point it is said to be the seat of "reason and discretion," yet a moment later it reappears, riding the wild id, who forces it into fears of separation, into parasitic relations, into a general secondary position. Again, that is, the id has begun to take over.

In the third chapter, this type of usage grows and subsides. This is where the superego is introduced by means of a long and abstract (and not altogether clear) discussion of how the ego is formed as a "precipitate" of abandoned object-incorporations. Freud does not talk about the origin of the idea of a superego; this, he says, was done elsewhere (e.g. in the work *Massenpsychologie und Ich-Analyse*, 1921, translated as *Group psychology and the analysis of the ego;* the term "mass psychology" is probably more accurate however). His main concern, rather, is to explain a new idea, namely that "this part of the I has a less firm connection with consciousness" (p. 256). This he does, of course, by surmising that the superego "is not simply a residuum of the first object-choices of the It [i.e., the parents], but also has the meaning of an energetic reaction-formation against these" (i.e., the Oedipus-complex] (p. 262). And further: "Its relation to the I is not exhausted in the warning: You *should* be and do this (like the father);

it also includes the prohibition: You *may not* be or do this (like the father), meaning you cannot do all that he does; some things remain reserved for him" (p. 262). Freud therefore makes of this third entity, the superego, a speaking (commanding) precipitate, a mixture of the ego's materiality and the id's animacy. Such organicism cannot but lead Freud to the portals of biology. And this is exactly what happens, for the discussion now turns to the "genesis of the superego" given in the terms of evolution and inheritance. The author claims the importance of "two most significant biological factors, the long period of childish helplessness and dependence, and the fact of the Oedipus Complex" (p. 263). Repeating a hypothesis made by Ferenczi, he describes the latency period as "an evolutionary inheritance imposed on human culture by the ice age," an idea whose acceptance argues that "the separating off of the ideal-I from the I is not a matter of chance [but] represents the most significant characteristics of the evolution of the individual and of the species" (p. 263).

This is all supposition and speculation, and is left unsupported by any evidence. Such does *not* make it unscientific, however; technical writing then and now remains often wedded to the "house of ifs" approach to theory building. What weakens any legitimacy of true science is the ice age idea itself, a form of environmental determinism (or exaggerated Lamarckism) that, by the 1920s, with the advent of genetic research, qualified as biological bunk. Not content with invoking an older, abandoned phase in technical thought, however, Freud goes yet further:

> What biology and the fate of the human species have created and left behind in the It are, through formation of the ideal-I, taken possession of by the I and reexperienced individually by it. The ideal-I, as a consequence of its history of formation, has the most abundant linkages with the phylogenetic acquirement, the archaic inheritance, of the individual. What belongs to the deepest part of the individual life of the soul becomes through the formation of the ideal-I the highest part of the human soul [*Menschenseele*] in the sense of our values. (pp. 264–265)

Scientific imagery is here blended into a type of language that verges on the mystical. Again, there is no explanation of what "phylogeny" might mean in the context of "soul," other than the phrase "archaic inheritance," which simply provides a kind of literary synonym. Freud is waxing in a fashion Jung and Neumann would perfect, in their own particular idiom, therefore in a manner that goes even beyond the hypothetical.

This entire evolutionary scheme reaches an even more heightened crescendo in the last two paragraphs of the chapter, where the id is finally brought into play again:

> We must grant the differentiation of the I and It not only in primitive humanity but even to much simpler forms of life, for it is the necessary expression of the influence of the external world. . . . It is here that the chasm between the real individual and the concept of species comes into view. . . . The experiences of the I at first seem lost to inheritance; yet when they repeat themselves frequently and strongly enough in numerous individuals, one after the other in successive generations, they transform themselves, so to speak, into experiences of the It, the impressions of which are retained through heredity. In this way does the It, which can be inherited, lodge in itself the remains of the existences of innumerable I's, and when the I creates its I-ideal out of the It, so may there arise only older forms of the I, given a resurrection. . . . The struggle that had raged at deeper levels and was not resolved by hasty sublimation and identification now continues in a higher region, as in Kaulbach's painting of the Battle of the Huns. (pp. 266–267)

As before, it is not the gingerbread logic of "if's," "could be's," and "why not's" that strike one so much in this passage. True, these do climb ever higher until even religious heights are attained ("resurrection"). But they still do not form the more remarkable aspect, which, as I have been saying, is the actual "science" called upon. Clearly, it is impossible to understand in exactly what sense Freud intends the concept of species inheritance here. How can the id be "inherited"? How, on the other hand, are "experiences"—which are only "so to speak" experiences—inheritable? Freud yields himself to a few genetic terms without any discussion of how the concepts involved might be applied (the id in particular is proposed as if commensurate with the human gene pool itself). Then there is the question of the recycling of past egos, something that bears closer affinity to the Oriental idea of karma and rebirth than it does to Lazarus or Jesus.

What kind of "science" is this? One should ask not out of incredulity or dismissal but instead with fascination and in response to decades of assumptions about "psychoanalytic science." Obviously, it is not the type of science then being written by chemists, physicists, or biologists in Europe and America. It is a discourse that must be made "scientific" by powers outside itself, by needs and ambitions beyond the page. Reading it a second time, one sees that the entire passage is really based on metaphor—on analogies that persist without explanation. In this book, even more than in other major writings of the same period (e.g., *Beyond* the Pleasure Principle), Freud has a tendency to swing in and out of this speculative–mystical intellectual plane, which exceeds the theoretical. At the beginning of the next chapter (Chapter 4), in which he discusses his "two types of instincts," he picks it up again, and though with less ardor and in more measured tones, applies it even well beyond the limits of the human.

These two instincts or drives (*Trieben*), of course, are Eros and the death instinct (Thanatos), which he has taken up in earlier writings. Freud describes them here as follows: "On the basis of theoretical considerations supported by biology, we surmise a death instinct, which has been given the task of leading organic life back into the lifeless state, while Eros, by effecting an ever-more far-reaching combination of the particles into which living substance is dispersed, pursues the objective of making life more complex and thereby, of course, preserving it" (pp. 268–269). Riding this logic further, Freud concludes:

> As a consequence of the binding together of elementary unicellular organisms into multicellular life forms, the death instinct of the single cell could succeed in being neutralized and the destructive impulses diverted into the external world by the agency of a special organ. This organ would be the musculature, and the death instinct would express itself—probably, however, only in part—as an *instinct of destruction* against the external world. (p. 269; italics in the original; not given in the Strachey translation)

One cannot but be struck by the ingenuity of these statements. The speculative house that Freud builds is full of many ornamented rooms. But one sees that the architecture of his reasoning remains airy, a series of assumptions, one afloat upon the other. Strictly speaking, we are back in the halls of half-mystical images. But by saying "half-mystical," I am speaking in a certain historical sense. Freud's notion of a single cell having instincts or drives is a holdover from the days when theories of vitalism were current, when organisms were spoken of as the physical manifestations of some "active principle" or even "self-will," rather than as physicochemical entities. In this brief passage he seems to be also suggesting that the physiology of higher life forms is at least partly the result of an "instinct" whose essence is to demand its own "expression" in organic form. Such an "instinct" is like a cosmological principle that precedes and supersedes life, using it for its own needs, forcing aggression upon it well beyond any Darwinian urges for survival.

Freud thus makes a call upon *Naturphilosophie*. Though vitalism had its powerful proponents as late as the 1880s (above all, Rudolph

Virchow), the entire weight of the Helmholtz School, and the general drift of Western science as a whole, had been thrown against it. By the 1920s, genetic explanations were the rule; the body's structure was viewed as a result of adaptation. The days were long gone when unaccountable forces, touched by teleology, directed the substance and movement of life. Calling Freud a "biologist of the mind" (Sulloway, 1979), is therefore to invoke, like it or not, a distinctly, historically defunct biology.

⑤⑤

From this point, the chapter follows an interesting course, and re-peats a pattern I have noted before. Introducing Eros and the death in-stinct, Freud goes on to speak of them in terms of "a great domain of facts" having to do with human behavior and the psyche. He discusses, briefly, such things as epilepsy ("we surmise that the epileptic fit is a product and sign of a mixing of these instincts"), homosexuality, and paranoia, and the change of love into hate, before going on to the ego and id again. Turning to the id once more, Freud claims that it some-times shows a "strange indifference" toward the objects of its erotic choices, even, in neurotic states, aiming its "revenge" against the "wrong persons." At this point, he inserts a joke: "This behavior of the uncon-scious inevitably makes one think of the well-known comic anecdote, in which one of the three village tailors had to be hanged because the only village blacksmith committed a capital offense. There must be punish-ment, even if it doesn't fall on the guilty" (p. 274).

This development in Freud's narrative shows another shift away from the more academic, and at times technical, style in use up to this point. Speaking of the id seems to have drawn the author once more in a different direction. From now on, in fact, Freud's ever more tenuous hold on resemblances to contemporary science—which he continues to implore through such terms as "energy," "neutral," "discharge," and so on—begins to snap, strand by strand. But a half-page later, images of violence and subjugation intrude. The ego is described as trying to "take possession" of the id's libido, to "force itself upon the It as a love-ob-ject" (p. 275). The death instinct, due to its relative invisibility, "is es-sentially mute," whereas "the clamor of life" comes mainly from Eros— "And from the struggle against Eros!" (p. 275). The exclamation point here, by itself, speaks a volume of historical distance from the contem-porary "scientific." Such an outburst, such a display of emotional at-tachment to the narrative itself, defines another mode of expression that had been largely banished from technical prose by this time. Here, in Freud, it is clearly more than a matter of rhetoric. It is importunate, a tug on the reader's sleeve ("let us not forget!"). Coming in the first line of the chapter's final, summarizing paragraph, it puts back before us Freud, the speaking person, the involved lecturer, looking out over his audience.

VI

In the fifth and final chapter of the book ("The Dependencies of the I"),[18] the imagery of violence and power takes over completely. The writing starts off reserved, yet with a very interesting humility: "The interwoven nature of the material [in this book] should excuse the fact that none of the headings entirely corresponds to the content of its chapter, and that we are always reaching back to matters already settled when we desire to study new connections" (p. 277).[19] Freud seems aware that his discourse has led him, if not really astray, then in directions he had not really planned or intended. The qualifier he has inserted here, at the very beginning of his final section, provides a total contrast to the declaration with which he began the book. Instead of expressing confidence, the book approaches its closure confessing something of the opposite. Somewhere along the way, in endeavoring upon "what is new," the author has fallen from grace: the rider has lost command of the horse.

Very soon, we are told that the superego, which derives its energy from the id, has "the capability to set itself against the I and to master it. It is a memorial to the former weakness and dependency of the I and it pursues its domination over the mature I as well. As the child stands under the compulsion to obey its parents, so does the I subject itself to the categorical imperative of its I-ideal" (pp. 277–278). A page later, and such imagery becomes yet more insistent. Freud is now speaking about obsessive neurosis and melancholia (depression), where the "feeling of guilt is excessively conscious; the I-ideal shows a unique severity and rages against the I, often in a cruel manner" (p. 280). In melancholia, particularly, Freud's sense is that "the I-ideal has seized hold of consciousness. But the I dares make no objection, instead admitting itself guilty and submitting itself to punishment" (p. 281). With regard to hysteria, on the other hand, "We know that the I ordinarily carries out repressions in the service and under the command of the I-ideal; here, however, is a case where it turns the same weapon against its harsh overlord" (p. 281).

The metaphorical frame of Freud's language in this last chapter has begun to gain focus. It is not merely a frame of violence and power but one of *state* violence and power, ruled entirely by military and arbitrary legal forces that are directed, in one form or another, against the ego (one can see, immediately, where Kafka may have acquired his ideas for such works as *The Trial* or "In the Penal Colony"). Indeed, in discussing melancholia and neurosis, Freud takes such imagery even further. In both illnesses, he says, the superego is "pitiless," because "what now reigns in the I-ideal is akin to a pure culture of the death instinct, and in truth it often succeeds well enough in driving the I toward death, if the I does not defend itself beforehand against its tyrant by a sudden change into mania" (p. 283). The id and the superego are both described as "cruel," the ego as "innocent." The id is "amoral," the ego "strives to be moral," and the superego can be "hyper-moral, and thus just as cruel as the It" (p. 284). In fact, it is exactly when Freud comes up with the term "innocent" for the ego that he says "our ideas about the I are [finally] beginning to clear themselves." He is right; this is his authorial sign that he has awakened to the central imagery he has been developing all along, for he then offers a simile that tries to soften everything: "With regard to its behavior, the I has something of the position of a constitutional monarch, without whose sanction nothing legal can occur, but who must consider very carefully before extending its veto against a proposal brought by Parliament" (p. 285). This is not the type of governmental or judicial state he has offered earlier on. It is in the Weimar mold, the mold of 1920s Germany, and it is not to last. Only a paragraph later and one happens upon this striking sentence: "Psychoanalysis is a tool which should render the I capable of a progressive conquest of the It" (p. 286). A mere few lines more, and, in a summation remarkable for its unleashed metaphorical energy, the reign of terror returns:

> Yet on the other hand we see this same I as a poor thing which stands under a triple servitude and, as a result, suffers under the threats from three dangers, from the external world, from the libido of the It, and from the harshness of the I-ideal. . . . As a border-entity, the I wishes to mediate between the world and the It, to make the It accommodate to the world and the world, by means of its muscular action, accord with the It's desires. The I behaves, actually, like the physician in an analytic cure, in that it offers itself, with its concern for the real world, as a libido object to the It and wants to attach itself to this libido. It is not only the helper of the It, but also its submissive servant, who seeks the love of his master. . . . In its position midway between the It and reality, it only too often gives in to

the temptation to become obsequious, opportunistic, and lying, rather like the politician, who, though possessing good judgment, seeks to maintain himself in the favor of public opinion. . . .

That the I's work of sublimation has, as its consequence, the unmixing of the instincts and the release of the aggressive instinct in the I-ideal, it delivers itself over . . . to the danger of maltreatment and death. When it suffers or even gives in under the aggression of the I-ideal, its fate is a counterpart to that of the protozoa, which are destroyed by the very products of decomposition that they themselves have created. (pp. 286–287)

Freud pours forth this mixture of statement, description, and metaphor with gathering clarity and enthusiasm. The ego, in particular, is rendered into an excited flood of images, a struggle without end. Its nature is that it is constantly in danger, constantly under threat both from the id and the superego, which can be not merely violent but seductive, capricious, and toxic. Whether as a doctor, a politician, or a protozoa, the ego is torn between forces beyond its control. Its situation—the situation of "morality" and "innocence"—seems embattled. Gone is the balance of powers, the Weimar interlude. If still a monarch, the ego is much closer to Charles I, with his nervous disabilities and his failure to rule, standing before a Parliament eager and ready to condemn him, sharpening the blade off-stage.

I take these images of violence to be the mark of a deep sympathy, a tragic compassion. Freud is drawing a picture of his own clinical experiences, his own witnessing of the struggles of his patients, and his narrative seems the result of his own struggle to project within these sufferers, and to then see evinced a "voice" of comprehension and understanding. It is this sympathy, normally quieted by the act of authorship, that at least partly directs his deepening attachment to the writing and that helps direct its searching, groping qualities. Certainly the metaphors in the passage just cited lack a total unity. Yet, at the same time, they are all representations of life being placed in dire servitude or peril—the analyst who gives himself up; the politician as slave to public opinion; the single-cell organism that kills itself in its blind acts of life. There is both despair and hope in such images. Such are the hallmarks, Charcot once said, of the medical mind. In this connection, one should note, too, that there are strong echoes in Freud's imagery to medical discourse itself, which by this time had become deeply immersed in military imagery (see Chapter 3). Freud's language overlaps this, but it is more specific too; in medicine, the ideas of "attack" and "defense" were central and involved "exterior agents" such as bacteria, viruses, and the like. Here, in Freud (who no doubt encountered this discourse in his own training), the agencies are all internal, native. The link therefore remains, as elsewhere, connotative.

In any case, if we follow the role of these metaphors in the larger text, they cease to hold, and quickly become contradictory to what was said earlier on. Does the analyst—who only a moment before was enlisted in the conquest over the id—really act as a "submissive servant," seeking "the love of his master"? Can the ego be called "moral" or "innocent" if it acts like a cunning politician, whose scruples involve "lying" and "opportunism"? Finally, does the ego, once upon a time comparable to the ovum, now reducible to a protozoa, follow a biochemical law of self-destruction instead of reproduction? Taken an inch beyond its surface appeal, Freud's discourse logically flies apart. But, to borrow a famous image from *Die Zerlegung*, it flies apart like a crystal thrown to the ground, along preexisting planes of meaning. One plane is the author's impulse to rely upon figuration as a means of searching out what he wants to say, a process that is often intuitive, not entirely consistent. Another cleavage is the conceit of violence, which ends up opposing almost everything positive claimed in the name of the ego. Finally, there is the compulsion on Freud's part to personify his three agencies, to treat them as living creatures, to such a point that he can no longer discuss them at all except in this completely analogical fashion, as if the analogy no longer existed but had become literal truth. All these surfaces of "weakness" have been present from the beginning of the book. It is their transformation into actual planes of breakage that we see in the above quotations. The force applied to them is Freud's own struggling attempt to come up with an adequate description of the ego and its relations. In the end, he still seems dissatisfied; the search for an adequate, standard image has retrieved only fragments.

The book ends with the id, as one feels it must. The id, after all, which propelled or inspired Freud to "lose control" and venture ever further into the realm of figuration, remains even more a resistant formation than the ego. It cannot even be described in the type of overt metaphors offered above. "The It," Freud writes, with projected sympathy again, "has no means to declare love or hate toward the I. It cannot say what it wants; it has brought about no unitary will. Eros and the death instinct struggle within it; we have mentioned with what means one instinct answers the call to arms against the other. We could portray things as if the It, held under the domination of the mute but powerful death instinct, might find a degree of peace and would desire to pacify the mischief-maker Eros at a signal from the pleasure principle, but we are apprehensive lest thereby the role of Eros be undervalued" (p. 289). Just as the id "cannot say what it wants," Freud cannot, finally, subdue it as a narrative subject.[20]

Though it begins with boredom, the book closes with anxiety. From "nothing new to be said," it progresses to a fear of not having said enough. From metaphors and images derived from science, it tumbles into those that call upon war. From the ego and its "coherent organization of processes in the soul," it descends to the id as having "no unified will" or ability to speak. Moreover, from the ego as the "court of the soul which exerts control over all its constituent processes," it declines to the ego as being "in the service and under the command of the I-ideal," and still more, "a poor thing," a "border entity" standing "under a triple servitude."

Together, then, the mind, its metaphors, and Freud's language in general all undergo a radical change from the first chapter of this work to the last. It is a change that skids away from the "scientific," or its attempted pretense, into an almost purely analogical domain, yet which looks back every so often, hoping to find remnants of its earlier ambition. This move is gradual at first, but becomes ever more rapid and irreversible. Wherever it draws the id directly into its reach, it receives an "impulse" to go further. Freud meanwhile, stumbling upon images of power (the horse-and-rider image first of all, those of violence later on), finds he cannot escape them. He must play them out, follow their logic, until he has so hyperanimated his three divisions of the psyche, so imbued them with fervid intentionality and conflict, that he seems able to pull away from them, from the book itself, only by an act of admitted failure or angst. It is not the ego, not the "conquest of the id," that finishes the book. Freud's discourse does not act as a mimesis for his larger hope and glory, the mission he has set for psychoanalysis. That final "but we are apprehensive" even has a despairing ring to it. In a strange, eerie, almost frightening way, the progress of Freud's narrative in *Das Ich und das Es* has given us a portent of 20th-century history: beginning with a picture of calm relations, welded by "science," this narrative plunges ever deeper into a swirl of endless strife, of aggressive drives and combative relations that bear one along in an urgent and seductive exile of logic toward some destined uncertainty. In this discourse, too, the Weimarian interlude is decisively short, the transfer of violence both terminable and interminable. We are left, at the very end, with a fateful struggle and with an attempt to comprehend it that may well be illusory. Just as the present century has thrown chilling doubt on the ancient belief that nation-states exist to make peace among peoples, so does this writing of Freud's disavow, in some part, the claim that psychoanalysis is not a matter of trained intuition but instead is essentially a "science" about restoring harmony and order, via language, an entirely rational means to deal with irrational forces.

Conclusion

Lacan noted some time ago, and critics like Ricoeur have repeated, that the question of language vis-à-vis Freud is central far beyond matters of style or expression. Before Freud, the somatic conception of neurosis and mental illness generally led to therapies that were speechless: quiet removal, rest, pills, massage, active restraint, denial, even punishment. It was Freud who placed the most human element of all at the center of everything, who therefore transformed the patient into a literary entity of sorts.

It is therefore all the more interesting to see Freud himself struggling to achieve adequate control. I say this not at all out of disparagement or frustration but out of a sense of what might be called realism. The innumerable superlatives lavished upon Freud seem to me often excessive and uncritical. Mahoney (1987) has advised us to appreciate "that Freud's writing *produces* knowledge rather than merely describ[es] it," a point he feels cannot be emphasized too strongly. But what, exactly, does this mean? How does a particular writing *create* knowledge, rather than (or in addition to), say, *contain* or *transfer* or merely *claim* it? I take Mahoney to mean that Freud, in many of his works, tries to pursue certain points in order to derive fresh conclusions, new insights, unseen implications and assumptions. This is certainly true; as we have seen above, Freud can be very persistent in following his own logic even into the irrational. But is this what we would call "knowledge"? Semantic problems rush in. Do Freud's texts, then, actively generate new understanding beyond the limits of the said? Do they do the epistemological work of training, of experience? Do they stand in for our own efforts at learning, or impose that learning upon us? Do they finally turn us, as readers, into psychoanalysts? In the end, Mahoney's appeal, it seems to me, simply poses another problem. A work like *The Ego and the Id*, as it goes along, makes a series of speculations, hypotheses, and, most of all, claims. Claims making, however, especially the type Freud employs, is not the same thing as knowledge production. Merely saying that "biology teaches us . . . " does not make it so, particularly when the biology being spoken of is no longer itself viewed as a legitimate part of scientific understanding.

With Mahoney and other skilled writers on Freud, I heartily agree on another point, namely, that Freud had no real contemporaries. He was powerfully, commandingly unique, a revolutionary both in his ideas and his discourse. But I would say, in terms of the latter, that his uniqueness stemmed from a kind of magnificent regression, as the image of Strachey's "English gentleman" suggests. Freud's corpus, his "science," is an expression of 20th-century ideas clothed in an early to mid-19th-century form. Among European scientists, those who employed a prose more closely allied to his were not Meynert or Wundt (as we have seen), nor Planck or Einstein, nor even, in the end, Charcot. They are more kin to Darwin and Lyell, a still older breed, who wrote at a time and in a field where materialist–determinist ideas had begun to reign but only with a tentative power and rarely as a guide to every expression. Strachey's gentleman, it seems to me, needs to be moved a few decades backward. He needs to be placed in the era when it was still common for scientific authors to weave their readers directly into their writing; to employ many types of literary technique; to be dramatic and metaphorical, and to admit as much; to employ humor and humility; to anthropomorphize forces or objects to a high degree; and to discuss on a regular basis such subjects as philosophy, history, literature, and art. Such was the era when scientists were, so to speak, among the most learned, cultivated, and eloquent writers of the time. In this and other things, Freud might even be said to bear certain affinities to a "geologist of the mind."

Indeed, beyond matters of style and expression, important analogies, possibly influences, exist between 19th-century ideas of earth history and Freudian concepts of the "layered" self. This has been pointed out before (though always in cursory fashion): Freud often used the term *Schichten* in referring to the mind, a term meaning "levels" or "layers," but as a technical word, "strata" (which is probably how he meant it, given his reading in archaeology and anthropology), this being how Strachey faithfully renders it nearly everywhere it appears. Biology may have been the far more frequent source of imagery for Freud, yet 19th-century biology, especially as affected by Darwin, was closely bound up with geology. Evolutionism, which permeated late-19th-century thought in so many areas, including history and politics, and was an undoubted *Anschauung* embedded in Freud's own training (see Ritvo, 1989), represented a fertile merging of these two disciplines, one that had existed for some time, at least from Lamarck onward. The idea of "struggle," so central to Freud's mental anatomy, long preceded Darwin and Spencer. Indeed, it was a concept no less ambient to the late 18th and early 19th centuries, codified in the theories of those such as Adam Smith and Thomas Malthus, in the philosophies of Hegel and Marx, and in a number of other manifestations. But in geology, early on, it takes a particular form:

> We defined geology to be the science which investigates the former changes that have taken place in the organic, as well as in the inorganic kingdoms of nature; and we now proceed to inquire what changes are now in progress in both these departments. Vicissitudes in the inorganic world are most apparent, and . . . they may claim our first consideration. We may divide the great agents of change in the inorganic world into two principal classes, the aqueous and the igneous. . . . Both these classes are instruments of decay as well as of reproduction; but they may be also regarded as antagonist forces. The *aqueous* agents are incessantly labouring to reduce the inequalities of the earth's surface to a level, while the *igneous*, on the other hand, are equally active in restoring the unevenness of the external crust, partly by heaping up new matter in certain localities, and partly by depressing one portion, and forcing out another of the earth's envelope. (Lyell, 1830/1990, p. 167)

Here is a type of imagery that would have served Freud extremely well. Geology was not the science to which Freud wished to try and attach his own, but it is as much geology to which some of his language and concepts return. Can we even read in this type of passage an anticipation of Freud's own antagonistic "dynamics of the soul"? But the larger point, no doubt, is how style and content necessitate one another here. Lyell is not a model for Freud, strictly speaking (Freud probably never read him); he is, instead, a kind of historical prototype, a stylistic colleague. According to his own testimony, and that of so many others who knew him, Freud sought to imitate Goethe in his writing (and when did Goethe write?). Yet it seems to me he lacks the precise mixture of romantic flourish and grandeur so characteristic of the *Farbenlehre* (*Theory of Colors*, 1810), Goethe's major sally into the precincts of science.[22]

I am not saying that a writer such as Lyell should have been Strachey's own model. That would have made little sense. It is more that the elements one can see in Lyell's writing, including the place of "struggle" as a generator of images, seem far more close to the discourse Freud was erecting a full three-quarters of a century later than do any of the works of Freud's own contemporaries. There is no doubt that Freud developed a "jargon," a nomenclature designed to specifically give flesh to the observations and interpretations of the new field he was devising.

But there is no doubt, too, that, with few exceptions, this was a jargon built out of ordinary words, just like Lyell's geology and Darwin's evolutionism. It was not a terminology coined from Greek and Latin roots like that of late-19th and early-20th-century medicine, psychology, biology, chemistry, and so on. Freud's language was older, earlier, simpler, more reflective of a period before "literature" and "science" were sundered.

One must therefore also point out many of the things that Freud's prose did *not* do, vis-à-vis the standards of scientific writing in the 1910s and 20s, for these absences are equally instructive. It did not favor observation over theory; it did not seek to quantify; it did not illustrate its ideas or "results" in the form of technical diagrams, charts, or maps; it did not suppress the I of authorship; it did not refrain from emotionalism or outright speculation; it did not make use of citations (except, as we have seen, in a literary manner); it did not include a bibliography; it was not divided into sections on "procedure," "data," "discussion," and "conclusions"; finally, it did not often present itself as a local type of "investigation," an inquiry employing empirical methods into a particular epistemological unit (it was much more often a working-out of universal conclusions on the level of reflection). Freudian discourse did not, in short, obey any of the new strictures and limits placed upon technical writing, as a special area of Western discourse. It did, however, follow those of an earlier period, before any of these tenets had grown up around "science."

In the sense I am defining here, Freud did not write a "borderline" or "uncategorizable" prose, as is so often claimed. Instead, he brought forward what was magnificent and successful in the not-so-recent scientific past, at a time when that past had been honed to ideas of "theory," "hypothesis," and "evidence." On a fundamental level, the reductionism of Helmholtz and others had not entirely eliminated this past. The idealistic tradition in Germany was too strong for such an erasure and continued on in such forms as Gestalt psychology, the psychiatry of "will" in the work of Karl Jaspers, and others. What Helmholtz had helped achieve was a practical redefinition of older ideas of "theory," such that, while never completely relinquishing its reflective fertility, it became more demanding of material evidence. The concept of "struggle" rendered into "survival" and "adaption," revealed through the composition and morphology of the fossil record, provides an excellent example.

Freud divided his total effort between Charcot's emphasis on clinical work and the Helmholtzian emphasis on theory. But it should never be in doubt that his emphasis was, in the end, on the latter taken in its broadest meaning, and that his writing reached for the sorts of inclusiveness that historically preceded Helmholtz, that Lyell expresses above. It was to this kind of discourse that Freud remained loyal; it was on this basis of "theory" too that Jones could so ardently claim Freud's writings as "science" and could succeed in getting them accepted as such, via Strachey's modifications, in America and then England.

Freud reaches back to pull forward the intellectual breadth and often imagistic reasoning of that earlier "past." In a manner of speaking, he extends back to the days of his own childhood, when writing like his own was still a style practiced in many parts of the Western world of science. In his discourse, he stands between the 19th century and the 20th, between an older era of "educated reflection" and a newer one of "knowledge" and "fact." To use an analogy he might have enjoyed, Freud seems like a painter such as Jan van Eyck, poised between worlds of sensibility, between a medieval universe of graceful color and scholarly universals and a more modern one fascinated by form, detail, and studied perception. His uniqueness, at the level of discourse, was at least partly in line with his own project: to reveal, through language, an earlier stage of development. It is my feeling, at the last, that one of his enduring attractions as a writer, and the source of the enigma surrounding this attraction, can be found in this idea: that he discovered a way to re-create, in a time of growing austerity, an earlier period of more lush and searching expression.

Notes

1. Much of the criticism that has been leveled at Bettelheim's book within the psychoanalytic community (see, e.g., Gilman, 1991; Ornston, 1985c, 1992b) seems beside the point. Usually, such critique takes the form of quibbling over precise usage or other details, over the author's failure as a "true expert" or scholarly knower of Freud's language and thought. As such, it makes itself appear the result of professional jealousy—an attempt to reclaim territory taken too far "outside." There is a sense here (and it is often true of insider-type professions) that a code was breached, that Bettelheim blew a whistle, did not obey the given channels, that an important issue was made public before it was allowed to become fully professional, and therefore that aspersions were cast upon this same profession, which had not had the foresight to study the matter already. In any case, Bettelheim's book was written as nothing more than an announcement of the problem; this explains its excesses and simplifications. Those who criticize it for this seem to be demanding, ironically, that Bettelheim should have exhausted the issues involved, therefore curtailing further work on the matter, which has proved so fertile and productive. Professions, how-

ever, do not like to feel that they can be moved so obviously by public announcements.

2. In this and all other translations given in this chapter, I have tried to remain as close to the original as possible, so as to give a better sense of how it was written at the time. This particular article was entitled *Neue Untersuchungen über die Mikroorganismen bei infektiösen Wundkrankheiten* (New investigations on microorganisms in surgical infections), and appeared in the October 26th edition (1878) of the *Deutsche Medizinische Wochenschrift*, pp. 1–4. The original German of the citation given is as follows:

> Vielfach sind bei infektiösen Wundkrankheiten Mikroorganismen gefunded. Gleichwohl berechtigen diese Befunde noch nicht zu der Annahme, dass die Wundinfektionskrankheiten lediglich durch das Eindringen der Mikroorganismen in den Körper und ihre Vermehrung in demselben bedingt werden, mit einem Worte also parasitare Krankheiten sind. Denn es wird mit Recht gegen die Beweiskraft jener Befunde geltend gemacht, dass gar nich selten in Fällen von unzweifelhaft infektiösen Wundkrankheiten die Mikroorganismen vermisst, in anderen ebensolchen Fallen in zu geringer Zahl gefunded werden, um die Krankheitssymptome oder den todlichen Ausgang der Krankheit zu erklären. (p. 1)

3. This type of structure was traditional in German higher level prose, having existed ever since the Renaissance period of vernacularization, when Greek and Hebrew—not merely Latin—forms had served as crucial models for writing, due to Luther's incredibly influential rendition of the Bible (based on these two older languages; see, e.g., Waterman, 1976). To a degree, Koch's type of narrative retains the outward cloth of this history, though the detailed patterns of its earlier expressiveness are now gone.

4. Bernfeld (1949, p. 175) lists the following titles, which were published contemporaneously with Brücke's own laboratory studies:

Fragments of a Theory of the Formative Arts (1877)

Training in the Classical Languages is Necessary for Physicians (1879)

The Metric Accentuation in Verses (1879)

Action in Painting and Sculpture (1887)

5. Both of these qualities contributed to a major contribution Charcot made to medicine: his pivotal defense of Pasteur before the French Academy of Medicine. Charcot appears to have been one of the only physicians willing to visit Pasteur's laboratory, be shown the experimental procedures, and read through the results.

6. For my own part, I am most persuaded by something Marthe Robert has said of Freud's essay on Charcot:

Nothing is more significant than the passage in which Freud, anticipating his own scientific achievement, brings together the great modern naturalist [Cuvier] and the Biblical Adam, both scientists, both creators of words, both poets by virtue of the supreme human privilege of distributing and naming the things of this world. He did not see the French naturalist's beautiful classifications or Charcot's verbal inventions as simple technical commodities, but as the model for the poetic science of which he already had the vision, if not the key . . . (1964, p. 59)

The passage to which Robert alludes, meanwhile, is the following:

For physicians it need only be hinted at what wealth of forms for neuropathology were won by [Charcot], what higher precision and certainty of diagnosis made possible by his observations. Yet for the student who spent hours accompanying him on his rounds through the wards of the Salpêtrière—that museum of clinical facts whose names and uniqueness emerged in large part from his own self—he would be reminded of Cuvier, whose statue, standing before the Jardin des Plantes, presents the great knower and describer of the animal world surrounded by a wealth of animal forms, or else he would be drawn to the myth of Adam, who, when God brought before him the living creatures of Paradise to be divided and named, must have felt to the highest degree the intellectual pleasure Charcot prized so deeply. (Freud, 1893, p. 23)

7. The French title translates as "Lectures on diseases of the nervous system"; Freud's own title is "New lectures on diseases of the nervous system, especially regarding hysteria."

8. For a range of essays that reveal the span of such accusations, see the two collections edited by Timms and Segal (1988) and Ornston (1992d).

9. I can't resist, in this connection, repeating a portion of a letter Strachey once wrote to his wife on this subject (quoted by Leupold-Lowenthal, 1991, p. 349; and also by Ornston, 1992, p. 108):

The little beast [Jones] . . . is really most irritating. . . . They [Jones and Rivière] want to call "das Es" "the Id." I said I thought everyone would say "the Yidd."

So Jones said there was no such word in English: "There's 'Yiddish,' you know. And in German 'Jude'. But there is no such word as 'Yidd.'"

"Pardon me, doctor, 'Yidd' is a current slang word for a Jew."

"Ah! A slang expression. It cannot be in very widespread use then."

10. As a single example of German scientific influence in the United States, one could note that since the 1840s, American professors, students, and science professionals had made regular pilgrimages to German laboratories—Liebig's, Helmholtz's, and Wundt's first among them—to gain a level of training and experience unavailable at home (see Bruce, 1987). German

science was seen as having no equal in Europe, in fields as diverse as physics, biology, and psychology. Indeed, the "new (read: scientific) psychology" in particular was the object of an attempted wholesale transplantation by G. Stanley Hall from Wundt's laboratory to Clark University. It is no accident, then, given Hall's general view of German superiority, that he was the one (and only one) to bring Freud to America in 1910. (See Chapter 5 for influence of German science in Japan.)

11. At the same time, meanwhile, Freud's brand of "metaphysical hypothesis" also put him afoul of German philosophy, which remained traditionally idealistic in the mold of Kant and Hegel, and whose most formidable representatives, Edmund Husserl and Karl Jaspers (a psychiatrist), continued to argue that "soul" and "mind" (*Sinn*) were unique, nonmaterialistic identities, and therefore not analyzable in universal terms. Jaspers especially felt that the entire Freudian system was a type of intellectual delusion. The human body was physiological; it could be studied scientifically. The human psyche was a different matter, something mysterious and unknowable, the essence of each individual. It could not be examined as an "organ," and to try and do so was to exceed the limits of both science and reason.

12. I speak here of three different types of authorship, the scientist, the scholar, and the artist, because these more clearly define the otherwise overly polar "scientific" versus "literary" scenario. An excellent example of one who combined all three in the mid-to late-19th century, from the "humanist" side, would be the famous historian Leopold von Ranke, who claimed to found a "scientific history" based on factual re-creation, yet whose works from the 1830s through the 1870s reveal a magnificent interplay of styles, including lyrical, dramatic, *romanesque*, philosophical and religious asides, as well as the more instrumental description of events. For a brief review of Ranke's style, see Gay (1974).

13. The fact that the Strachey edition is actually titled *The Standard Edition of the Complete Psychological Works of Sigmund Freud* simply rephrases the problem.

14. The closest the *Standard Edition* and the *Gesammelte Werke* come to offering examples of any of this early work is their inclusion of a list of abstracts Freud wrote and had privately printed in 1897, part of a vita he assembled as an application for a higher academic position (Professor extraordinarus). The total list includes 38 separate publications. Those articles on this list that were published after 1892 and which are of direct relevance to psychoanalysis (e.g., those on neurosis and hysteria) are included in full in the same volume of each standard edition. Those dealing with other topics, e.g., neurological disease and disorder in children, are omitted. The policy of selectively deleting Freud's "science" is therefore even more clearly laid out in this case.

15. In what follows, I will use the established terms—ego, id, and superego—in my own commentary but change these to "I," "It," and/or "Ideal-

I" in my translations of Freud's own text. By this, I hope to create an added distance with regard to Freud's language and to suggest its qualities to a German speaker unfamiliar with psychoanalytic discourse.

16. It is interesting, too, that *Das Ich und das Es* is one of the few Freudian works in which the original German is significantly easier to read and comprehend than Strachey's English version, a result I attribute to the innate resistance of this complicated text to being smoothed out and rendered consistent, both in its style and imagery. Strachey applied no little effort to this task. He imposed a technical continuity on Freud's nomenclature; he tried to even the tone; he changed many meanings; he deleted many italicizations and inserted some of his own. He did all this and more, but his effort, in the end, could not eliminate the essential difficulties of the original.

17. As this is the only diagram Freud ever drew to graphically illustrate his concepts of the mind, and is one of the very few that appear anywhere in the whole of his psychoanalytic work, it has been much discussed (it first appeared in a different and earlier form in a letter to W. Fliess, dated December 6, 1896, then in *The Interpretation of Dreams* (1900), and again, in highly modified form, in the lecture *Die Zerlegung* mentioned earlier— Lecture 31 in the 1933 collection). I don't wish to rehearse the contents of what has already been said. Only a few comments might be made: what strikes the eye is a difference in orthography, the words "Ich" and "Es" being both larger and in cursive type, with "Es" slightly larger and its shape closely mimicking that of the lower part of the diagram. These aspects retain a suggestive quality; "Ich" and "Es," in their form, give the appearance of being more endemic to the curvilinear diagram as a whole. The whole is hardly "artistic," but it has a decided nontechnical aspect. All this is lost in the Strachey version, which puts all words in print, equalizes and reduces the size of "Ego" and "Id," and writes out the entire word "Repressed." The result is more "scientific" in appearance. Moreover, Strachey's translation of Freud's own qualifying remarks about the drawing are especially unfaithful at this point. Compare:

> We can sketch these relationships as a drawing, whose contours serve only for purposes of description and should claim no special meaning. (p. 252)

> The state of things which we have been describing can be represented diagrammatically (Fig. 1); though it must be remarked that the form chosen has no pretensions to any special applicability, but is merely intended to serve for purposes of exposition. (Freud, 1962, p. 24)

My own impression is that Strachey's verbosity was, in one way or another, an attempt to cover over Freud's own casualness about the figure. Labeling it as "Fig. 1" seems ridiculous, given that there never was a "Fig. 2," here or anywhere else in Freud's entire corpus. But the title helps formalize the drawing, turning it into an "illustration" or "diagram" (Freud's own term was *Zeichnung*, which is clearly vernacular), just as such phrases as "represented diagrammatically," "no pretensions to any special applicability," and "purposes of exposition." Here, in other words, is an undeniable example of the larger policy to scientize Freud.

18. This is rendered by Strachey as "The Dependent Relationships of the Ego." The title in German is *Die Abhängigkeiten des Ichs*.

19. Note that Strachey alters the sense of this, trying to keep Freud on his pedestal of control: "The complexity of our subject-matter must be an excuse for the fact that none of the chapter-headings of this book quite correspond to their contents, and that in turning to new aspects of the topic we are constantly harking back to matters that have already been dealt with" (Freud, 1962, p. 48).

20. This "failure," however, was anything but true of the later, essay version of the book, "The Dissection of the Psychic Personality." Ten years and more than a dozen new works gave Freud much time, space, and effort to work his ideas into better shape. In the essay, he is fully in command and is quite able to provide the id with an adequate imagery (he calls it, for example, "a chaos, a cauldron seething with agitations"). There is, moreover, a host of new images employed to describe the relations between the three agencies—one of which, the most dominant of all, is that of a political-territorial system in which the ego emerges as "leader." The final paragraph of this work is worth quoting in full, to contrast it with that just cited of *Das Ich und das Es*:

> In the end, we must admit that the therapeutic endeavors of psychoanalysis have chosen a similar point of approach. For their intention is to strengthen the I, to make it more independent from the Higher-I, to broaden its perceptual field and extend its organization so that it is able to annex for itself new pieces of the It. Where It was, so I shall be. It is the work of civilization, like the draining of the Zuyder Sea.

The essay therefore ends with the victorious ego, on an entirely different note from its predecessor. Indeed, the phrase "where It was . . ." is one of the most famous and oft-quoted from all of Freud's works—yet it is nearly opposite to the sense of the last lines of *Das Ich und das Es*, where we are left "apprehensive" at the potential power of the "mischief-maker," Eros, within the id.

Most writers who have touched on the subject treat the book of 1923 and the essay of 1933 as if they were closely similar, as if the latter had adopted most of the imagery of the former and offered a parallel treatment. To my mind, this is wholly untrue. These two segments of Freud's corpus are very different pieces of writing, on almost any plane one might choose to describe.

21. One should perhaps note, too, that in the very beginning of this work, Goethe writes "Fruitless is it for us to try and describe the character of a man; instead, one assembles his actions, his deeds together, and an image of his character then comes forward" (1963, p. 5).

References

Amacher, P. (1965). *Freud's Neurological Education and its Influence on Psychoanalytic Theory. Psychological Issues, 4*(4). Monograph 16.

Anzieu, D. (1986). The Place of German Language and Culture in Freud's Discovery of Psychoanalysis Between 1895 and 1900. *International Journal of Psychoanalysis, 67,* 219–226.

Bass, A. (1985). On the History of a Mistranslation and the Psychoanalytic Movement. In J. F. Graham (Ed.), *Difference in Translation* (pp. 102–141). Ithaca, NY: Cornell University Press.

Bernfeld, S. (1944). Freud's Earliest Theories and the School of Helmholtz. *Psychoanalytic Quarterly, 13,* 341–362.

Bernfeld, S. (1949). Freud's Scientific Beginnings. *American Imago* 6(3), 163–196.

Bernfeld, S. (1951). Sigmund Freud, 1882–1885. *International Journal of Psychoanalysis, 32,* 204–217.

Bettelheim, B. (1982). *Freud and Man's Soul.* New York: Knopf.

Bourguignon, A., P. Cotet, J. Laplanche, and F. Robert. (1989). *Traduire Freud.* Paris: Presses Universitaires de France.

Brücke, E. (1881). *Vorlesungen über Physiologie.* 3d ed. Vienna: Wilhelm Braumüller.

Bruce, R. (1987). *The Launching of Modern American Science, 1846–1876.* New York: Knopf.

Charcot, J. M. (1881–1892). *Oeuvres Complètes.* 13 vols. Paris: Lecrosnier et Babe.

Charcot, J. M., and A. Pitres. (1883). *Étude Critique et Clinique de la Doctrine des Localisations Motrices dans l'Ecorce des Hémispheres Cérébraux de l'Homme.* Paris: Germer Bailliere.

Charcot, J. M., and P. Richer. (1887). *Les Demoniaques dans l'Art.* Paris: Delahaye et Lecrosnier.

Charcot, J. M., and P. Richer (1889). *Les Difformes et les Malades dans l'Art.* Paris: Lecrosnier et Babe.

Charcot, J. M. (1877–1889). *Lectures on the Diseases of the Nervous System.* 3 vols. Transl. by G. Sigerson. London: New Sydenham Society.

Codet, H., and R. Laforgue. (1925). L'influence de Charcot sur Freud. *Progrès Medical, 22,* 801–802.

Coleman, W., and F. L. Holmes. (Eds.). (1988). *The Investigative Enterprise: Experimental Physiology in Nineteenth-Century Medicine*. Berkeley and Los Angeles: University of California Press.

Decker, H. S. (1977). *Freud in Germany: Revolution and Reaction in Science, 1893–1907. Psychological Issues*, 11(1). Monograph 41.

Federn, E. (1988). Psychoanalysis—The fate of a science in exile. In E. Timms and N. Segal (Eds.), *Freud in Exile* (pp. 156-162). New Haven, CT: Yale University Press.

Fisher, S., and R. P. Greenberg. (1977). *The Scientific Credibility of Freud's Theory and Therapy*. New York: Basic Books.

Freud, S. (1877a). Über den Ursprung der hinteren Nervenwürzeln im Rückenmarke von Ammocoetes. *Sitzungsberichte der mathematisch-naturwissenschaftlichen Classe der Kaiserlichen Akademie der Wissenschaften, 75*(1), 15–27.

Freud, S. (1879). Notiz über eine Methode zur anatomischen Praparation des Nervensystems. *Centralblatt für die medicinischen Wissenschaften, 17*, 468–469.

Freud, S. (1882). Über den Bau der Nervenfasern und Nervenzellen beim Flusskrebs. *Sitzungsberichte der mathematisch-Naturwissenschaftlichen Classe der Kaiserlichen Akademie der Wissenschaften, 85*, 9–46.

Freud, S. (1884). A New Histological Method for the Study of Nerve-Tracts in the Brain and Spinal Chord. *Brain: A Journal of Neurology, 7*, 86–88.

Freud, S. (trans). (1886). *Neue Vorlesungen über die Krankheiten des Nervensystems, Insbesondere uber Hysterie*. (Translation of *Leçons sur les Maladies du Système Nerveux*. Leipzig: Toeplitz & Deuticke.

Freud, S. (1893). Charcot. In *Gesammelte werke*, 1: 19–37.

Freud, S. (1899). Inhaltsangaben der Wissenschaftlichen Arbeiten des Privatdozenten Dr. Sigmund Freud (1877–1897). In *Gesammelte werke*, 1: 461–488.

Freud, S. (1923). *Das Ich und das Es*. In *Gesammelte werke*, 13: 237–289.

Freud, S. (1924–1934). *Gesammelte Schriften*. 13 vols. Vienna: Internationaler Psychoanalytischer Verlag.

Freud, S. (1940–1968). *Gesammelte werke*. 18 vols. London: Imago.

Freud, S. (1933). *Neue Folge der Vorlesungen zur Einführung in die Psychoanalyse*. In *Gesammelte werke*, 15.

Freud, S. (1953–1960). *The Standard Edition of the Complete Psychological Works of Sigmund Freud*. 24 vols. Transl. and ed. by J. Strachey. London: Hogarth Press and Inst. of Psychoanalysis.

Freud, S. (1962). *The Ego and the Id*. Transl. by J. Strachey. New York: Norton. (This work also appears in *Standard Edition*).

Freud, S. (1965). *New Introductory Lectures on Psycho-Analysis*. Transl. by J. Strachey. New York: Norton. (Also in *Standard Edition*, 22: 3–182)

Gay, P. (1974). *Style in History*. New York: McGraw-Hill.

Gay, P. (1988). *Freud: A Life for Our Time*. New York: Norton.

Gedo, J. E., and G. H. Pollock. (Eds.). (1976). *Freud: The Fusion of Science and Humanism*. New York: International Universities Press.

Gilman, S. L. (1988). Constructing the image of the appropriate therapist: The struggle of psychiatry with psychoanalysis. In E. Timms and N. Segal (Eds.), *Freud in Exile* (pp. 15–36). New Haven, CT: Yale University Press.

Gilman, S. L. (1991). Reading Freud in English: Problems, paradoxes, and a solution. *International Review of Psychoanalysis, 18*, 331–344.

Goethe, J. W. (1963). *Zur Farbenlehre: Didaktischer Teil*. Munich: Deutscher Taschenbuch.

Gradmann, C. (1993). Naturwissenschaft, Kulturgeschichte und Bildungs-begriff bei Emil Du Bois-Reymond. *Tractrix, 5*, 1–16.

Gray, H. (1948). Bibliography of Freud's Pre-analytic Period. *Psychoanalytic Review, 35*, 403–410.

Grubrich-Simitis, I. (1986). Reflections on Sigmund Freud's Relationship to the German Language and to Some German-Speaking Authors of the Enlightenment. *International Journal of Psychoanalysis, 67*, 287–294.

Grunbaum, A. (1984). *The Foundations of Psychoanalysis: A Philosophical Critique*. Berkeley and Los Angeles: University of California Press.

Guillain, G. (1959). *J.-M. Charcot: His Life, His Work*. New York: Paul B. Hoeber.

Hale, N. H., Jr. (1971). *Freud in America*. Vol. 1. New York: Oxford University Press.

Helmholtz, H. (1847). *Über die Erhaltung der Kraft*. Leipzig: L. Voss.

Helmholtz, H. (1896). *Vorträge und Reden*, 2 vols. 4th ed. Braunschweig: Friedrich Vieweg und Sohn.

Holder, A. (1988). Reservations about the *Standard Edition*. In E. Timms and N. Segal (Eds.), *Freud in Exile* (pp. 210–214). New Haven, CT: Yale University Press.

Holder, A. (1992). A Historical-Critical Edition. In D. G. Ornston, Jr. (Ed.), *Translating Freud* (pp. 75–96). New Haven, CT: Yale University Press.

Holt, R. (1965). A Review of Some of Freud's Biological Assumptions and Their Influence on His Theories. In N. Greenfield and W. Lewis, (Eds.), *Psychoanalysis and Current Biological Thought* (pp. 93–124). Madison: University of Wisconsin Press.

Hook, S. (1960). *Psychoanalysis, Scientific Method, and Philosophy*. New York: Grove Press.

Junker, H. (1987). General Comments on the Difficulties of Re-translating Freud into English. *International Review of Psychoanalysis, 14,* 317–320.

Junker, H. (1992). Standard Translation and Complete Analysis. In D. G. Ornston, Jr. (Ed.), *Translating Freud* (pp. 48–62). New Haven, CT: Yale University Press.

Lacan, J. (1966) *Écrits.* Paris: Seuil.

Laplanche, J., P. Cotet, and A. Bourguignon. (1992). Translating Freud. In D. G. Ornston, Jr. (Ed.), *Translating Freud* (pp. 135–190). New Haven, CT: Yale University Press.

Laplanche, J. and J.-B. Pontalis. (1973). *The Language of Psychoanalysis.* New York: Norton.

Leupold-Lowenthal, H. (1991). The impossibility of making Freud English. *Int. Rev. Psychoanal., 18,* 345–350.

Liebig, J. (1840). *Die Organische Chemie in Ihrer Anwendung auf Agrikultur und Physiologie.* Leipzig: L. Voss.

Lyell, C. (1830). *Principles of Geology.* Vol. 1. Chicago: University of Chicago Press, 1990. (Facsimile reprint of 1st ed.)

Mahoney, P. J. (1982). *Freud as a Writer.* New York. International University Press.

Mahoney, P. J. (1987). *Freud as a Writer.* Rev. ed. New Haven, CT: Yale University Press.

Mahoney, P. J. (1989). *On Defining Freud's Discourse.* New Haven, CT: Yale University Press.

Mahoney, P. J. (1992). A Psychoanalytic Translation of Freud. In D. G. Ornston, Jr. (Ed.), *Translating Freud* (pp. 24–47). New Haven, CT: Yale University Press.

Mann, T. (1929). Die Stellung Freuds in der Modernen Geistesgeschichte. *Pschoanalytische Bewegung, 1,* 3–32.

Marcus, S. (1984). *Freud and the Culture of Psychoanalysis.* New York: Norton.

McIntosh, D. (1986). The Ego and the Self in the Thought of Sigmund Freud. *International Journal of Psychoanalysis, 67,* 429–448.

Meyers, G. E. (1986). *William James: His Life and Thought.* New Haven, CT: Yale University Press.

Meynert, T. (1890). *Klinische Vorlesungen über Psychiatrie auf wissenschaftlichen Grundlagen.* Vienna: Wilhelm Braumuller.

Muschg, W. (1930). Freud als Schriftsteller. *Die Psychoanalytische Bewegung, 2,* 467–509.

Nagel, T. (1974). Freud's anthropomorphism. In R. Wollheim (Ed.), *Freud: A Collection of Critical Essays* (pp. 11–24). Garden City, NY: Anchor

Books.

Ornston, D. G., Jr. (1982). Strachey's Influence: A Preliminiary Report. *International Journal of Psychoanalysis, 63,* 409–426.

Ornston, D. G., Jr. (1985a). Freud's Conception is Different From Strachey's. *Journal of the American Psychoanalytic Association, 33,* 379–412.

Ornston, D. G., Jr. (1985b). The invention of "cathexis" and Strachey's strategy. *International Review of Psychoanalysis, 12,* 391–399.

Ornston, D. G., Jr. (1985c). Review of Bettelheim's *Freud and man's soul. Journal of the American Psychoanalytic Association, 33* (book suppl.), 189–200.

Ornston, D. G., Jr. (1988). How standard is the Standard Edition? In E. Timms and N. Segal (Eds.), *Freud in Exile* (pp. 196–209). New Haven, CT: Yale University Press.

Ornston, D. G., Jr. (1992a). Improving Strachey's Freud. In D. G. Ornston, Jr. (Ed.), *Translating Freud* (pp. 1–23). New Haven, CT: Yale University Press.

Ornston, D. G., Jr. (1992b). Bruno Bettelheim's *Freud and Man's Soul.* In D. G. Ornston, Jr. (Ed.), *Translating Freud* (pp. 63–74). New Haven, CT: Yale University Press.

Ornston, D. G., Jr. (1992c). Alternatives to a Standard Edition. In D.G. Ornston, Jr. (Ed.), *Translating Freud* (pp. 97–113). New Haven, CT: Yale University Press.

Ornston, D. G., Jr. (Ed.). (1992d). *Translating Freud* (pp. 97–113). New Haven, CT: Yale University Press.

R. A. Paskauskas. (1988). The Freud–Jones Era. In E. Timms and N. Segal (Eds.), *Freud in Exile* (pp. 109–123). New Haven, CT: Yale University Press.

Pasteur, L. (1923–1929). *Oeuvres.* 7 vols. Paris: Libraires de l'Académie de Médecine.

Peters, U. (1988a). The Psychoanalytic Exodus: Romanti Antecedents, and the Loss of German Intellectual Life. In E. Timms and N. Segal (Eds.), *Freud in Exile* (pp. 54–64). New Haven, CT: Yale University Press.

Pines, M. (1988). The question of revising the Standard Edition. In E. Timms and N. Segal (Eds.), *Freud in exile* (pp. 177–180). New Haven, CT: Yale University Press.

Pollak-Cornillot, M. (1986). Freud Traducteur: Une Contribution à Traduction de ses Propres Oeuvres. *Revue Française de Psychoanalyse, 53,* 1235-1246.

Ricoeur, P. (1970). *Freud and Philosophy: An Essay on Interpretation.* New Haven, CT: Yale University Press.

Ritvo, L. B. (1965). Darwin as the source of Freud's neo-Lamarckism. *Journal of the American Psychoanalytic Association, 13.*

Ritvo, *L. B. (1990). Darwin's Influence on Freud.* New Haven, CT: Yale University Press.

Roazen, P. (Ed.). (1973). *Sigmund Freud.* Englewood Cliffs, NJ: Prentice-Hall.

Robert, M. (1964). *La Revolution Psychoanalytique.* Paris: Payot.

Robinson, P. (1993). *Freud and His Critics.* Berkeley: University of California Press.

Rycoff, C. (1973). *A Critical Dictionary of Psychoanalysis.* Totowa, NY: Littlefield Adams.

Schlessinger, N., J. E. Gedo, J. A. Miller, G. H. Pollock, M. Sabshin, and L. Sadow. (1967). The Scientific Styles of Breuer and Freud and the Origins of Psychoanalysis. *Journal of the American Psychoanalytic Association, 15,* 404–422.

Schnitzler, A. (1887, 12 December). Review of *Neue Vorlesungen über die Krankheiten des Nervensystems, insbesondere über Hysterie,* by J. M. Charcot. Transl. by Sigmund Freud. *Internat. Klinische Rundschau,* columns 19–20.

Schönau, W. (1968). *Sigmund Freuds Prosa.* Stuttgart: Metzlersche.

Solms, M., and M. Saling. (1986). On Psychoanalysis and Neuroscience: Freud's Attitude to the Localizationist Tradition. *International Journal of Psychoanalysis, 67,* 397–416.

Spence, D. P. (1987). *The Freudian Metaphor.* New York: Norton.

Stanton, M. (1991). *Laissez-faire*: James Strachey and Freud's French. *International Review of Psychoanalysis, 18,* 393–400.

Steiner, R. (1987). A World-Wide international trade mark of genuineness? *International Review of Psychoanalysis, 14,* 33–102.

Steiner, R. (1988). Die Weltmachtstellung des Britischen Reichs. In E. Timms and N. Segal (Eds.), *Freud in Exile* (pp. 180–195). New Haven, CT: Yale University Press.

Steiner, R. (1991). To Explain Our Point of View to English Readers in English Words. *International Review of Psychoanalysis, 18,* 351–400.

Sulloway, F. J. (1979). *Freud: Biologist of the Mind.* New York: Basic Books.

Swanson, D. R. (1977). A Critique of Psychic Energy as an Explanatory concept. *Journal of the American Psychoanalytic Association, 25,* 603–633.

Ticho, E. A. (1986). The influence of the German-language culture on Freud's Thought. *International Journal of Psychoanalysis, 67,* 227–234.

Timms, E. (1988). Freud's Library and His Private Reading. In E. Timms and N. Segal (Eds.), *Freud in Exile* (pp. 65–79). New Haven, CT: Yale University Press.

Trilling, L. (1953). Freud and Literature. In *The Liberal Imagination* (pp. 32–54). Garden City, NY: Anchor.

Tuchman, A. (1993). Helmholtz and the German Medical Community. In Cahan, D. (Ed.), *Hermann von Helmholtz and the Foundations of Nineteenth-Century Science* (pp. 17–49). Berkeley and Los Angeles: University of California Press.

Vermorel, M. and H. Vermorel. (1986). Was Freud a Romantic? *International Review of Psychoanalysis, 13,* 15–37.

Watson, A. S. (1958). Freud the Translator. *International Journal of Psychoanalysis, 39,* 326–329.

Waterman, J. T. (1976). *A History of the German Language.* Rev. ed. Prospect Heights, IL: Waveland Press.

Wilson, E. (1987). Did Strachey Invent Freud? *International Review of Psychoanalysis, 14,* 299–316.

Wundt, W. (1893). *Grundzüge der Physiologischen Psychologie.* Vol. 1. 4th ed. Leipzig: Wilhelm Engelmann.

Wundt, W. (1904). *Principles of Physiological Psychology.* Vol. 1. Transl. by E. B. Titchener. New York: Macmillan.

Wyatt, F. (1988). The Severance of Psychoanalysis from its Cultural Matrix. In E. Timms and N. Segal (Eds.), *Freud in Exile* (pp. 145–155). New Haven, CT: Yale University Press.

Whyte, L. (1960). *The Unconscious Before Freud.* New York: Basic Books.

Concluding Remarks

Science, in great measure, is a matter of language. It is much else besides, of course: people, labor, equipment, institutions, capital, education, and so forth. But as knowledge, as a collection of formal understandings that aim at communality and communal power, science must begin and end in words and images, for it is here that literate societies demand all effort and thought find their material embodiment.

This book began with the question of human discourse and its changing direction. This direction can be charted, not simply in traditional linguistic terms, but in larger cultural and historical terms. Today, the maps that might seek to orient the present would inevitably highlight scientific discourse as an ultimate, long-term model for all professional forms of speech. Any study of this discourse of science, therefore—*any* study whatsoever—has an imminent political dimension. It is a study of power, as has been suggested by such diagnosticians as Michel Foucault and Jürgen Habermas. The true penetrative influence of scientific speech is completely commensurate with that of scientific knowledge, with government policies, with military weaponry and strategy, with the university and its major sources of funding, with all of medicine, with a great deal of international economics, with "nature" and "the universe" as public ideas, with myriad institutional sources involved

430

in the distribution of money, materials, and belief, and finally with the deepest, most private ideas about illness, the self, the possible. Scientific discourse is more penetrative, more urgent as an everyday force in human society, than one commonly recognizes. Shedding light on its character and history, therefore, is more than mere "scholarship."

Even within the realm of academic discussion, language in science is far more than a structuralist's daydream; nor is it a literary critic's vision of flatland or of hell. It can't be analyzed away; reduced to a series of formulae, narrative moves, or tactics of persuasion; or dismissed as a tale of agents blindly or cleverly employing convention to achieve some particular end. If scientists today are less individual in the discourse they use, if they are more often the servants of their language than in the past (and they are), if the speech of their chosen intellectual striving has suffered the loss of certain sophistications for the gain of others (and it has), this is a historical result that cannot be understood simply as a consequence of "rhetoric." Throughout its history, authorship in the various fields of science has not evolved independently, like an island of self-reference, but as part of the greater field of speech and writing. It has done this, moreover, both by drawing directly on words and images from other areas of culture and also by adopting a great many of the stylistic forms of expression found in literary, philosophical, historical, and other types of authorship. Now, as in the past, scientists and their writing do not exist as an insider region populated by a multitude of obedient scribblers and a few heroic originals. The language of science today retains many voices, many traces and tracings, of the real-world evolution of both itself and society. A number of these voices have been weakened, silenced, even systematically suppressed during the past century and a half, for reasons that have everything to do with political realities as well as epistemological ones. Many others, however, have not; and it remains only to step around that older, more monolithic and intimidated view of "science" in order to find them still very much alive and insistent. Science is as much a history of writers as of texts, of writing as of the written; it is enormously more than its physical remains. Its voices, therefore, deserve to be reclaimed. And the work of such reclamation should not be viewed merely as a victory over former prejudice or provincialism. It is also one way—a very central and necessary way—to frame and define "science" in terms of the powers it has achieved in the contemporary world.

This same science after all, in its infancy, actually began partly out of arguments over language itself. The prose style of "experimental philosophy," in England, for example, developed as a concerted literary endeavor within a larger conflict over proper and improper writing, good and bad mental work. One sees, today, that the common distinctions made between "science" and "literature"—so often invoked in attempts to trace various influences back and forth (e.g., Galilean astronomy in Milton's poetry)—are themselves a fiction when imposed upon this time period. The early battles for natural philosophy were fought over the expressive force, debilities, excesses, and needed reforms not of "language" generally but of *writing* in particular. And these battles did not merely involve the "new philosophy" alone, but all of literature at once— poetry, drama, the sermon, satire, fiction, all forms of written discourse. One finds that the arguments of the day were aimed at making the written word less ornamental, rhetorical, and courtly, and more controlled and obedient—more a model for a hoped-for, postmedieval, expanding social order. A more exacting and disciplined word would temper men's minds into a similar substance, and would therefore create a new society of harmony, loyalty, and consolidated power. If this sounds especially well tuned to an age of civil war, dynastic conflict, and increasing colonial ambitions, this is no accident.

One tends to look upon the arguments of that time as naive, or, as Foucault might say, circumscribed by belief in the ultimate power of the sign. The complaints of Bacon and Wilkins about the power of words to rule and deform the mind seem situated irrevocably in their time, a topic for distanced study. Yet to stop here would be shortsighted. These writers, deeply influenced by their own classical learning, understood something crucial and of continual relevance, namely, that writing is a mode of conceiving as well as of expressing, a force for creating, pursuing, solidifying, and imposing knowledge, not merely offering it. We see this at present profoundly revealed in the entire psychology of disease, health, and death—indeed, in the entire edifice of our institutional structures erected upon concepts of illness and its cure. Such structures are so deeply implicated by medical language, in its most recent forms, as to seem the very evocation of this speech, its fearful crystallizations. There has been much talk in recent years of "medical metaphors," of medicine's "allegorical imagery of the body," and so forth. What this type of thinking fails to comprehend—and it is an enormous failure, when one considers what is at stake—is that, often enough, what is being discussed are not metaphors at all but scientific knowledge as it has come to be hardened and finalized. In the case of war and computer imagery, the relevant language lies at the very core of medical science and has long stopped operating on anything approaching a figurative level but instead now helps dictate the very process and direction of research, therapy, technology, instrumentation, the doctor–patient relationship, training, indeed, all aspects of the "medical encounter." Talk of metaphors, as metaphors, is therefore critical with regard to historical origins, but otherwise beside the point. It is disingenuous to treat a knowledge of this type, with its vast practical powers, as if it had quotation marks around it. The question, rather, becomes one of examining the particular species of logic inherent in such imagery, what aspects of the past it bears within it, the guiding rationale of this logic with regard to conception, the type of present and future it has helped chart out for inquiry and healing, for public and private emotion, for "medicine" as a central aspect of modern society.

In a letter to his friend, Gerhard Scholem, the young critic Walter Benjamin wrote of an idea he had come upon, that "philology" (the study of human language as a vehicle for literature and as an expression of culture) be reconceived as "the *history of terminology* at its deepest level." Such a history, Benjamin felt, might be a type of chronicle, whose interpretation would reveal "the intention of the content, since its content interpolates history." One doesn't have to share Benjamin's idealism regarding content as possessing a life of its own to comprehend the implications of such a history. If we expand (as we must) the idea of terminology to include names and visual images, for example, we can see what Benjamin's concept might mean for the sciences. Titles and pictures represent acts of attribution and definition; they are means through which scientists have been able to discover and interact with the historical moment, to impose a personal will whose contours of possession, subtly or otherwise, reach far beyond their own. The Moon, as we have seen, reveals this to a magnificent degree, in the several phases of creation it underwent as an object of observation, study, and mapping, and in the naming schemes it attracted. The close interplay between Europe and the New World at this time, particularly involving the collapse of Spanish power, paralleled by a similar shift within astronomy itself which brought the "downfall" of the ancients, was played out again upon the lunar surface as an avid consciousness of history. And this staged consciousness, for all intents and purposes, will be rehearsed forever across the plains, seas, and mountains of this same surface, which therefore constitute a geography commensurate with Benjamin's notion of interpolating content.

Finally, with regard to the chapters in this book, one comes to the issue of translation and therefore to the reality that scientific discourse is itself a fiction, inasmuch as it remains spoken of as a single, unified phenomenon of human linguistic effort. Like any other human effort, scientific understanding cannot escape its forms, and its forms are multiple, changing, not always stable, and certainly not universal across language boundaries. As the examples of Freud in English and Japanese science make plain, the movement of knowledge between languages does not take place without profound and critical transformations—and with equally profound cultural effects. This can be guided by ideology, as in the case of Freud, or, in the case of Japan, by a range of complex and competing interests, both private and national in scope, evolving over centuries and deeply involved in definitions of the "foreign" and the "indigenous." That issues of translation must be viewed as central to the history of science is amply shown by the adaptation Freud underwent to the reigning demands of the "scientific" in the early 20th century—in effect, the creating of a scientific Freud in the Anglo-American idiom. In the case of Japan, meanwhile, the record shows that "science" and "translation" were not merely inseparable at an early stage, but actually identical: to do "science" meant to translate texts, to establish a discourse and vocabulary in Japanese that could make "research" possible. The creation of this discourse, then and thereafter, reveals how the choices made in this process have been at every step guided by allegiances to political and cultural beliefs, those having everything to do with struggles regarding the formation of modern Japanese identity and its place in world history. Both these cases, Freud and Japan, tell us that scientific knowledge has never been merely "transmitted" or "adopted" from one culture to another. Translation has involved many processes of nativization which have until now gone almost entirely unstudied by scholars. The universality of scientific discourse is another grand and obfuscating myth—perhaps the last of the great legends of positivistic science to remain in place.

5.5.

There are an infinite number of histories and evaluations that might be written about science and its languages, just as there are about any other major area of long-term human effort. I have offered only a few possibilities here, which, I hope, suggest this breadth, not only in subject and point of view but in style of expression as well. A full history of scientific discourse would be an enormous contribution by itself, a gateway to a new field. The overwhelming and yet still growing centrality of medical knowledge to modern society, throughout the globe, begs that one consider, in far more detail than I have been able to do here, the institutional, psychological, and even cognitive effects of linguistic choice—where our ideas of illness and health have come from and where they have taken us. The power of names to create and possess "territory" both for science and for Western culture pervades so much of technical knowledge as to comprise another realm of major inquiry. Equally vast and untouched is the subject of translation, the degree to which modern science has depended upon it from the 12th century onward (when most of Greek learning, which was not yet "Western," came into Europe), and still more, how this science has been spread throughout the world's cultures by various processes of nativization. The issues here are pressing today in an entirely concrete, political–economic sense: nations seeking to outdo each other technologically are engaged in bringing to market not "science" per se, as a simple universalism, but "science" in their own national dialect, in direct competition with other such encompassing dialects, a "science" in each case adapted by language, tradition, and culture to a specific group, to a specific set of institutional and cognitive circumstances, and therefore to a specific set of possibilities for success. The myth of universality cannot hold when faced with this evidence; a new view is required.

Technical discourse is not—to drag an old cliché to the whipping post—a mansion of many rooms, but a vast urbanity of linguistic activities sharing all the ordinary aspects of human discourse generally. The many realities that exist with regard to writing in human culture, past and present, exist within the domains of science as well. Indeed, there is no real division. Human beings have always been scribblers of details that have gone into the particularity of their time. Indeed, this is part of what makes them "human," as writers of the lived and the felt, not merely of the "observed" and the "concluded." It is also what makes, and has always made, science a human enterprise. If scholarship too often focuses on lives for the sake of their products, it is perhaps not merely due to convention, but because of the lurking danger of an unwanted reflection.

One of Oscar Wilde's fairy tales, "The Star Child," opens with several animals arguing about how cold and snowy it is and who should be blamed. "'Nonsense!,'" says the wolf to all other statements of the debate. "'I tell you that it is all the fault of the Government, and if you don't believe me I shall eat you.'" The Wolf, Wilde informs us, "had a thoroughly practical mind, and was never at a loss for a good argument." Science, perhaps, has been too often cast as the wolf at the door of modernism, whether in its pre- or postmodern forms. It seems time, more than ever, to find in it a menagerie of new areas for study and inquiry that might expand beyond present boundaries, and that might reflect back, more largely, on what we have acquired, culturally speaking, from this enormously plural activity, and what we might rightly expect and desire from it in the future.

Index